AutoCAD® 2006 Tutorial
Second Level: 3D Modeling

Randy H. Shih
Oregon Institute of Technology

ISBN:
1-58503-229-8 (Text only)

1-58503-231-X (Text bundled with 180-day license AutoCAD 2006)

SDC
PUBLICATIONS

Mission, Kansas

Schroff Development Corporation
P.O. Box 1334
Mission KS 66222
(913) 262-2664
www.schroff.com

Trademarks
The following are registered trademarks of Autodesk, Inc.: 3D Studio, ADI, Advanced Modeling Extension, AME, AutoCAD, AutoCAD Mechanical Desktop, AutoCAD Development System, AutoCAD LT, Autodesk, Autodesk Animator, AutoLISP, AutoShade, AutoVision, and Heidi.
The following are trademarks of Autodesk, Inc.: ACAD, Autodesk Device Interface, AutoCAD DesignCenter, AutoTrack, Heads-up Design, ObjectARX and Visual LISP.
Microsoft, Windows are either registered trademarks or trademarks of Microsoft Corporation.
All other trademarks are trademarks of their respective holders.

Copyright © 2005 by Randy Shih

All rights reserved. No part of this book may be reproduced, stored in a retrieval system, or transcribed in any form or by any means -electronic, mechanical, photocopying, recording, or otherwise - without the prior written permission of Schroff Development Corporation.

Shih, Randy H.
   AutoCAD® 2006 Tutorial / Second Level: 3D Modeling

Randy H. Shih

ISBN 1-58503-229-8 (Text Only)
ISBN 1-58503-231-X (Text bundled with AutoCAD 2006)

The author and publisher of this book have used their best efforts in preparing this book. These efforts include the development, research and testing of the material presented. The author and publisher shall not be liable in any event for incidental or consequential damages with, or arising out of, the furnishing, performance, or use of the material.

Printed and bound in the United States of America.

Preface

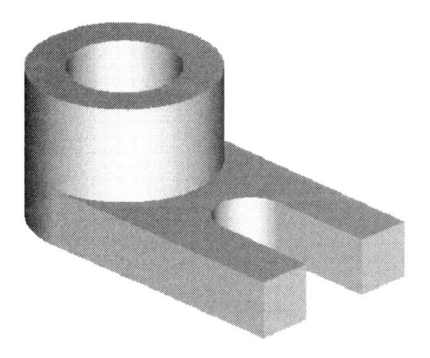

The primary goal of *AutoCAD® 2006 Tutorial: 3D Modeling* is to introduce the aspects of **Computer Three Dimensional Modeling**. This text is intended to be used as a training guide for students and professionals. This text covers *AutoCAD® 2006* and the chapters proceed in a pedagogical fashion to guide you from constructing 3D wireframe models, 3D surface models and 3D solid models to making multiview drawings. This text takes a hands-on, exercise-intensive approach to all the important 3D modeling techniques and concepts. This textbook contains a series of ten tutorial style chapters designed to introduce CAD users to 3D modeling with **AutoCAD® 2006**. This text is also helpful to AutoCAD users upgrading from a previous release of the software. The new improvements and enhancements of the software are incorporated into the tutorials. The 3D modeling techniques and concepts discussed in this text are also designed to serve as the foundation to the more advanced feature-based CAD/CAE packages such as AutoCAD® Mechanical Desktop, AutoCAD® Architectural Desktop, and Autodesk Inventor®. The basic premise of this book is that the more 3D designs you create using AutoCAD® 2006, the better you learn the software. With this in mind, each tutorial introduces a new set of commands and concepts, building on previous chapters. This book does not attempt to cover all of AutoCAD® 2006's features, only to provide an introduction to the software. It is intended to help you establish a good basis for exploring and growing in the exciting field of Computer Aided Engineering.

Acknowledgments

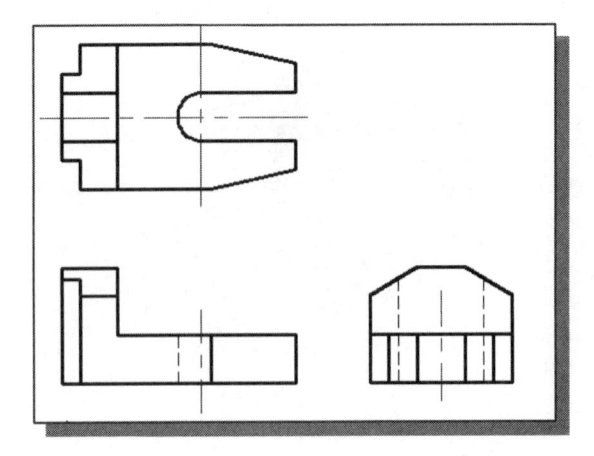

This book would not have been possible without a great deal of support. First, special thanks to two great teachers, Prof. George R. Schade of University of Nebraska-Lincoln and Mr. Denwu Lee, who taught me the fundamentals, the intrigue, and the sheer fun of Computer Aided Engineering.

The effort and support of the editorial and production staff of Schroff Development Corporation is gratefully acknowledged. I would especially like to thank Stephen Schroff and Mary Schmidt for their support and helpful suggestions during this project.

I am grateful that the Mechanical and Manufacturing Engineering Technology Department of Oregon Institute of Technology has provided me with an excellent environment in which to pursue my interests in teaching and research. I would especially like to thank Professor Brian Moravec and Emeritus Professor Charles Hermach for helpful comments and encouragement.

Finally, truly unbounded thanks are due to my wife Hsiu-Ling and our daughter Casandra for their understanding and encouragement throughout this project.

Randy H. Shih
Klamath Falls, Oregon
Spring, 2005

Table of Contents

Preface
Acknowledgments

Chapter 1
Getting Started

Introduction	1-2
Development of Computer Geometric Modeling	1-2
Why use AutoCAD® 2006?	1-5
Getting Started with AutoCAD® 2006	1-7
AutoCAD Screen Layout	1-8
Pull-down Menus	1-9
Standard Toolbar	1-9
Object Properties Toolbar	1-9
Graphics Window	1-9
Graphics Cursor or Crosshairs	1-9
Command Prompt Area	1-9
Draw Toolbar and Modify Toolbars	1-10
Mouse Buttons	1-10
[Esc] - Canceling commands	1-11
On-Line Help	1-11
Leaving AutoCAD	1-12
Creating a CAD files folder	1-13

Chapter 2
User Coordinate System and the Z-Axis

Introduction	2-2
The Floor Plan design	2-2
Starting up AutoCAD	2-3
Drawing Units Setup	2-4
Drawing Area Setup	2-5
GRID and *SNAP* intervals Setup	2-7
Drawing Polylines	2-8
Creating an offset polyline	2-9
Creating interior walls	2-10
Completing the doorway using the *TRIM* command	2-13
User Coordinate System – It is an XY CRT, but an XYZ World	2-15
Viewing the 2D Design in 3D space	2-16
Adding the 3rd dimension to the Floor Plan Design	2-17

Viewing the design using the Hide option	2-19
Adding New Layers	2-20
Moving entities to a different layer	2-21
Moving the UCS	2-21
Creating the roof	2-23
Rotating the UCS	2-25
Sketching on the Rotated UCS	2-26
Viewing the design using the Hidden option	2-27
Questions	2-28
Exercises	2-29

Chapter 3
3D Wireframe Modeling

Introduction	3-2
The Locator Design	3-3
Starting up AutoCAD	3-4
Using the Startup Options	3-4
Creating the rectangular base of the design	3-6
Create a 3D box	3-7
Object Snap Toolbar	3-10
Using the Snap options to locate the top corners	3-10
Using the Copy option to create additional edges	3-12
Using the *TRIM* Command	3-14
Using the Trim-Project-View option	3-16
Using the View Toolbar	3-18
Dynamic Rotation – 3D orbit	3-18
Using the *OFFSET* command to create parallel edges	3-19
Creating a circle above the UCS sketch plane	3-21
Completing the wireframe model	3-22
Questions	3-24
Exercises	3-25

Chapter 4
UCS, Viewports and Wireframe Modeling

Introduction	4-2
The V-block design	4-2
Starting Up AutoCAD	4-3
Layers setup	4-4
Creating the rectangular base of the design	4-5
Creating a wireframe Box	4-6
Using the View and the UCS Toolbars	4-7
Creating construction lines in the front view	4-8

Copying in the negative Z direction   4-10
Creating an Inclined Line at the Base of the Model   4-11
Creating object lines   4-12
Multiple Viewports   4-14
Using the *MIRROR* command   4-15
Turn *OFF* the Construction Lines   4-17
Creating a NEW UCS   4-18
Creating a New Named View   4-20
Creating the V-Cut feature on the inclined plane   4-22
Extend the Cut and GRIP Editing   4-23
Questions   4-25
Exercises   4-26

Chapter 5
3D Surface Modeling

Introduction   5-2
Starting up AutoCAD   5-4
Using the UCS and Surfaces Toolbars   5-5
Creating a surface using the *2D SOLID* command   5-5
Using the Shade Toolbar   5-8
Creating a surface using the *3D FACE* command   5-10
Creating a surface of irregular shape   5-12
Using the *Invisible Edge* option   5-14
The *Locator* wireframe model   5-16
Moving Objects to a different Layer   5-16
Predefined Surface models   5-18
Advanced Surface Modeling commands   5-20
Using the *Tabulated Surface* option   5-23
Using the *Ruled Surface* option   5-24
Questions   5-28
Exercises   5-29

Chapter 6
Solid Modeling - Constructive Solid Geometry

Introduction   6-2
The Guide-Block Design   6-2
Constructive Solid Geometry Concept   6-3
Binary Tree   6-4
The Guide-Block CSG Binary Tree   6-5
Starting Up AutoCAD® 2006   6-6
Creating the First 3-D object   6-7
Viewing the 3-D Block   6-8

Creating the Second Solid Feature	6-9
Boolean Operation - *UNION*	6-10
Creating the Second Cylinder Feature	6-11
Boolean Operation - *SUBTRACT*	6-12
Creating Another Solid Feature	6-13
Shaded Options	6-14
Creating the Final Feature	6-15
Rotating the Rectangular Block	6-16
Moving the Rectangular Block	6-17
Questions	6-18
Exercises	6-19

Chapter 7
Regions, Extrude and Solid Modeling

Introduction	7-2
The V-blockS design	7-2
Starting Up AutoCAD	7-3
Layers setup	7-4
Setting up a 2D sketch	7-5
Defining the Front edges of the design	7-7
Creating a Region	7-10
Solids Toolbar	7-11
Creating a 2D sketch at the Base of the Model	7-13
Creating a Copy of the 2D sketch	7-14
Creating the cutter solids	7-15
Boolean operation - *Subtract*	7-16
Mass properties of the solid model	7-17
Align the UCS to the inclined face	7-18
Creating the V-cut	7-19
Questions	7-22
Exercises	7-23

Chapter 8
Multiview Drawings From 3D Models

Introduction	8-2
The V-block design	8-2
Starting Up AutoCAD	8-3
AutoCAD Paper Space	8-4
Deleting the displayed viewport	8-5
Adding borders and title block in the layout	8-6
Setting up Viewports inside the Titleblock	8-7
Setting up the standard views	8-8

Determining the necessary 2D views	8-10
Establishing an auxiliary view in Model Mode	8-11
Adding a viewport for an Auxiliary view	8-13
Using the *DVIEW* command	8-16
Adjusting the Viewport Scale	8-17
Locking the Base view	8-18
Aligning the 2D views	8-18
Creating 2D projected entities - SOLPROF	8-21
Completing the 2D drawing	8-23
Questions	8-25
Exercises	8-26

Chapter 9
Symmetrical Features in Designs

Introduction	9-2
A Revolved Design: PULLEY	9-2
Modeling Strategy - A Revolved Design	9-3
Starting Up AutoCAD	9-4
Layers Setup	9-5
Setting up a 2D sketch for the Revolved feature	9-6
Perform 2D Boolean Operations	9-8
Creating the Revolved Feature	9-12
Mirroring Part	9-13
Combining Parts	9-14
3D Array	9-15
Position and perform the Cut	9-18
Questions	9-20
Exercises	9-21

Chapter 10
Advanced Modeling Tools & Techniques

Introduction	10-2
A Thin-walled Design: OIL SINK	10-2
Modeling Strategy	10-3
Starting Up AutoCAD	10-4
Layers Setup	10-5
The First Extruded Feature	10-6
Create an Offset geometry from an Extracted surface	10-8
Extrude with Draft Angle	10-10
Aligning the parts	10-11
Create another Extracted Surface	10-12
Combining parts – Boolean UNION	10-15

Creating 3D Rounds and Fillets	10-16
Creating a Shell Feature	10-18
Creating a Rectangular Array Cut feature	10-19
Creating Another Rectangular Array cut feature	10-21
Conclusion	10-23
Questions	10-24
Exercises	10-25

Index

Chapter 1
Getting Started

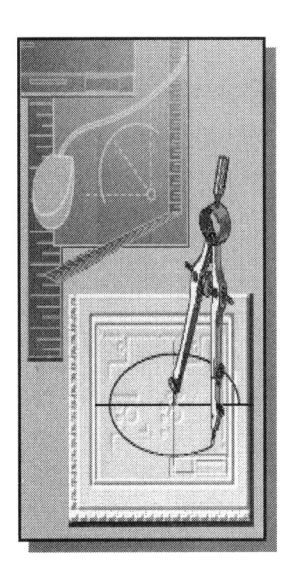

Learning Objectives

- **Development of Computer Aided Design**
- **Why Use AutoCAD 2006?**
- **Getting Started with AutoCAD 2006**
- **The AutoCAD Startup Dialog Box and Units Setup**
- **AutoCAD 2006 Screen Layout**
- **Mouse Buttons**

Introduction

Computer Aided Design (CAD) is the process of doing designs with the aid of computers. This includes the generation of computer models, analysis of design data, and the creation of the necessary drawings. **AutoCAD® 2006** is a computer aided design software developed by *Autodesk Inc*. The **AutoCAD® 2006** software is a tool that can be used for design and drafting activities. The two-dimensional and three-dimensional models created in **AutoCAD® 2006** can be transferred to other computer programs for further analysis and testing. The computer models can also be used in manufacturing equipment such as machining centers, lathes, mills, or rapid prototyping machines to manufacture the product.

The rapid changes in the field of **computer aided engineering** (CAE) have brought exciting advances in industry. Recent advances have made the long-sought goal of reducing design time, producing prototypes faster, and achieving higher product quality closer to a reality.

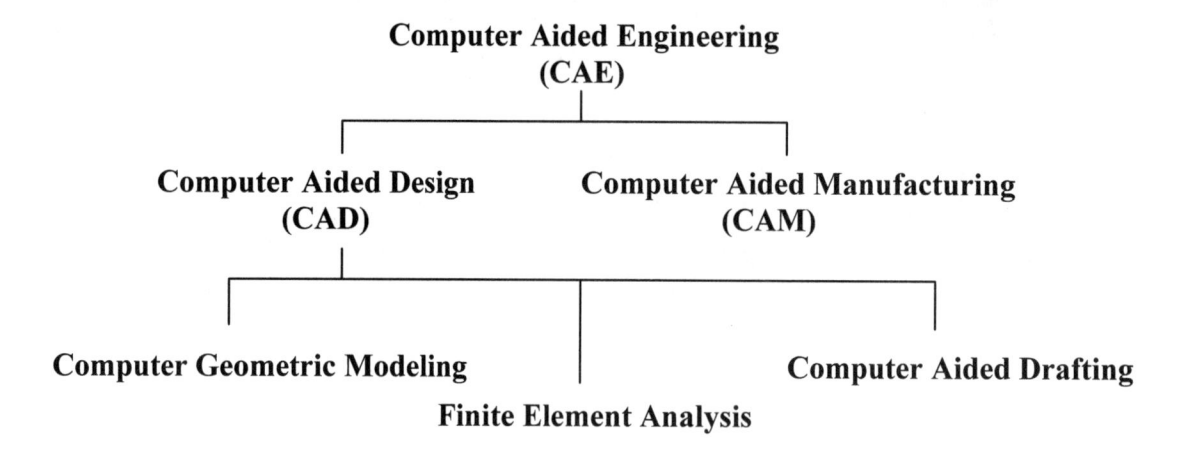

```
              Computer Aided Engineering
                       (CAE)
            ┌────────────┴────────────┐
     Computer Aided Design    Computer Aided Manufacturing
            (CAD)                     (CAM)
     ┌──────┴──────┐
Computer Geometric Modeling        Computer Aided Drafting
              Finite Element Analysis
```

Development of Computer Geometric Modeling

Computer Aided Design is a relatively new technology and its rapid expansion in the last fifty years is truly amazing. Computer modeling technology advanced along with the development of computer hardware. The first generation CAD programs, developed in the 1950s, were mostly non-interactive; CAD users were required to create program codes to generate the desired two-dimensional (2D) geometric shapes. Initially, the development of CAD technology occurred mostly in academic research facilities. The Massachusetts Institute of Technology, Carnegie-Mellon University, and Cambridge University were the lead pioneers at that time. The interest in CAD technology spread quickly and several major industry companies, such as General Motors, Lockheed, McDonnell, IBM, and Ford Motor Co., participated in the development of interactive CAD programs in the 1960s. Usage of CAD systems was primarily in the automotive industry, aerospace industry, and government agencies that developed their own programs for their specific needs. The 1960s also marked the beginning of the development of finite element analysis methods for computer stress analysis and computer aided manufacturing for generating machine toolpaths.

The 1970s are generally viewed as the years of the most significant progress in the development of computer hardware, namely the invention and development of **microprocessors**. With the improvement in computing power, new types of 3D CAD programs that were user-friendly and interactive became reality. CAD technology quickly expanded from very simple **computer aided drafting** to very complex **computer aided design**. The use of 2D and 3D wireframe modelers was accepted as the leading edge technology that could increase productivity in industry. The developments of surface modeling and solid modeling technology were taking shape by the late 1970s; but the high cost of computer hardware and programming slowed the development of such technology. During this time period, the available CAD systems all required extremely expensive room-sized mainframe computers.

In the 1980s, improvements in computer hardware brought the power of mainframes to the desktop at less cost and with more accessibility to the general public. By the mid-1980s, CAD technology had become the main focus of a variety of manufacturing industries and was very competitive with traditional design/drafting methods. It was during this period of time that 3D solid modeling technology had major advancements, which boosted the usage of CAE technology in industry.

In the 1990s, CAD programs evolved into powerful design/manufacturing/management tools. CAD technology has come a long way, and during these years of development, modeling schemes progressed from two-dimensional (2D) wireframe to three-dimensional (3D) wireframe, to surface modeling, to solid modeling and, finally, to feature-based parametric solid modeling.

The first generation CAD packages were simply 2D **Computer Aided Drafting** programs, basically the electronic equivalents of the drafting board. For typical models, the use of this type of program would require that several to many views of the objects be created individually as they would be on the drafting board. The 3D designs remained in the designer's mind, not in the computer database. The mental translation of 3D objects to 2D views is required throughout the use of the packages. Although such systems have some advantages over traditional board drafting, they are still tedious and labor intensive. The need for the development of 3D modelers came quite naturally, given the limitations of the 2D drafting packages.

The development of the 3D wireframe modeler was a major leap in the area of computer modeling. The computer database in the 3D wireframe modeler contains the locations of all the points in space coordinates and it is sufficient to create just one model rather than multiple models. This single 3D model can then be viewed from any direction as needed. The 3D wireframe modelers require the least computer power and achieve reasonably good representation of 3D models. But because surface definition is not part of a wireframe model, all wireframe images have the inherent problem of ambiguity.

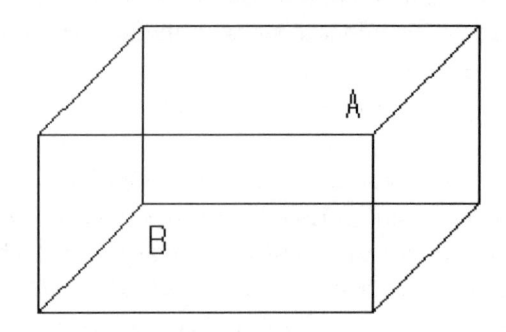

Wireframe Ambiguity: Which corner is in front, A or B?

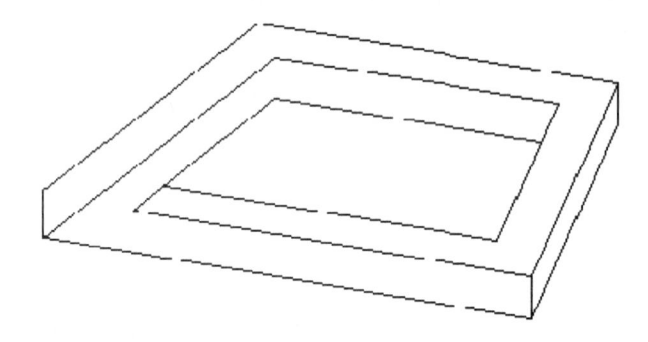

A non-realizable object: Wireframe models contain no surface definitions.

Surface modeling is the logical development in computer geometry modeling to follow the 3D wireframe modeling scheme by organizing and grouping edges that define polygonal surfaces. Surface modeling describes the part's surfaces but not its interiors. Designers are still required to interactively examine surface models to insure that the various surfaces on a model are contiguous throughout. Many of the concepts used in 3D wireframe and surface modelers are incorporated in the solid modeling scheme, but it is solid modeling that offers the most advantages as a design tool.

In the solid modeling presentation scheme, the solid definitions include nodes, edges, and surfaces, and it is a complete and unambiguous mathematical representation of a precisely enclosed and filled volume. Unlike the surface modeling method, solid modelers start with a solid or use topology rules to guarantee that all of the surfaces are stitched together properly. Two predominant methods for representing solid models are **constructive solid geometry** (CSG) representation and **boundary representation** (B-rep).

The CSG representation method can be defined as the combination of 3D solid primitives. What constitutes a "primitive" varies somewhat with the software but typically includes a rectangular prism, a cylinder, a cone, a wedge, and a sphere. Most solid modelers allow the user to define additional primitives, which can be very complex.

In the B-rep representation method, objects are represented in terms of their spatial boundaries. This method defines the points, edges, and surfaces of a volume, and/or issues commands that sweep or rotate a defined face into a third dimension to form a solid. The object is then made up of the unions of these surfaces that completely and precisely enclose a volume.

By the 1990s, a new paradigm called *concurrent engineering* had emerged. With concurrent engineering, designers, design engineers, analysts, manufacturing engineers, and management engineers all work closely right from the initial stages of the design. In this way, all aspects of the design can be evaluated and any potential problems can be identified right from the start and throughout the design process. Using the principles of concurrent engineering, a new type of computer modeling technique appeared. The technique is known as the *feature-based parametric modeling technique*. The key advantage of the *feature-based parametric modeling technique* is its capability to produce very flexible designs. Changes can be made easily and design alternatives can be evaluated with minimum effort. Various software packages offer different approaches to feature-based parametric modeling, yet the end result is a flexible design defined by its design variables and parametric features.

In this text, we will follow a logical order, parallel to the development of computer geometric modeling, in learning the fundamental concepts and commands of **AutoCAD® 2006**. We will begin with creating 2½D models by modifying the entities properties of 2D designs and then move toward creating 3D wireframe and surface models. We will also discuss and demonstrate several procedures for creating three-dimensional solid models. The techniques presented in this text will also serve as the foundation for entering the world of advanced three-dimensional solid modeling using packages such as **AutoCAD Mechanical Desktop, AutoCAD Architectural Desktop** and **Autodesk Inventor**.

Why use AutoCAD® 2006?

AutoCAD® was first introduced to the public in late 1982, and was one of the first CAD software products that were available for personal computers. Since 1984, **AutoCAD®** has established a reputation for being the most widely used PC-based CAD software around the world. By 2003, it is estimated that there are over 3.5 million **AutoCAD®** users in more than 150 countries worldwide. **AutoCAD® 2006** is the twentieth release, with many added features and enhancements, of the original **AutoCAD®** software produced by *Autodesk Inc*.

CAD provides us with a wide range of benefits; in most cases, the result of using CAD is increased accuracy and productivity. First of all, the computer offers much higher accuracy than the traditional methods of drafting and design. Traditionally, drafting and detailing are the most expensive cost element in a project and the biggest bottleneck. With CAD systems, such as **AutoCAD® 2006,** the tedious drafting and detailing tasks are simplified through the use of many of the CAD geometric construction tools, such as *grids*, *snap, trim,* and *auto-dimensioning*. Dimensions and notes are always legible in CAD drawings, and in most cases, CAD systems can produce higher quality prints compared to traditional hand drawings.

CAD also offers much-needed flexibility in design and drafting. A CAD model generated on a computer consists of numeric data that describe the geometry of the object. This allows the designers and clients to see something tangible and to interpret the ramifications of the design. In many cases, it is also possible to simulate operating conditions on the computer and observe the results. Any kind of geometric shape stored in the database can be easily duplicated. For large and complex designs and drawings, particularly those involving similar shapes and repetitive operations, CAD approaches are very efficient and effective. Because computer designs and models can be altered easily, a multitude of design options can be examined and presented to a client before any construction or manufacturing actually takes place. Making changes to a CAD model is generally much faster than making changes to a traditional hand drawing. Only the affected components of the design need to be modified and the drawings can be plotted again. In addition, the greatest benefit is that, once the CAD model is created, it can be used over and over again. The CAD models can also be transferred into manufacturing equipment such as machining centers, lathes, mills, or rapid prototyping machines to manufacture the product directly.

CAD, however, does not replace every design activity. CAD may help, but it does not replace the designer's experience with geometry and graphical conventions and standards for the specific field. CAD is a powerful tool, but the use of this tool does not guarantee correct results; the designer is still responsible for using good design practice and applying good judgment. CAD will supplement these skills to ensure that the best design is obtained.

CAD designs and drawings are stored in binary form, usually as CAD files, to magnetic devices such as diskettes and hard disks. The information stored in CAD files usually requires much less physical space in comparison to traditional hand drawings. However, the information stored inside the computer is not indestructible. On the contrary, the electronic format of information is very fragile and sensitive to the environment. Heat or cold can damage the information stored on magnetic storage devices. A power failure while you are creating a design could wipe out the many hours you spent working in front of the computer monitor. It is a good habit to save your work periodically, just in case something might go wrong while you are working on your design. In general, one should save one's work onto a storage device at an interval of every 15 to 20 minutes. You should also save your work before you make any major modifications to the design. It is also a good habit to periodically make backup copies of your work and put them in a safe place.

Getting Started with AutoCAD® 2006

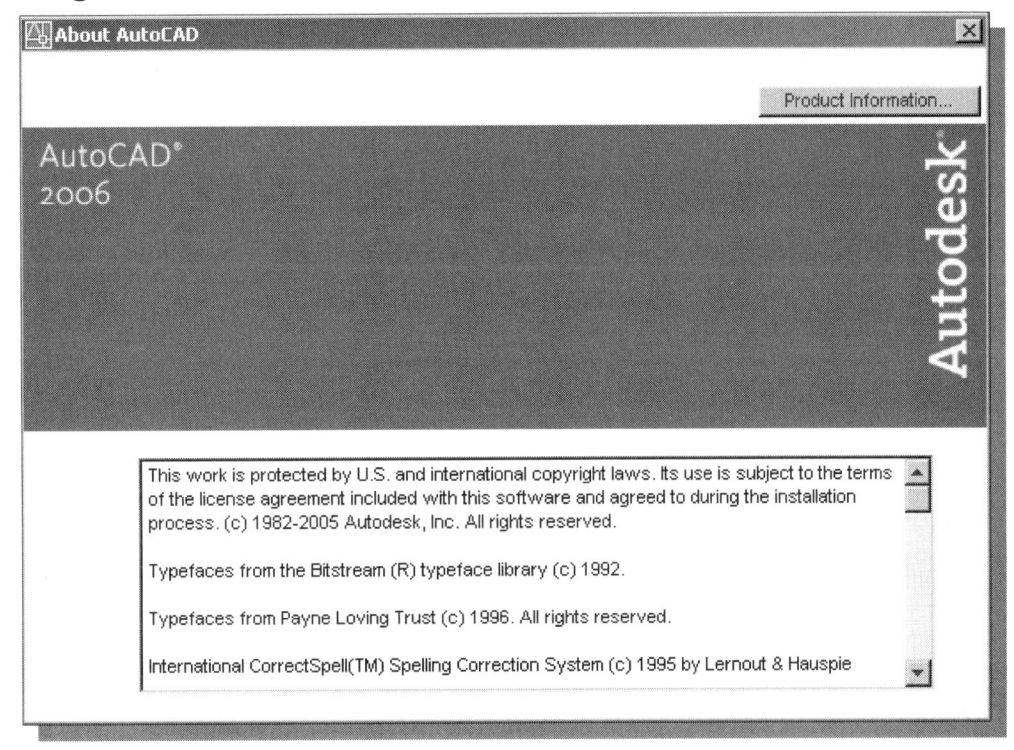

How to start **AutoCAD® 2006** depends on the type of workstation and the particular software configuration you are using. With most *Windows* systems, you may select the **AutoCAD 2006** option on the *Start* menu or select the **AutoCAD 2006** icon on the *Desktop*. Consult with your instructor or technical support personnel if you have difficulty starting the software.

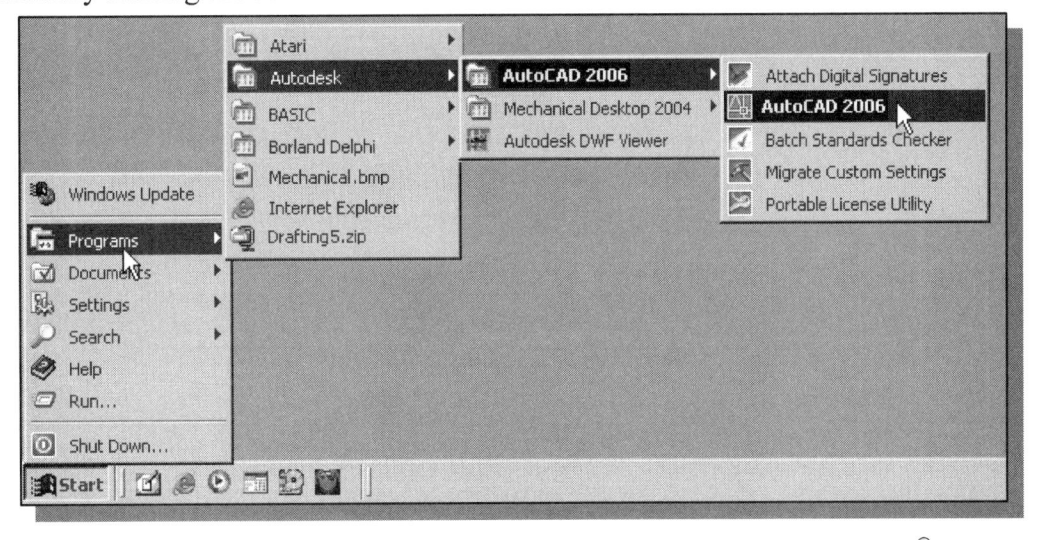

The program takes a while to load, so be patient. Eventually the **AutoCAD® 2006** main *drawing screen* will appear on the screen. The tutorials in this text are based on the assumption that you are using **AutoCAD® 2006**'s default settings. If your system has been customized for other uses, some of the settings may not work with the step-by-step instructions in the tutorials. Contact your instructor and/or technical support personnel to restore the default software configuration.

AutoCAD® 2006 Screen Layout

The default **AutoCAD® 2006** *drawing screen* contains the *pull-down* menus, the *Standard* toolbar, the *Object Properties* toolbar, the *Draw* toolbar, the *Modify* toolbar, the *command prompt area*, the *Status Bar* and the *Tool Palettes window*. A line of quick help text appears at the bottom of the window as you move the *mouse cursor* over different icons. You may resize the **AutoCAD® 2006** drawing window by click and drag at the edges of the window, or relocate the window by click and drag at the window title area.

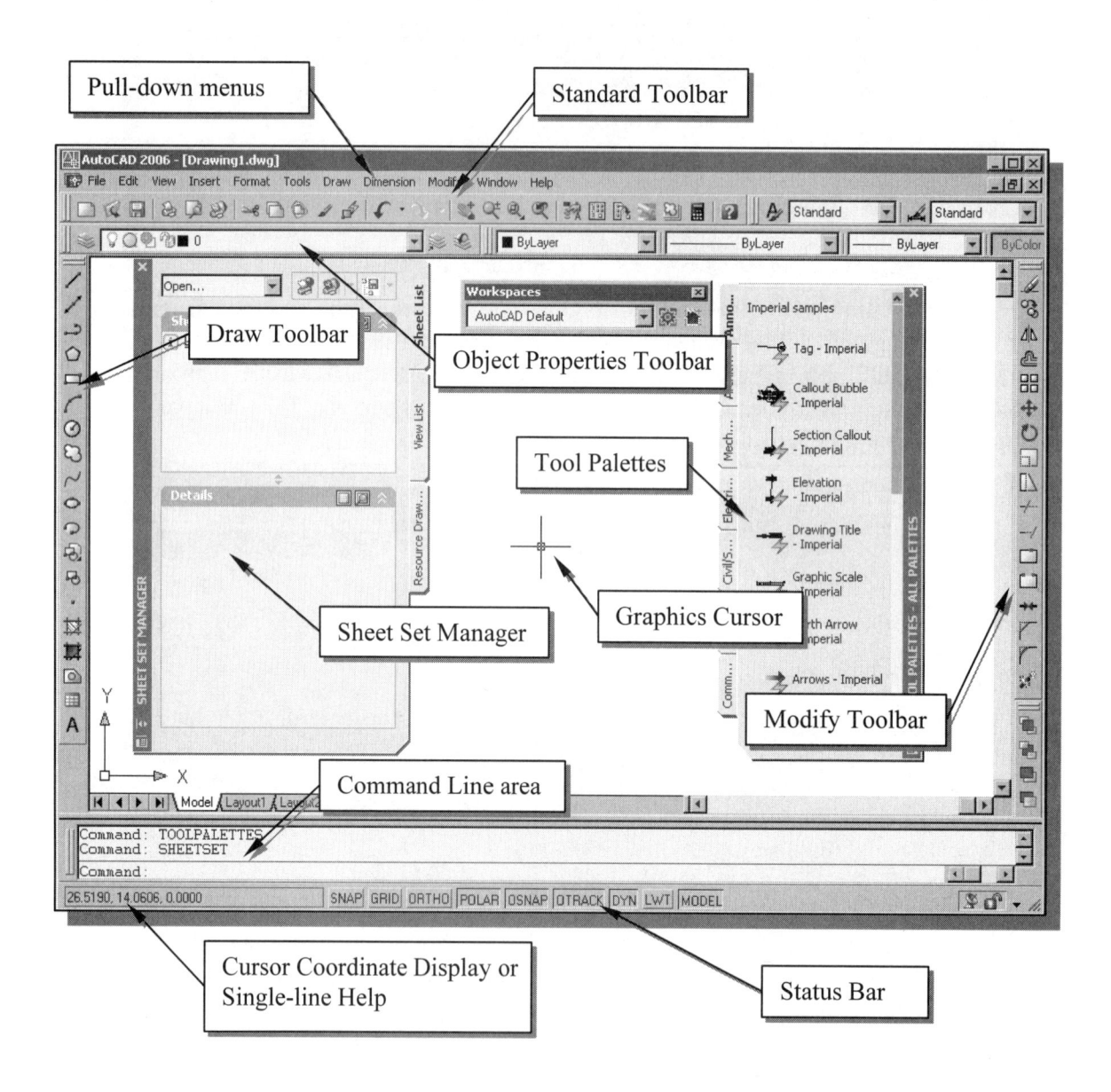

- **Pull-down Menus**
  The *pull-down* menus at the top of the main window contain operations that you can use for all modes of the system.

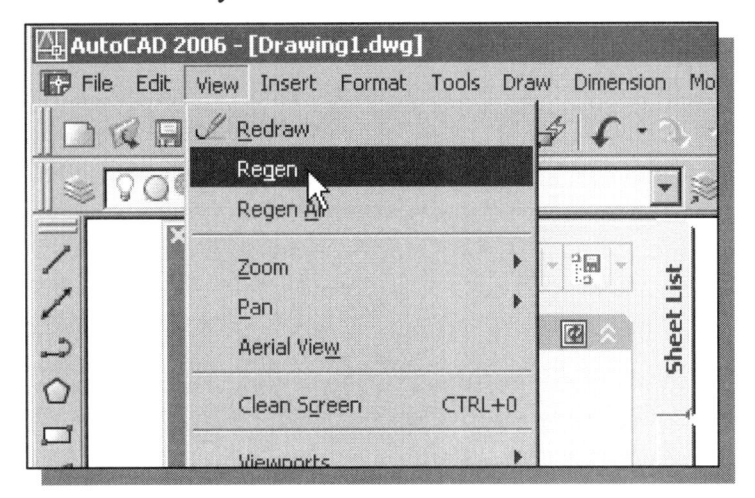

- **Standard Toolbar**
  The *Standard* toolbar at the top of the *AutoCAD* window allows us quick access to frequently used commands. We can customize the toolbar by adding and removing sets of options or individual commands.

- **Object Properties Toolbar**
  The *Object Properties* toolbar contains tools to help manipulate the graphical object properties, such as color, line type, and layer.

- **Graphics Window**
  The *graphics window* is the area where models and drawings are displayed.

- **Graphics Cursor or Crosshairs**
  The *graphics cursor*, or *crosshairs*, shows the location of the pointing device in the graphics window. The coordinates of the cursor are displayed at the bottom of the screen layout. The cursor's appearance depends on the selected command or option.

- **Command Line Area**
  The bottom section of the screen layout provides status information for an operation and it is also the area for data input.

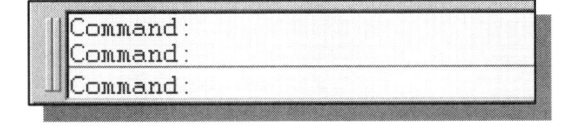

- **Draw Toolbar and Modify Toolbar**
  Additional toolbars are available in **AutoCAD® 2006**, and contain groups of buttons that allow us to pick commands quickly, without searching through a menu structure. The *Draw* toolbar and *Modify* toolbar contain icons for basic draw and modify commands.

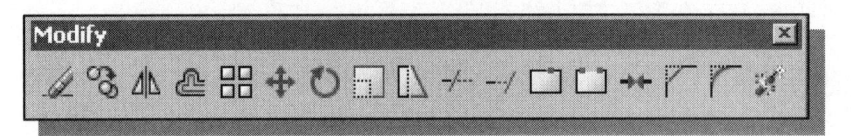

Mouse Buttons

AutoCAD® 2006 utilizes the mouse buttons extensively. In learning **AutoCAD® 2006**'s interactive environment, it is important to understand the basic functions of the mouse buttons. It is highly recommended that you use a mouse or a tablet with **AutoCAD® 2006** since the package uses the buttons for various functions.

- **Left mouse button**
  The **left-mouse-button** is used for most operations, such as selecting menus and icons, or picking graphic entities. One click of the button is used to select icons, menus and form entries, and to pick graphic items.

- **Right mouse button**
  The **right-mouse-button** is used to bring up additional available options. The software also utilizes the **right-mouse-button** as the same as the **ENTER** key, and is often used to accept the default setting to a prompt or to end a process.

- **Middle mouse button/wheel**
  The middle mouse button/wheel can be used to Pan (hold down the wheel button and drag the mouse) or Zoom (rotate the wheel) realtime.

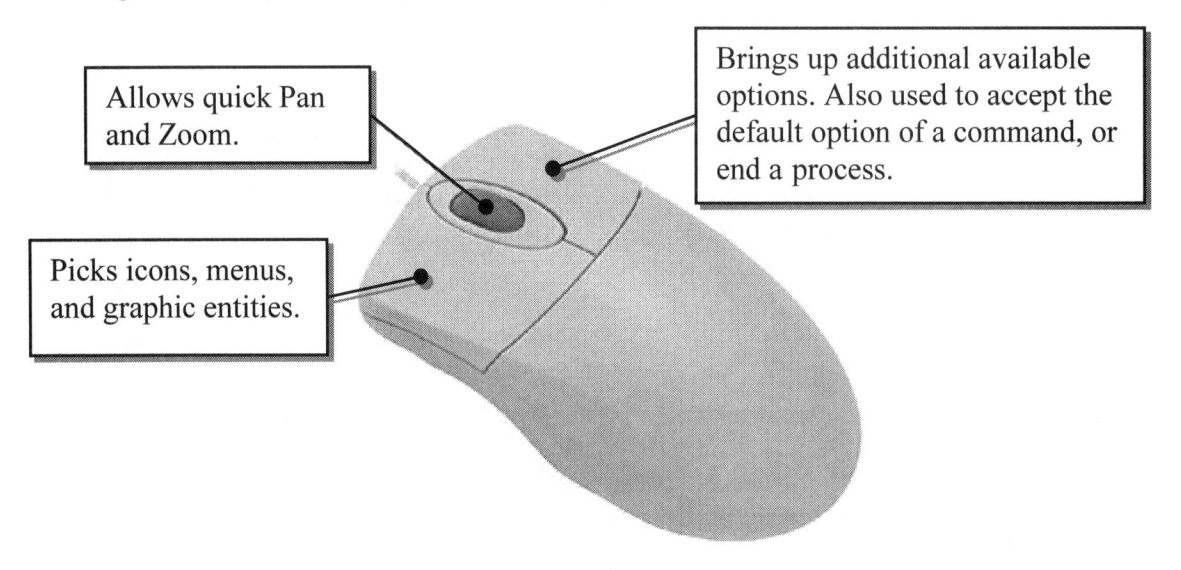

Allows quick Pan and Zoom.

Brings up additional available options. Also used to accept the default option of a command, or end a process.

Picks icons, menus, and graphic entities.

[Esc] – Canceling commands

The [**Esc**] key is used to cancel a command in **AutoCAD® 2006**. The [**Esc**] key is located near the top-left corner of the keyboard. Sometimes, it may be necessary to press the [**Esc**] key twice to cancel a command; it depends on where we are in the command sequence. For some commands, the [**Esc**] key is used to exit the command.

On-Line Help

- ❖ Several types of on-line help are available at any time during an **AutoCAD® 2006** session. The **AutoCAD® 2006** software provides many on-line help options:

- **Info Palette**:

The **Info Palette** option provides an instant help that dynamically displays information on the activated command. The guidance from the **Info Palette** system enables users to quickly get started on performing desired tasks. In the **Info Palette** window, links to additional information are also available. Clicking a blue-text-link expands the current **Info Palette** topic.

- To *close* the **Info Palette**, left-click the [**X**] icon located in the upper left corner of the *Info Palette* window as shown.

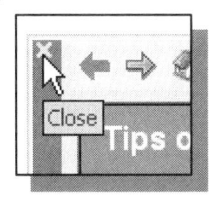

- **Additional Resources**: Click on the **HELP** option in the pull-down menu to access the **AutoCAD® 2006 Additional Resources**. Notice the different online resources available in the pull-down list.

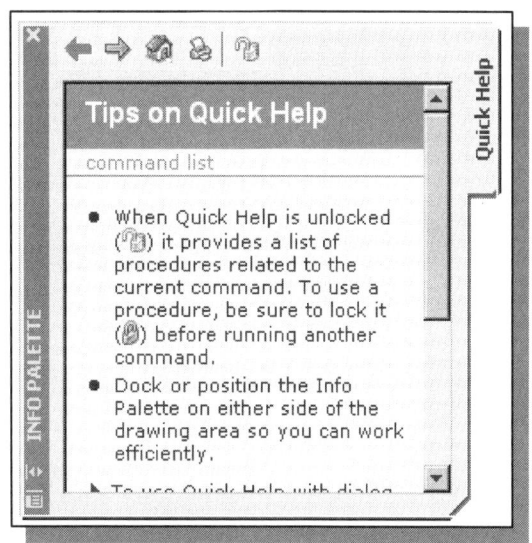

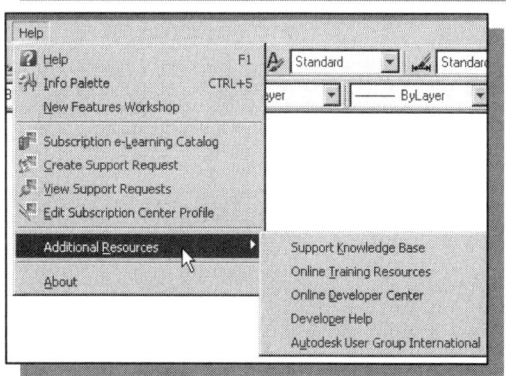

- **Standard toolbar**: Click on the **[?]** icon in the *Standard* toolbar to access Autodesk On-line Help: User Documentation.

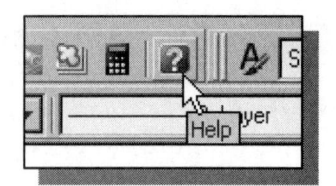

- **Command line and function key [F1]**: Press the **[F1]** key or enter a question mark **[?]** at the command prompt to access the **AutoCAD On-line Help system**.

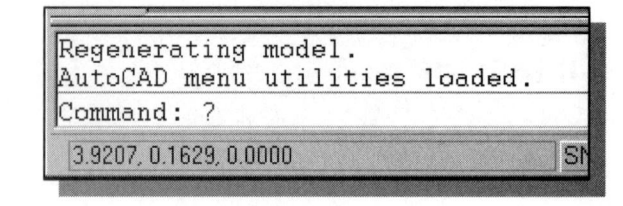

Leaving AutoCAD® 2006

➢ To leave **AutoCAD® 2006**, use the left-mouse-button and click on **File** at the top of the **AutoCAD® 2006** screen window, then choose **Exit** from the pull-down menu or type **QUIT** in the command prompt area.

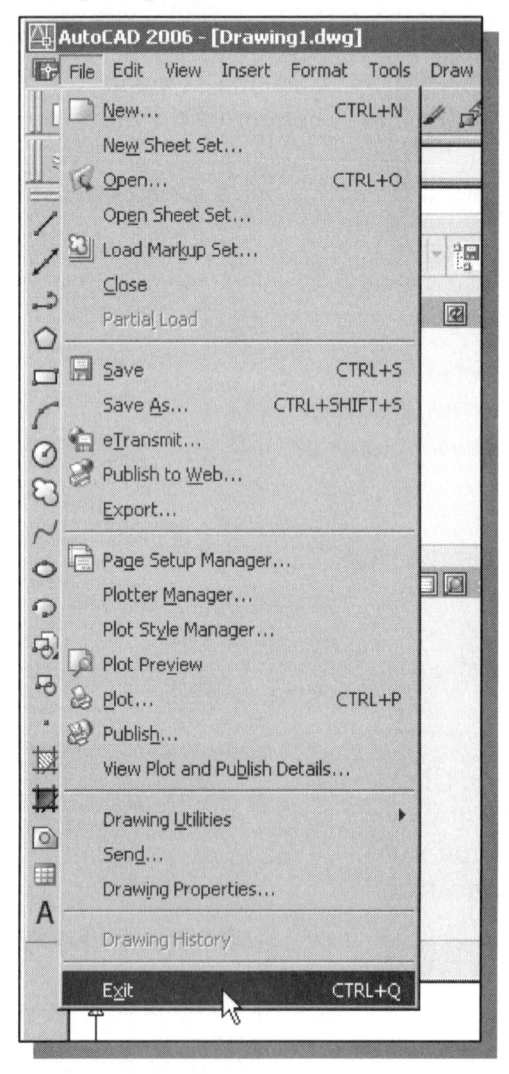

Creating a CAD File Folder

❖ It is a good practice to create a separate folder to store your CAD files. You should not save your CAD files in the same folder where the **AutoCAD® 2006** application is located. It is much easier to organize and backup your project files if they are in a separate folder. Making folders within this folder for different types of projects will help you organize your CAD files even further. When creating CAD files in **AutoCAD® 2006**, it is strongly recommended that you *save* your CAD files on the hard drive. However, if you do want to save your files on a floppy drive, be sure to exit the **AutoCAD® 2006** program before removing the diskette from the drive. The better alternative is to save the files on the hard drive and then copy the files onto a floppy diskette under the operating system.

➢ To create a new folder in the *Windows* environment:

1. In *My Computer*, or start the ***Windows Explorer*** under the *Start* menu, open the folder in which you want to create a new folder.

2. On the **File** menu, point to **New**, and then click **Folder**. The new folder appears with a temporary name.

3. Type a name for the new folder, and then press **ENTER**.

NOTES:

Chapter 2
User Coordinate Systems and the Z-Axis

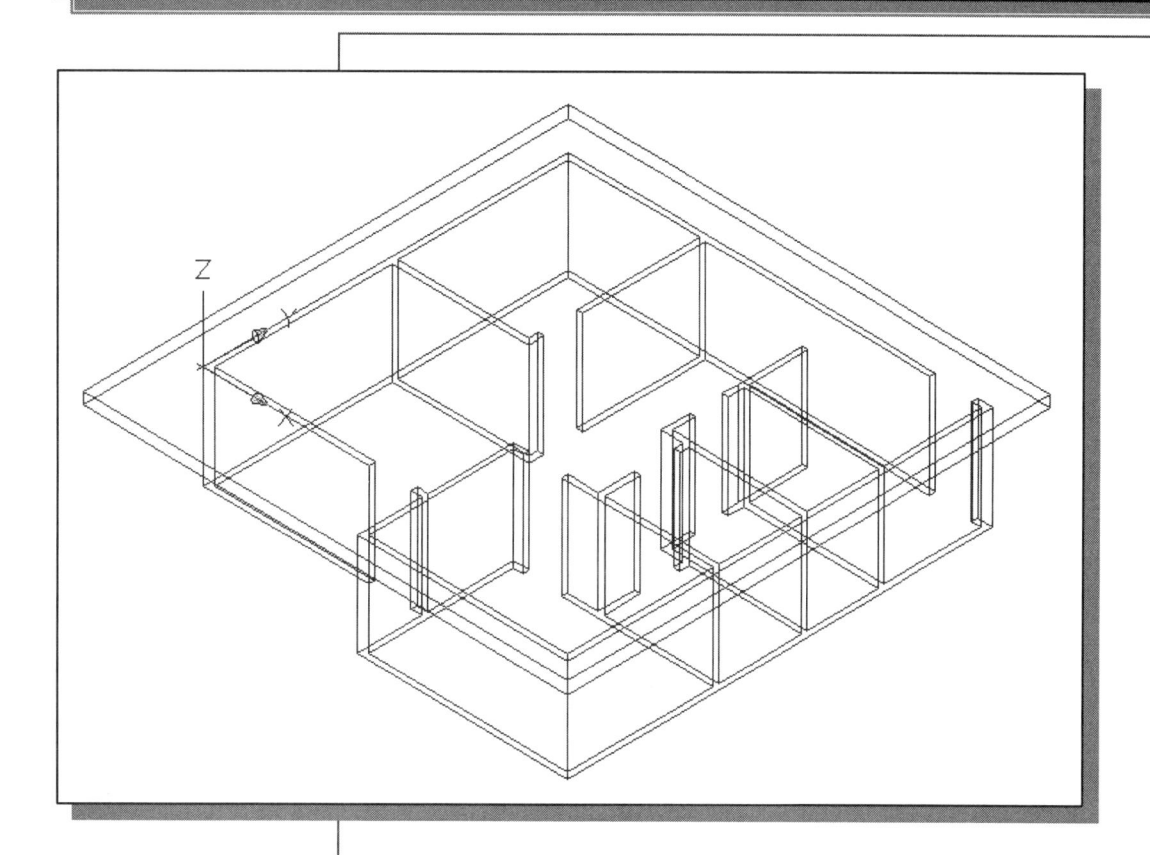

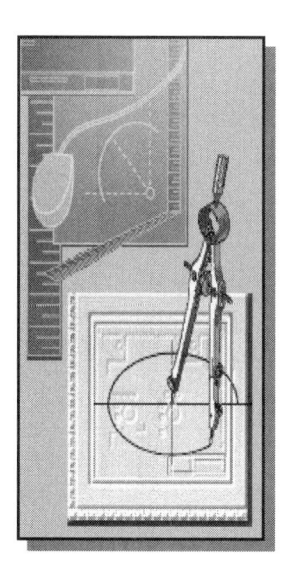

Learning Objectives

- Using the Drawing Units and Drawing Limits Commands
- Understand and Use the THICKNESS Option
- Understand the AutoCAD UCS
- Be able to Switch to Predefined Views
- Pre-selection of Objects
- Controlling Object Properties
- Viewing with the Hidden Option

Introduction

Creating three-dimensional models on a CAD system provides us a means to check the integrity of a design. Creating realistic three-dimensional models helps us visualize our final design much more clearly than we can with 2D representations of the design. Several different approaches are available in AutoCAD that allow us to quickly create 3D models. The simplest approach is to use the *Thickness* variable to perform an extrusion in the third dimension of 2D geometries. In AutoCAD, the *Thickness* variable can make certain 2D objects look like 3D objects. It is important to realize that this approach is somewhat limited and that the objects created are not true solids. This approach allows us to start considering the complexity involved in creating 3D models.

In this chapter, we will explore the virtual 3D environment in AutoCAD. Understanding how to maneuver in the virtual 3D environment and the use of a *User Coordinate System* (UCS) are the foundations and perhaps the most difficult parts of 3D modeling. We will also examine the basic viewing and rendering options that are available in AutoCAD. Specifically, we will use the Hide options to create hidden-line-removed images. The hidden-line-removed image makes it easier to visualize the model because the back faces are not displayed.

In this chapter, we will begin with the construction of a 2D floor plan design. This tutorial will also serve as a review of some of the 2D CAD construction techniques. The main goal of this chapter is to provide you with a basic understanding of the 3D environment of AutoCAD.

The *Floor Plan* Design

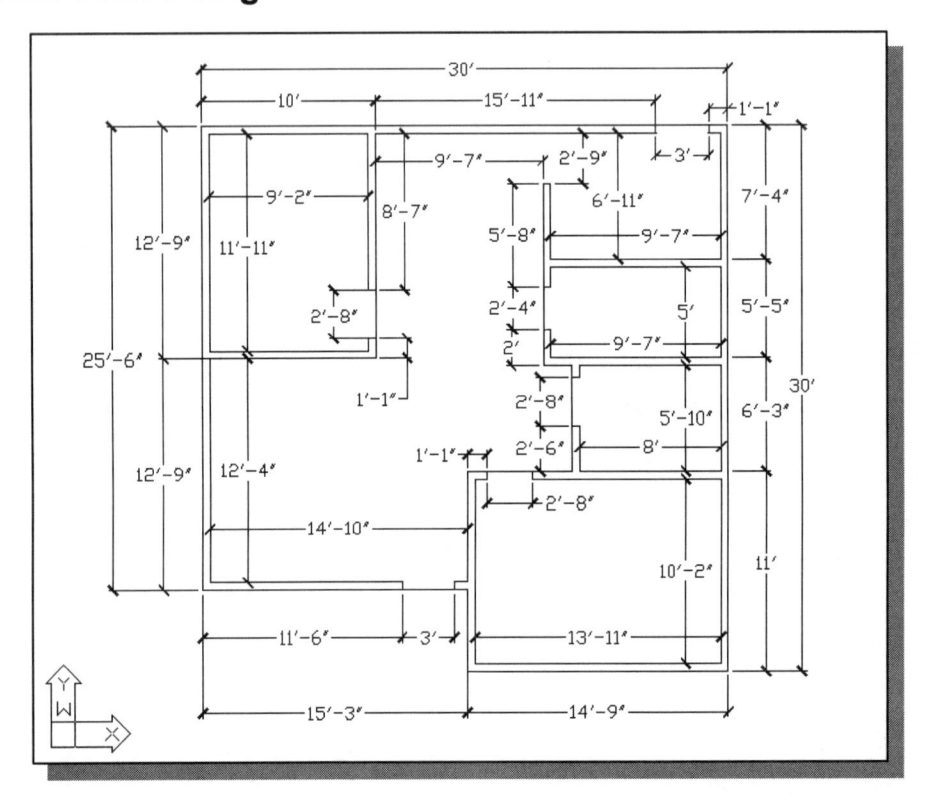

Starting Up AutoCAD® 2006

1. Select the **AutoCAD 2006** option on the *Program* menu or select the **AutoCAD 2006** icon on the *Desktop*. Once the program is loaded into memory, the **AutoCAD® 2006** main drawing screen will appear on the screen.

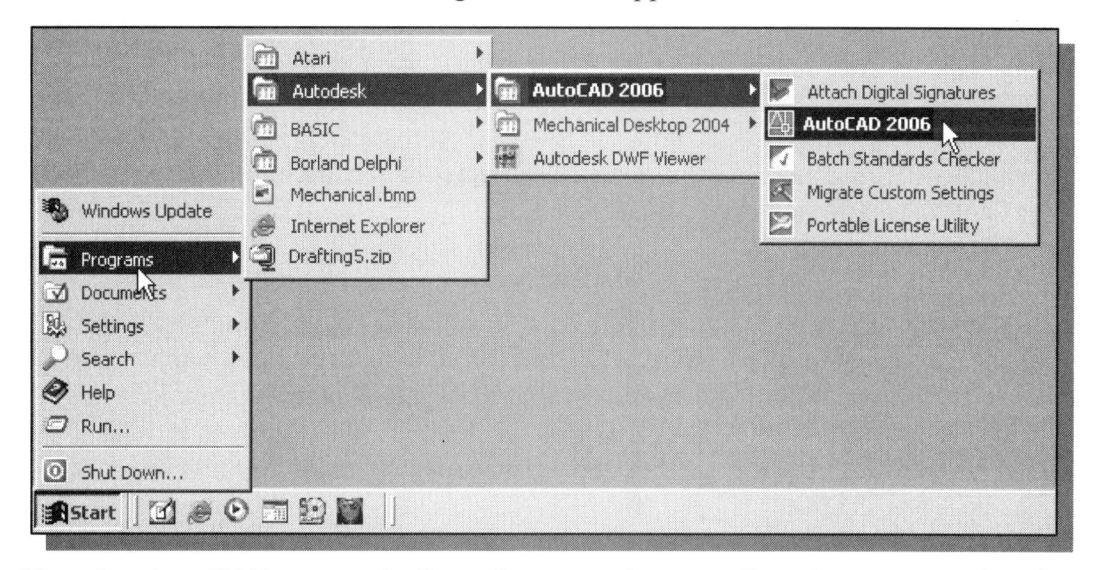

➢ Note that AutoCAD automatically assigns generic name, *Drawing x*, as new drawings are created. In our example, AutoCAD opened the graphics window using the default system units and assigned the drawing name *Drawing1*.

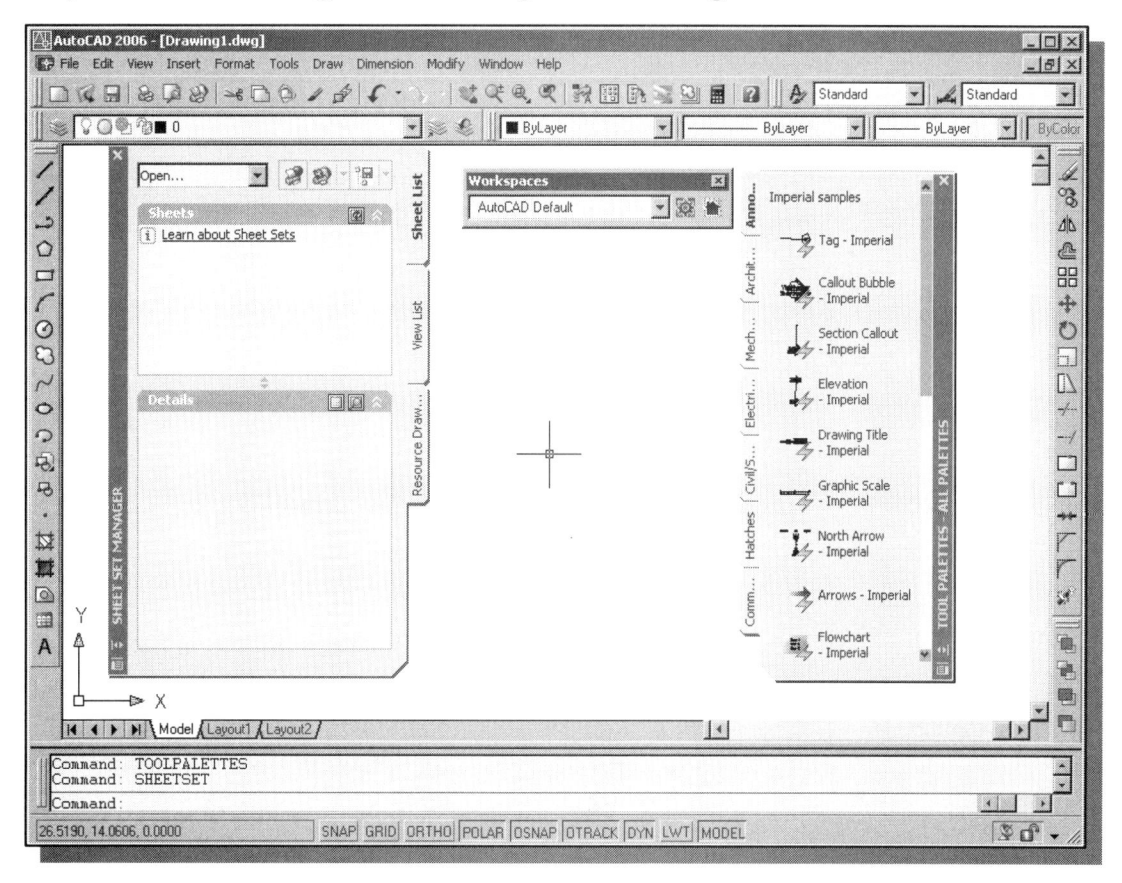

Drawing Units Setup

> Every object we construct in a CAD system is measured in **units**. We should determine the value of the units within the CAD system before creating the first geometric entities.

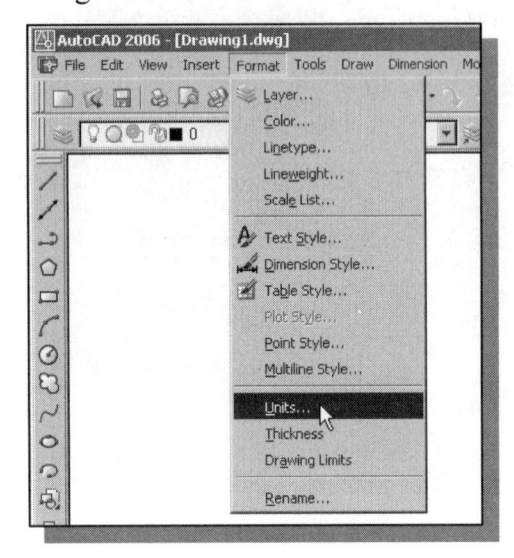

1. In the *pull-down* menus, select:

   [Format] → [Units]

2. In the *Drawing Units* dialog box, set the *Length Type* to **Architectural**. This will set the measurement to the default Architectural units, feet and inches.

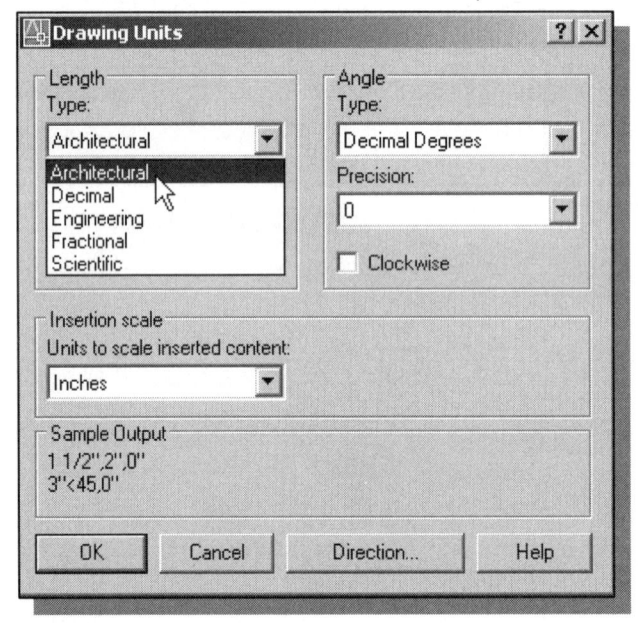

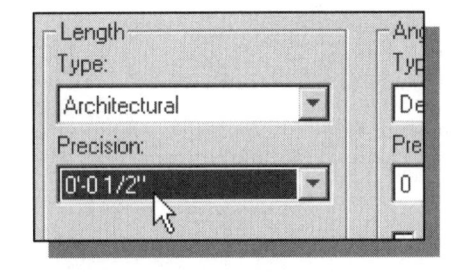

3. Set the *Precision* to **half-an-inch** as shown in the figure.

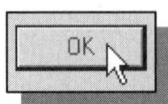

4. Pick **OK** to exit the *Drawing Units* dialog box.

Drawing Area Setup

❖ Next, we will setup the **Drawing Limits**; setting the Drawing Limits controls the extents of the display of the *grid*. It also serves as a visual reference that marks the working area. It can also be used to prevent construction outside the grid limits and also as a plot option that defines an area to be plotted/printed. Note that this setting does not limit the region for geometry construction.

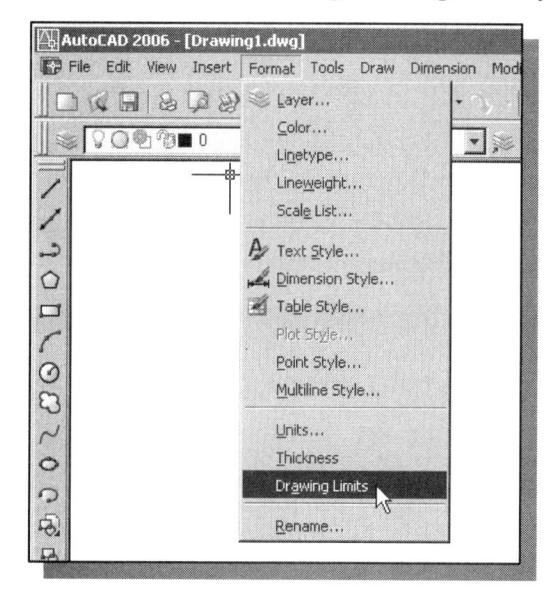

1. In the *pull-down* menus, select:

   [Format] → [Drawing Limits]

2. In the *command prompt* area, near the bottom of the AutoCAD drawing screen, the message "*Reset Model Space Limits: Specify lower left corner or [On/Off] <0'-0",0'-0">:*" is displayed. Press the **ENTER** key once to accept the default coordinates <**0'-0",0'-0"**>.

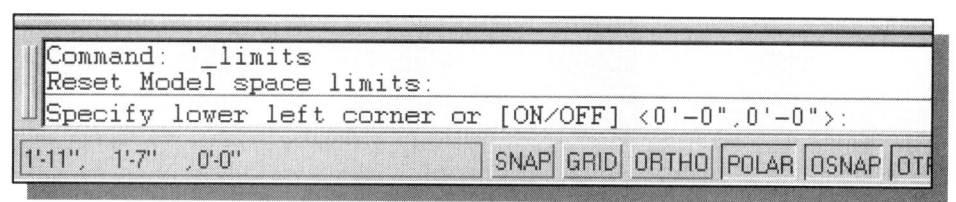

3. In the *command prompt* area, the message "*Specify upper right corner <1'-0", 0'-9">:*" is displayed. Enter **60',40'** to change the upper right coordinates to <**60',40'**>.

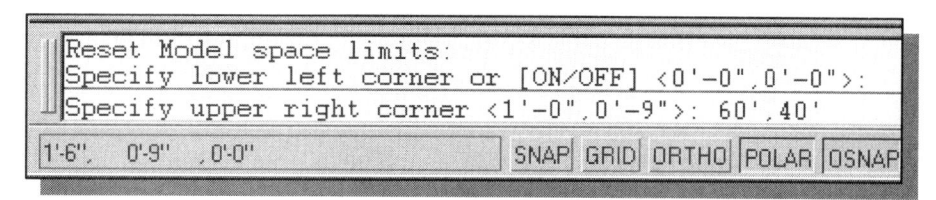

4. On your own, move the graphic cursor near the upper right corner inside the drawing area and note that the drawing area is unchanged. (The drawing limits command is used to set the drawing area; but the display will not be adjusted until a display command is used.)

5. In the pull-down menus, select:

   [View] → [Zoom] → [All]

   ❖ The **Zoom All** command can be used to adjust the display so that all objects in the drawing are displayed to be as large as possible. If no objects are constructed, the **Drawing Limits** are used to adjust the current viewport.

6. Move the graphic cursor near the upper right corner inside the drawing area and note that the display area is updated.

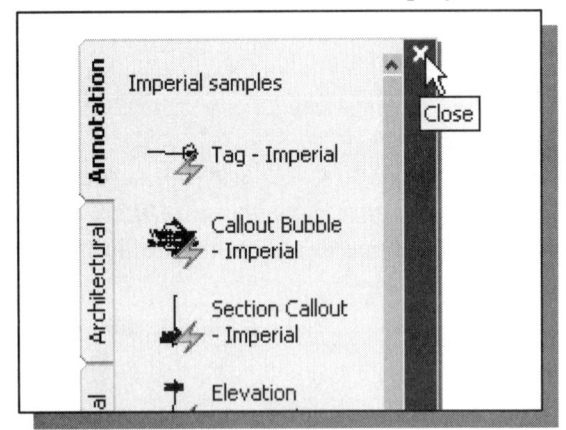

7. Close the *Tool Palettes* by clicking once on **Close** button located at the upper right corner of the window as shown.

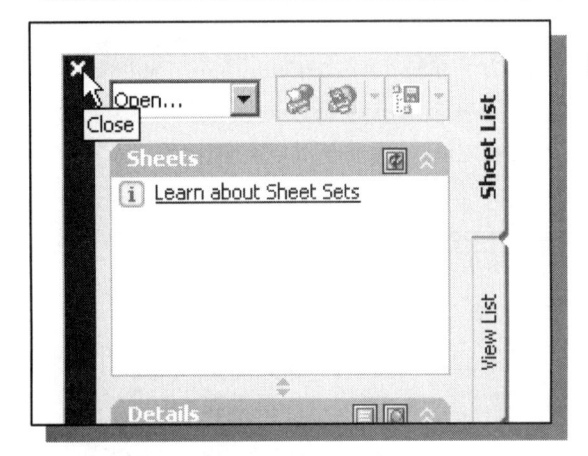

8. Close the *Sheet Set Manager* by clicking once on the **Close** button located at the upper right corner of the window as shown.

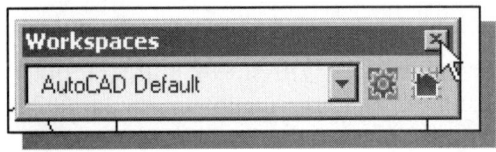

9. Close the *Workspaces* toolbar by clicking once on the **Close** button located at the upper right corner of the window as shown.

GRID and SNAP Intervals Setup

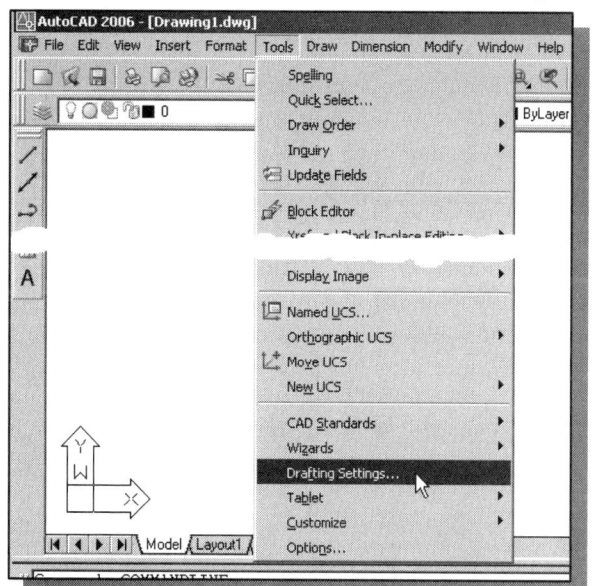

1. In the pull-down menus, select:
   [Tools] → **[Drafting Settings]**

2. In the *Drafting Settings* dialog box, select the **SNAP and GRID** tab if it is not the page on top.

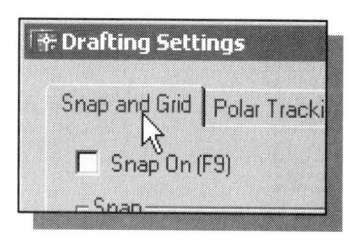

3. Change *Grid Spacing* to **6"** for both X and Y directions.

4. Also adjust the *Snap Spacing* to **6"** for both X and Y directions.

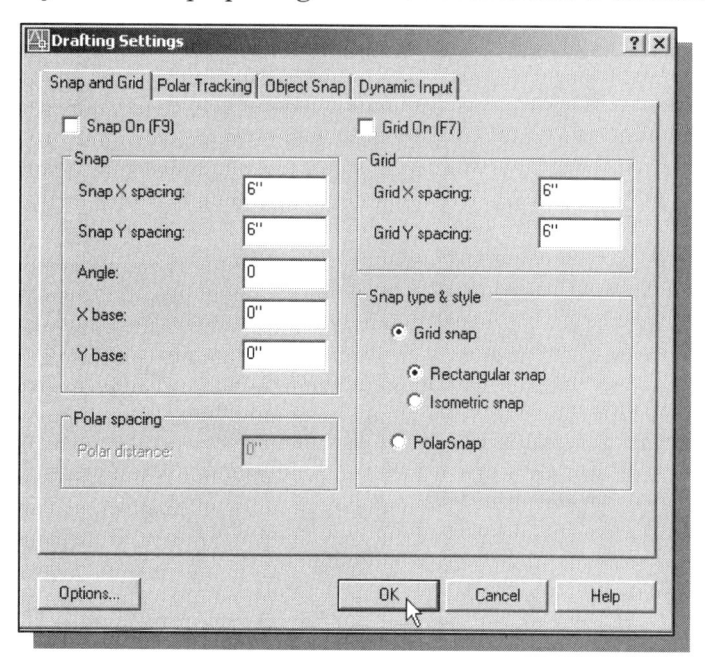

5. Pick **OK** to exit the *Drafting Settings* dialog box.

6. In the *Status Bar* area, reset the option buttons so that only *SNAP, GRID* and *MODEL* are switched *ON*.

Drawing Polylines

❖ In AutoCAD, a **polyline** is a connected sequence of line segments created as a single object. A polyline can contain straight line segments, arc segments, or a combination of the two.

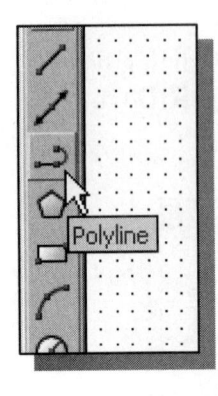

1. Select the **Polyline** command icon in the *Draw* toolbar.

2. In the command prompt area, the message "*Specify start point:*" is displayed. AutoCAD expects us to identify the starting location. Select a location that is near the coordinates (**25'-0", 10'-0"**) as the **starting point** of the polyline by left-clicking the mouse.

3. In the command prompt area, create a horizontal line by using the *relative rectangular coordinates entry method*, relative to the last point we specified
   Specify next point: **@-11'6",0 [ENTER]**

4. In the command prompt area, create a vertical line by using the *relative rectangular coordinates entry method*, relative to the last point we specified
   Specify next point: **@0,25'6" [ENTER]**

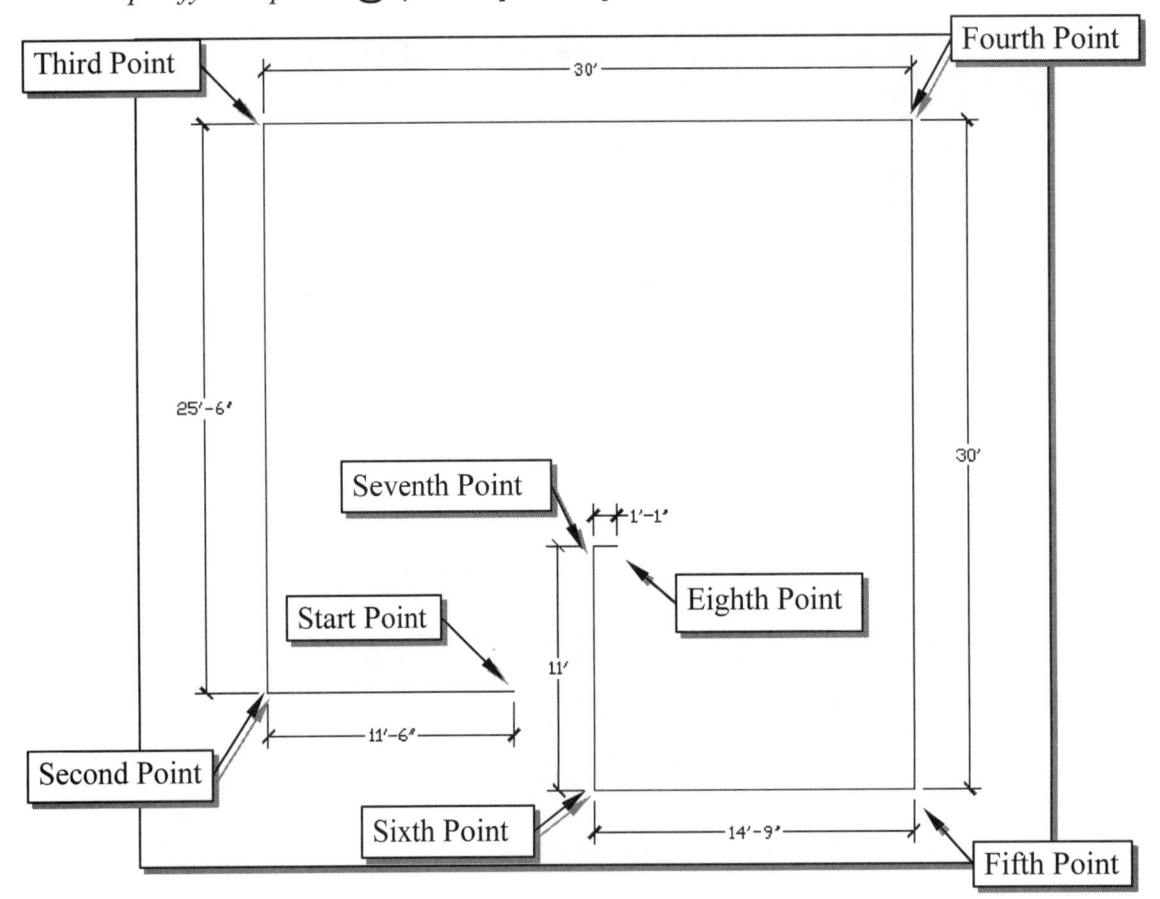

5. On your own, complete the polyline by specifying the rest of the points, *point four* through *point eight*, as shown in the above figure.

6. Inside the graphics window, right-mouse-click and select **Enter** to end the Polyline command.

Creating an Offset Polyline

- The **Offset** command creates a new object at a specified distance from an existing object or through a specified point.

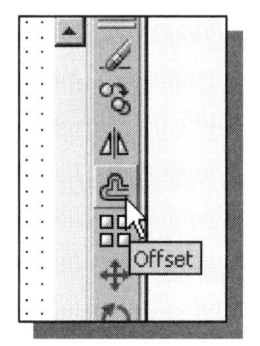

1. Select the **Offset** command icon in the *Modify* toolbar. In the command prompt area, the message "*Specify offset distance or [Through]:*" is displayed.

   Specify offset distance or [Through]: **6"** [ENTER]

2. In the command prompt area, the message "*Select object to offset or <exit>:*" is displayed. Pick the polyline on the screen. (Note that all line segments of the polyline are automatically selected.)

3. AutoCAD next asks us to identify the direction of the offset. Pick a location that is *inside* the polyline.

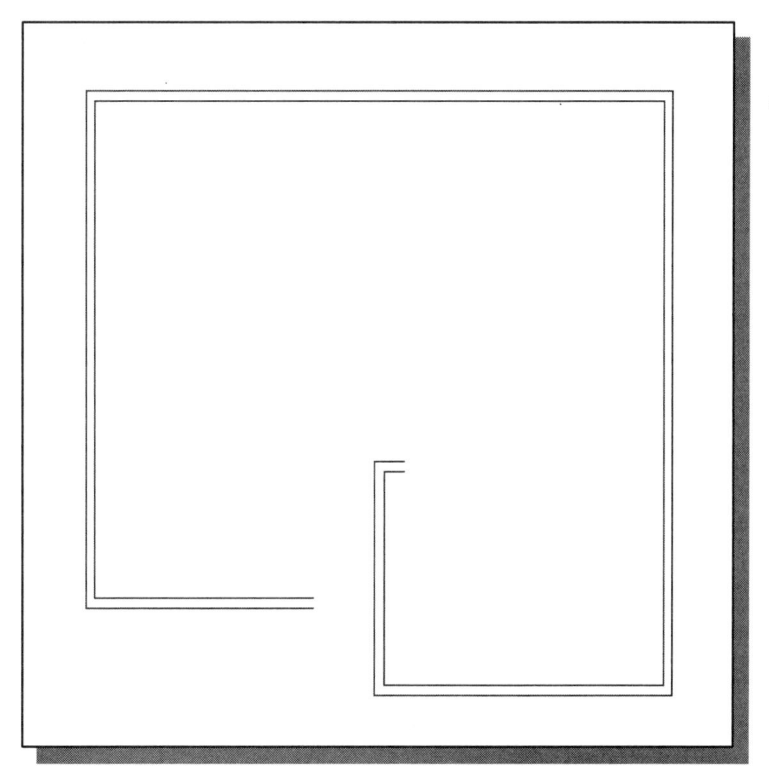

4. Inside the graphics window, **right-mouse-click** and select **Enter** to end the Offset command.

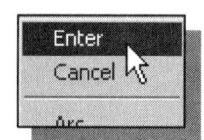

Creating Interior Walls

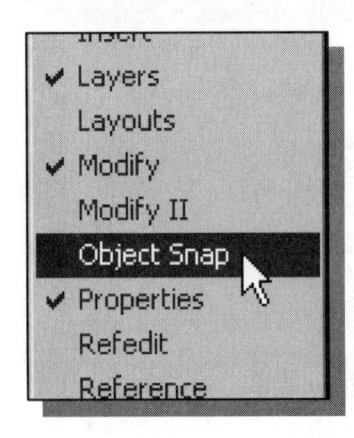

1. Move the cursor on any icon in the *Standard* toolbar area and **right-mouse-click** once to display a list of toolbar menu groups.

2. Select **Object Snap**, with the left-mouse-button, to display the *Object Snap* toolbar on the screen to assist the construction of the floor plan.

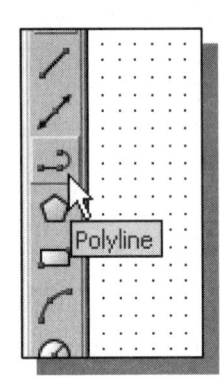

3. Select the **Polyline** command icon in the *Draw* toolbar. In the command prompt area, the message "*Specify start point or [Justification/Scale/Style]:*" is displayed.

4. In the *Object Snap* toolbar, pick **Snap to Midpoint**. In the command prompt area, the message "*_mid of*" is displayed. AutoCAD now expects us to select a geometric entity on the screen.

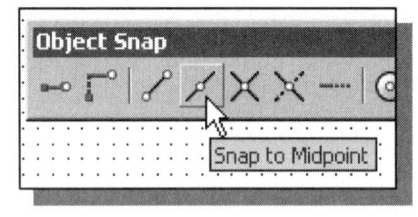

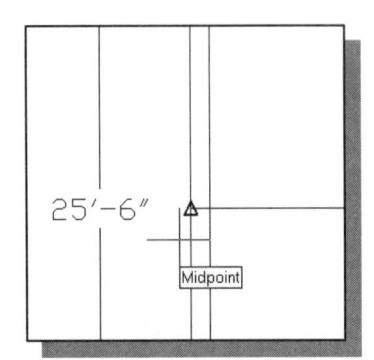

5. Select the vertical line on the left as shown.

6. At the command prompt, enter **@10',0 [ENTER]**.

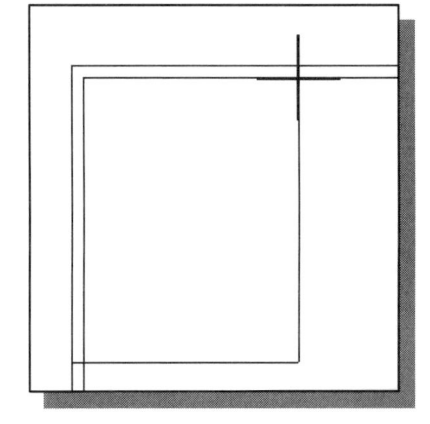

7. On your own, define a vertical line at the location as shown in the figure.

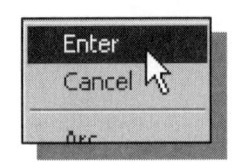

8. Inside the graphics window, **right-mouse-click** and select **Enter** to end the Polyline command.

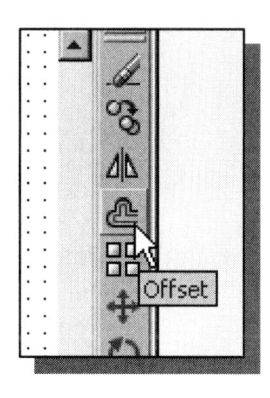

9. Select the **Offset** command icon in the *Modify* toolbar. In the command prompt area, the message "*Specify offset distance or [Through]:*" is displayed.

   Specify offset distance or [Through]: **6"** **[ENTER]**

10. In the command prompt area, the message "*Select object to offset or <exit>:*" is displayed. Pick the polyline we just created.

11. AutoCAD next asks us to identify the direction of the offset. Pick a location that is toward the upper left corner of the design.

12. Inside the graphics window, **right-mouse-click** to end the Offset command.

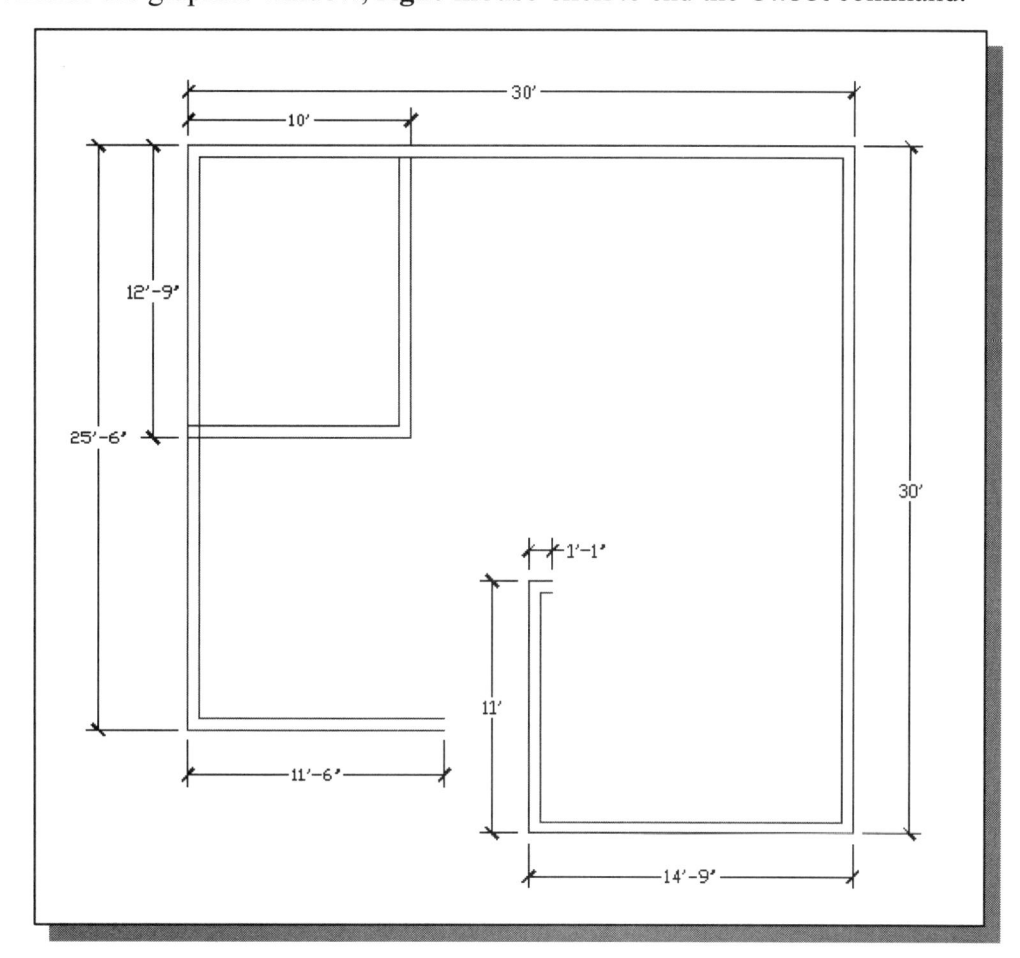

➢ In the *Status Bar* area, reset the option buttons so that only the *MODEL* button is switched *ON*.

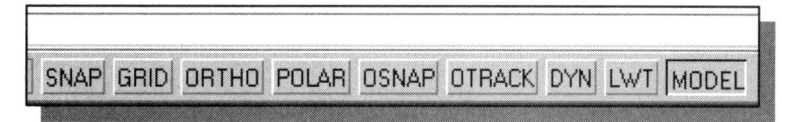

- Next, we will create a 2'-8" doorway.

13. Select the **Line** command icon in the *Draw* toolbar. In the command prompt area, the message "*_line Specify first point:*" is displayed.

14. In the *Object Snap* toolbar, pick **Snap From**. In the command prompt area, the message "*_from Base point*" is displayed. AutoCAD now expects us to select a geometric entity on the screen.

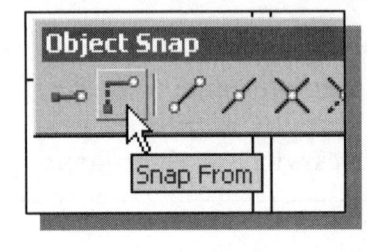

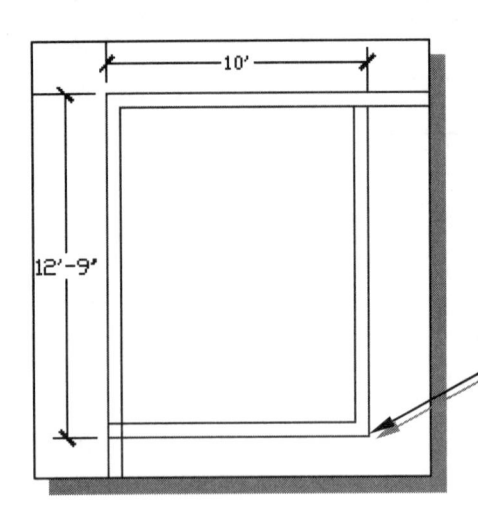

15. We will measure relative to the lower left corner of the interior wall. In the *Object Snap* toolbar, pick **Snap to Endpoint**. In the command prompt area, the message "*_from Base point:_endp of*" is displayed. Pick the lower right corner as shown.

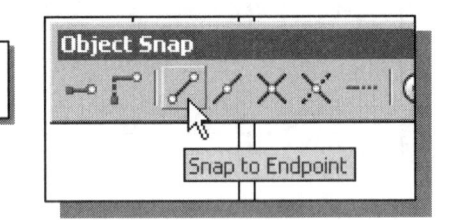

16. At the command prompt, enter **@0,1'-1"** [ENTER].

17. At the command prompt, enter **@-6",0** [ENTER].

18. Inside the graphics window, right-mouse-click once and select **Enter** to end the Line command.

19. On your own, use the **Offset** command to create a parallel line that is **2'-8"** above the line we just created.

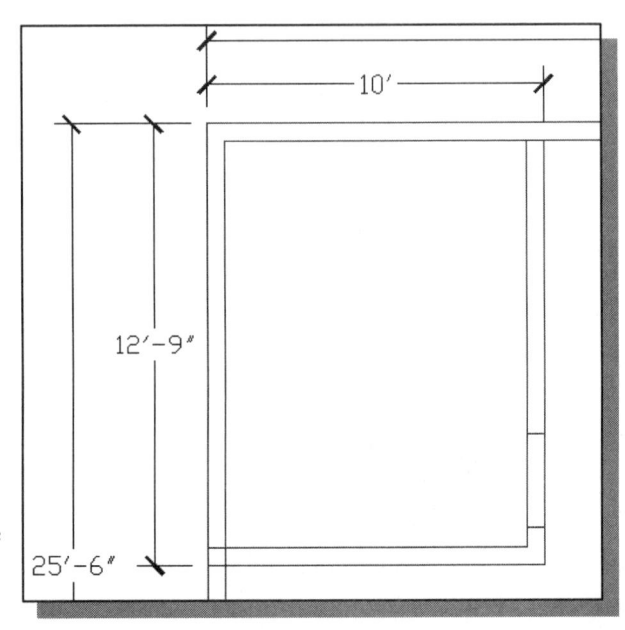

Completing the Doorway Using the *Trim* Command

- The **Trim** command shortens an object so that it ends precisely at a selected boundary.

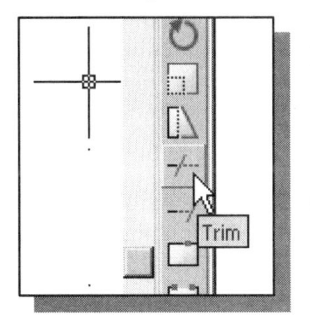

    1. Select the **Trim** command icon in the *Modify* toolbar. In the command prompt area, the message "*Select boundary edges... Select objects:*" is displayed.

    - First we will select the objects that define the boundary edges to which we want to trim the object.

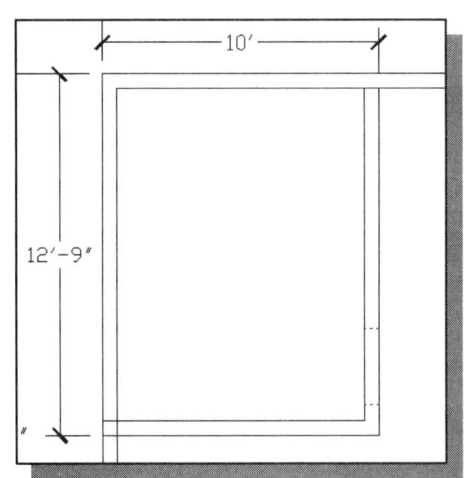

    2. Pick the **two horizontal lines** we just created as the *boundary edges*.

    3. Inside the graphics window, right-mouse-click to proceed with the Trim command.

    4. The message "*Select object to trim or shift-select object to extend or [Project/ Edge/ Undo]:*" is displayed in the command prompt area. Pick the vertical sections bound by the two selected horizontal lines.

    5. Inside the graphics window, right-mouse-click to activate the option menu and select **Enter** with the left-mouse-button to end the Trim command.

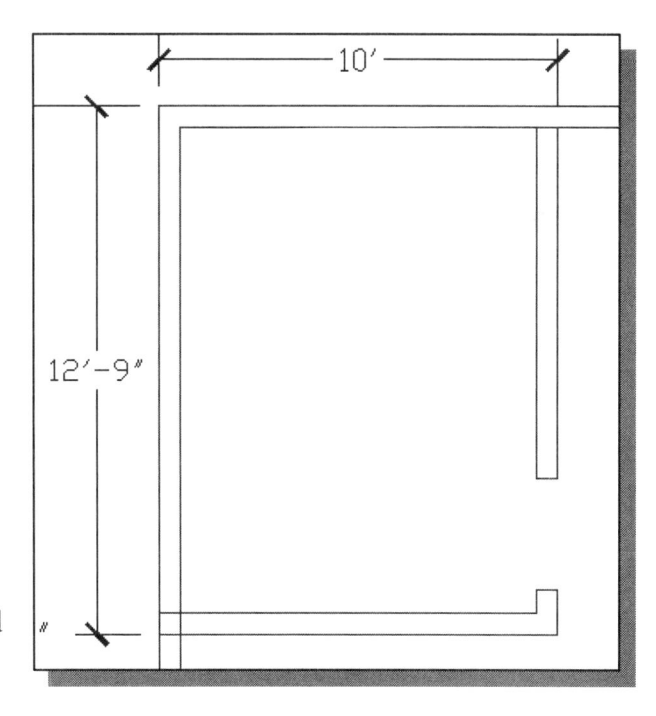

- Using the **Polyline** and **Line** commands, create the additional walls and doorways as shown.

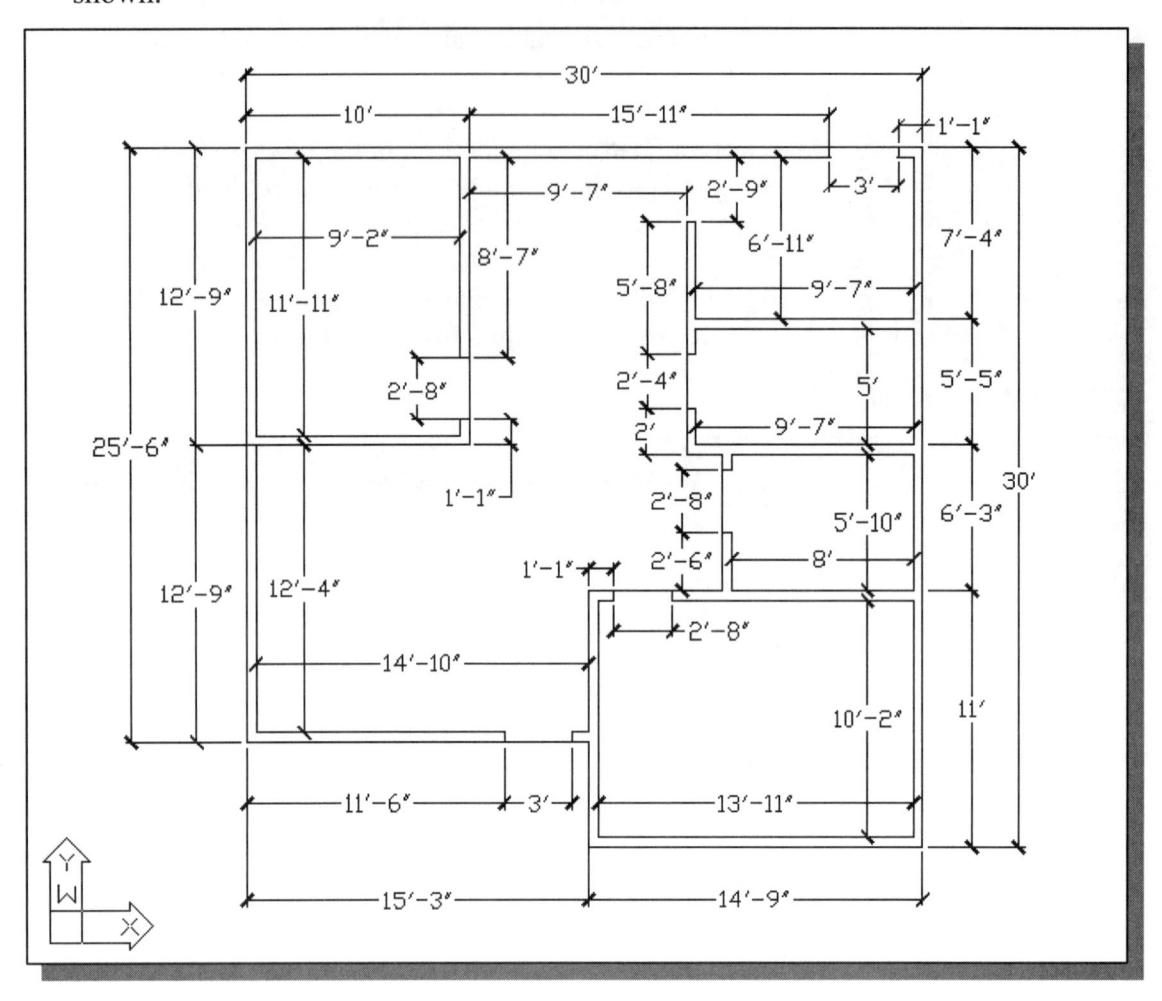

❖ Hints: Use the **Trim/Extend**, **Undo**, and **Erase** commands to assist the construction of the floor plan.

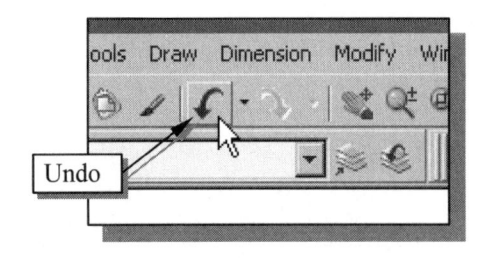

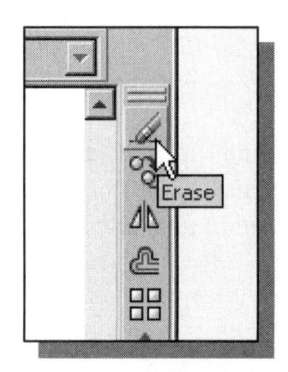

◆ Now is a good time to save the design. Select **[File]** → **[Save As]** in the pull-down menu and use *FloorPlan* as the *File name*.

User Coordinate System – It is an XY CRT, but an XYZ World

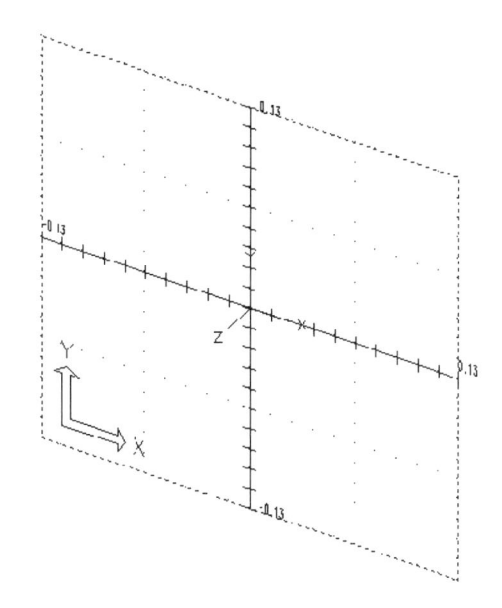

Design modeling software is becoming more powerful and user friendly, yet the system still does only what the user tells it to do. When using a geometric modeler, we therefore need to have a good understanding of what the inherent limitations are. We should also have a good understanding of what we want to do and what to expect, as the results are based on what is available.

In most 3D geometric modelers, 3D objects are located and defined in what is usually called **world space** or **global space**. Although a number of different coordinate systems can be used to create and manipulate objects in a 3D modeling system, the objects are typically defined and stored using the world space. The world space is usually a **3D Cartesian coordinate system** that the user cannot change or manipulate.

In most engineering designs, models can be very complex, and it would be tedious and confusing if only the world coordinate system were available. Practical 3D modeling systems allow the user to define **Local Coordinate Systems (LCS)** or **User Coordinate Systems (UCS)** relative to the world coordinate system. Once a local coordinate system is defined, we can then create geometry in terms of this more convenient system.

Although objects are created and stored in 3D space coordinates, most of the geometry entities can be referenced using 2D Cartesian coordinate systems. Typical input devices such as a mouse or digitizers are two-dimensional by nature; the movement of the input device is interpreted by the system in a planar sense. The same limitation is true of common output devices, such as CRT displays and plotters. The modeling software performs a series of three-dimensional to two-dimensional transformations to correctly project 3D objects onto a 2D picture plane.

The AutoCAD's **User Coordinate System (UCS)** is a special construction tool that enables the planar nature of the 2D input devices to be directly mapped into the 3D coordinate system. The *UCS* is a local coordinate system that can be aligned to the world coordinate system, an existing face of a part, or a predefined plane. By default, the UCS is aligned to the XY plane of the world coordinate system.

Think of the UCS as the surface on which we can sketch the 2D profiles of designs. It is similar to a piece of paper, a white board, or a chalkboard that can be attached to any planar surface. In the previous sections, we created the 2D design of a floor plan using the default settings where the UCS is aligned to the XY plane of the world coordinate system.

Viewing the 2D Design in 3D Space

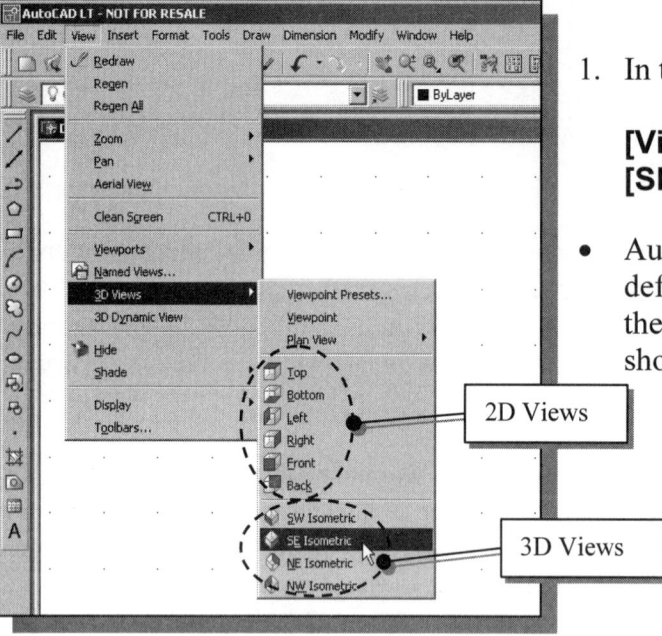

1. In the pull-down menus, select:

   [View] → [3D Views] → [SE Isometric]

   - AutoCAD provides a set of pre-defined views, which contains most of the standard 2D and 3D views as shown in the figure.

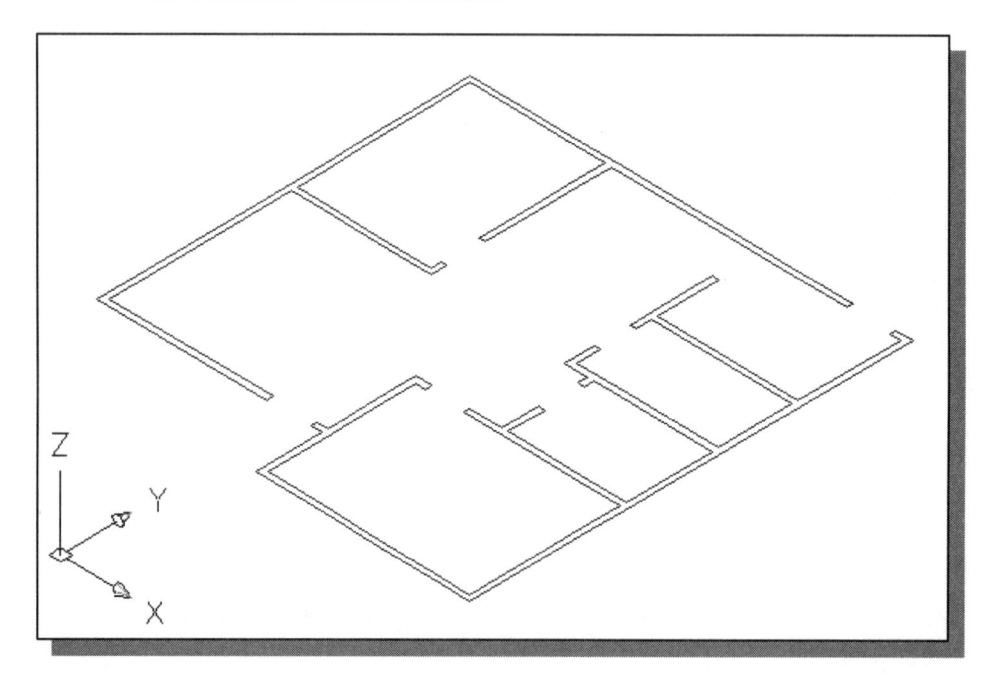

- The orientations of the preset 2D and 3D views are based on the world coordinate system as shown below. Note that the positive Y-axis direction is also identified as pointing toward **North** and the preset 3D views are oriented using the **North**, **East**, **South**, and **West** directions.

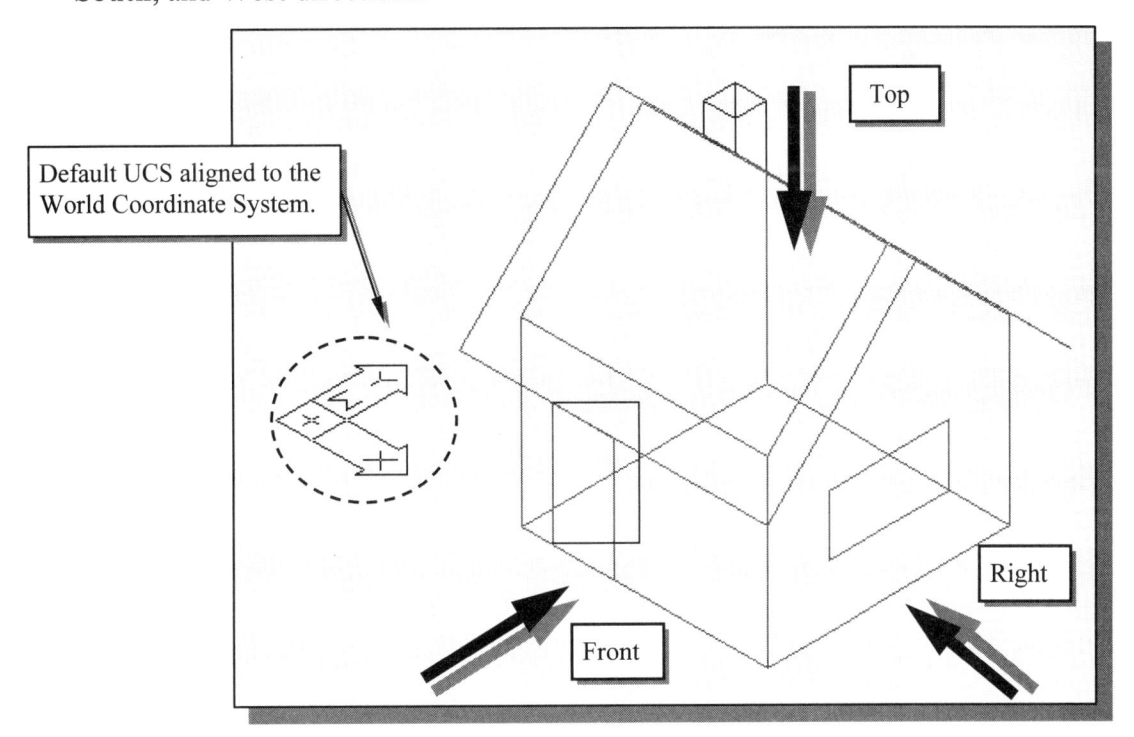

Adding the 3rd dimension to the *Floor Plan* Design

❖ **AutoCAD® 2006** provides a flexible graphical user interface that allows users to select graphical entities BEFORE the command is selected (*Pre-selection*), or AFTER the command is selected (*Post-selection*). We can preselect one or more objects by clicking on the objects at the command prompt (**Command:**). To deselect the selected items, press the [**Esc**] key twice.

1. Inside the *graphics window*, pre-select all objects by enclosing all objects inside a **selection window** as shown.

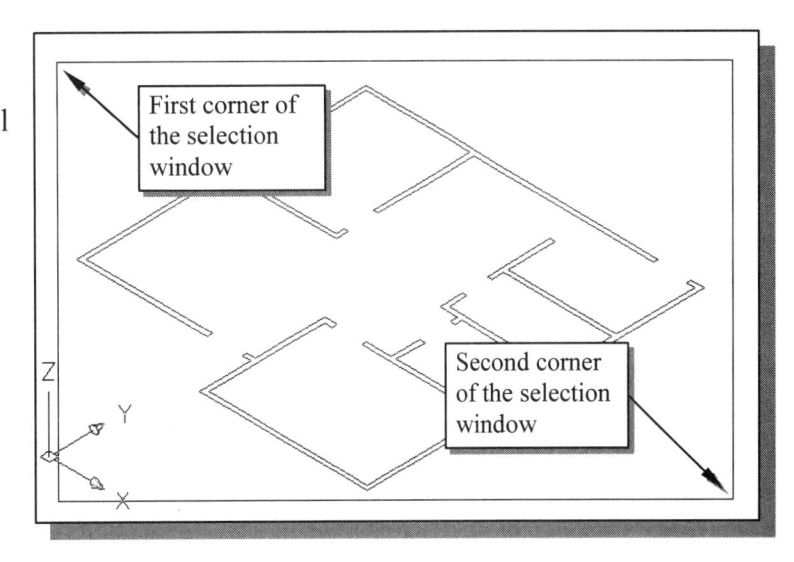

2. In the *Standard* toolbar, click on the **Properties** icon.

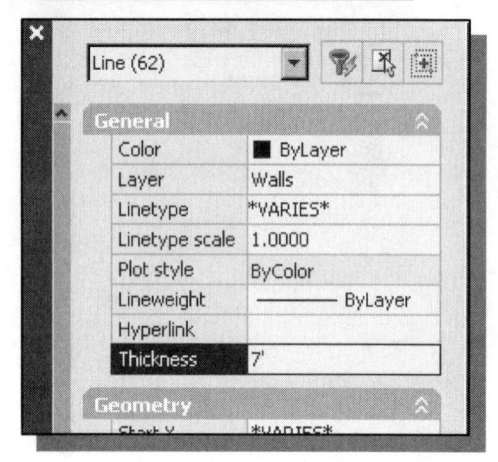

3. In the *Properties* dialog box, properties of the highlighted entities are displayed. Notice the current *Thickness* is set to 0".

4. **Left-click** the *Thickness* box and enter a new value: **7' [ENTER]** as the new thickness value.

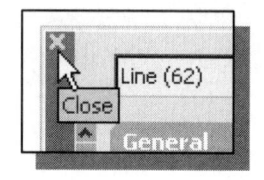

5. Click on the [**X**] button to exit the *Properties* dialog box.

In AutoCAD, the *thickness* of an object determines the distance that the object is extruded above or below its current Z-elevation. Positive thickness will extrude the objects upward (positive Z), negative thickness extrudes downward (negative Z), and zero thickness means no extrusion. The Z direction is determined by the orientation of the UCS at the time the object was created.

Changing the *thickness* of objects changes the appearance of certain geometric objects, such as circles, lines, polylines, arcs, 2D solids, and points. Note that the *Thickness* property is not available on certain geometric objects, such as *Splines* and *Multilines*. We can set the *thickness* of an object with the **Thickness** variable. AutoCAD applies the extrusion uniformly to an object. A single object cannot have different thickness for its various points.

Changing the thickness of objects simulates a simple Z-extrusion of creating solid objects in AutoCAD. The main advantage of using the object thickness to create the Z-elevation instead of creating a true solid is that the operation is quick and easy. In AutoCAD, objects with thickness can be hidden, shaded, and rendered as if they are three-dimensional solids. Once an object's thickness is set, we can visualize the results in any view other than the plan view.

This method is also known as the **2½D solid modeling approach**. This modeling technique was first invented in the 1970s and was adopted by several of the first-generation solid-modeling packages. This approach requires no additional 3D construction tools and only minimum computing power is needed. Note that this modeling approach does not create true 3D solids and objects can only be extruded in one direction. This modeling approach is considered to be less flexible and its application is somewhat limited.

Viewing the Design Using the Hide Option

- By default, AutoCAD produces a wireframe image of the constructed 2D/3D design on the screen. All geometric entities are present, including those that should be hidden by other objects. Several options are available in AutoCAD to create more realistic images of 3D designs. In this section, we will introduce the **Hide** option, which can be used to adjust the display of the design; the **Hide** option eliminates the hidden lines on the screen.

1. Deselect any objects by hitting the [**Esc**] key once.

2. In the pull-down menus, select:
   [View] → [Hide]

3. To display the wireframe image, select:
   [View] → [Regen]

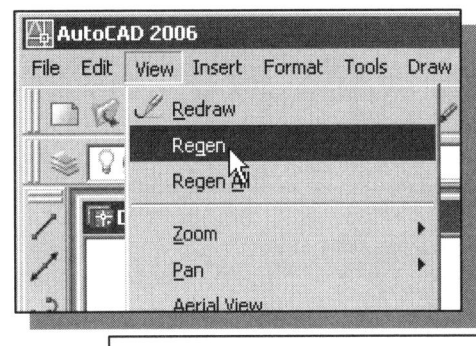

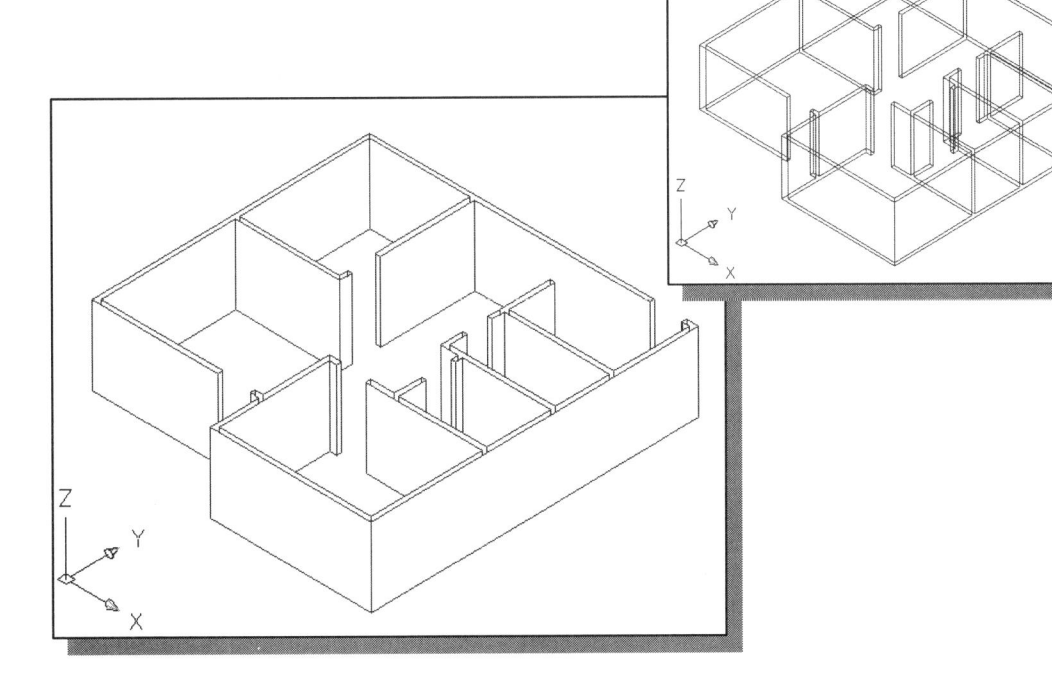

Adding New Layers

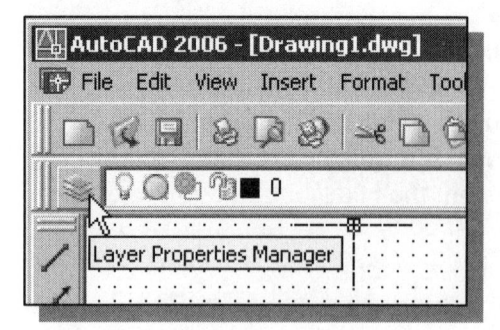

1. Pick **Layer Properties Manager** in the *Object Properties* toolbar. The *Layer Properties Manager* dialog box appears. AutoCAD creates a default layer, *Layer 0*, which we cannot rename or delete. *Layer 0* has special properties used by the system.

In AutoCAD, we always construct entities on a layer. It may be the default layer or a layer that we create. Each layer has associated properties such as the visibility setting, color, linetype, lineweight, and plot style.

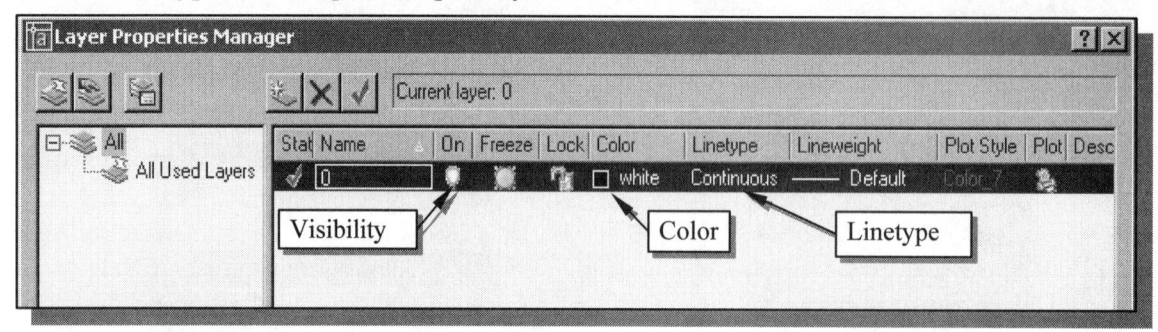

2. Click on the **New** button. Notice a layer is automatically added to the list of layers.

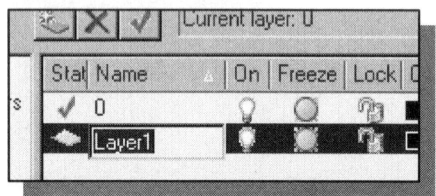

3. AutoCAD will assign a generic name to the new layer (*Layer1*). Enter **Walls** as the name of the new layer and change the layer color to **Green** as shown in the below figure.

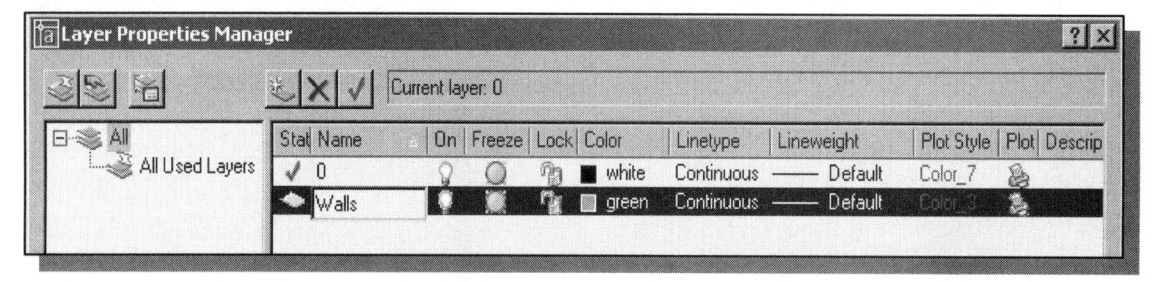

4. On your own, create another layer (layer name: **Roof**) and change the layer color to **Blue**.

5. Click on the **OK** button to accept the settings.

Moving Entities to a Different Layer

1. Inside the graphics window, pre-select all objects by enclosing all objects inside a **selection window**.

2. On the *Object Properties* toolbar, choose the ***Layer Control*** box with the left-mouse-button.

❖ Notice the layer name displayed in the *Layer Control* box is the selected object's assigned layer and layer properties.

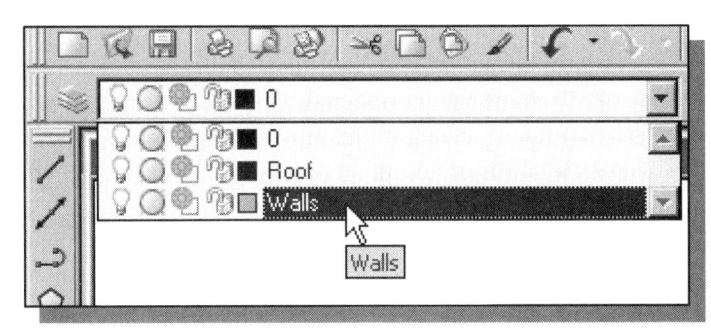

3. In the *Layer Control* box, click on the *Walls* layer name. The selected objects are now moved to the *Walls* layer.

4. Left-click once on the *Roof* layer name to set it as the ***Current Layer***.

Moving the UCS

- AutoCAD's **User Coordinate System (UCS)** is a special construction tool that enables the planar nature of the 2D input devices to be directly mapped into the 3D coordinate system. The **UCS** is a local coordinate system that can be repositioned and/or reoriented to an existing face of a part, or a predefined plane. By default, the UCS is aligned to the XY plane of the world coordinate system. Several options are available to reposition and reorient the UCS in 3D space.

1. In the pull-down menus, select:

   [Tools] → [Move UCS]

2. We will use the *Object Snap* and *Object Tracking* options to aid the repositioning of the UCS. In the *Status Bar* area, reset the option buttons so that the *POLAR, OSNAP, OTRACK, DYN* and *MODEL* options are switched *ON*.

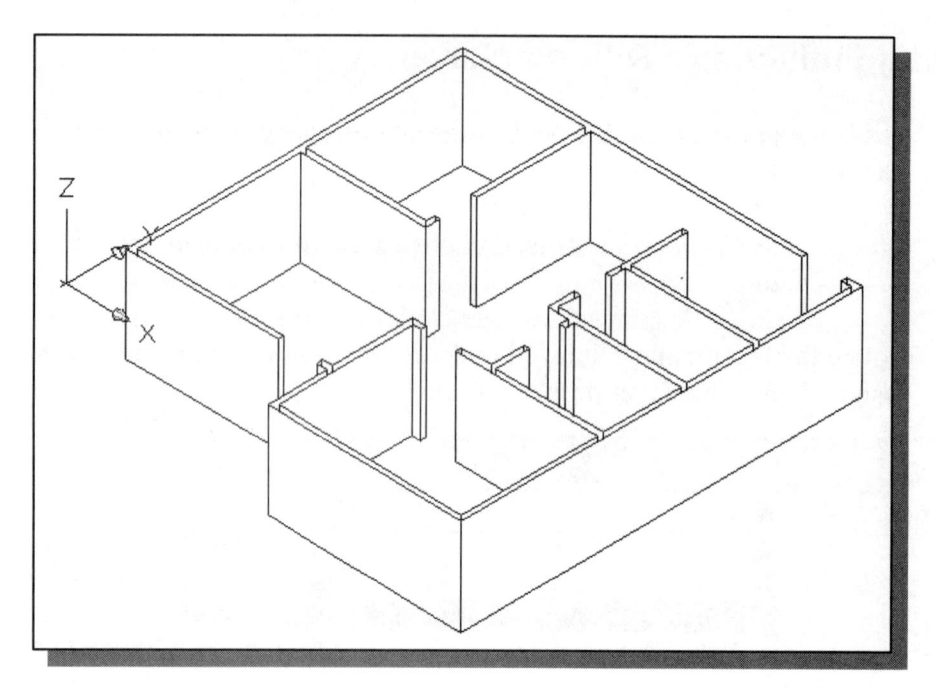

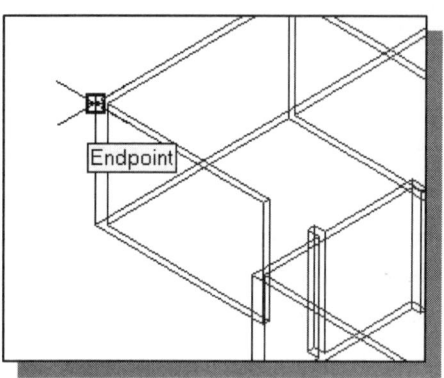

- We will reposition the UCS at the intersection of the left wall and the front wall; the intersection can be found by using the AutoCAD object tracking options.

3. Move the cursor on the top left corner of the external wall as shown; pause for a few seconds to activate the *OTRACK* option.

4. Move the cursor on the top front corner of the external wall as shown; pause for a few seconds to activate the *OTRACK* option.

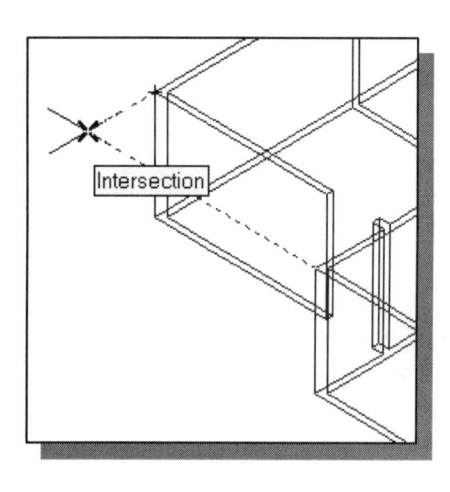

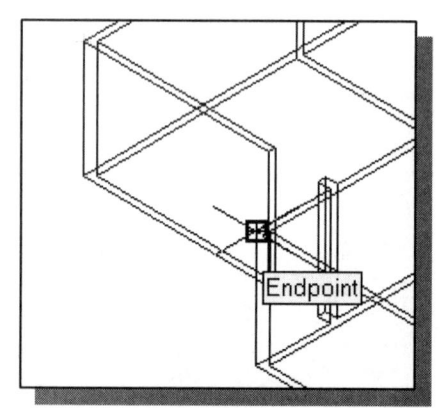

5. Left-click once when the cursor is aligned to the intersection of the left wall and the front wall as shown.

- The UCS is repositioned to the new location as shown. This location is the new origin for any geometric constructions.

Creating the Roof

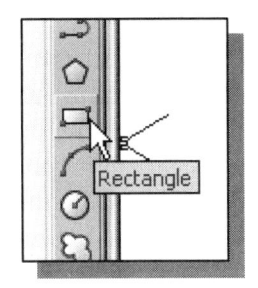

1. Select the **Rectangle** command icon in the *Draw* toolbar as shown.

2. In the command prompt area, the message "*Specify first corner point or [Chamfer/Elevation/Fillet/Thickness/Width]:*" is displayed. Move the cursor inside the graphics window and right-mouse-click to display the option menu.

- The thickness of certain geometry entities can be set as they are being created.

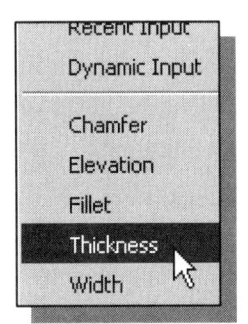

3. In the option menu, select **Thickness** by clicking once with the left-mouse-button.

4. In the command prompt area, the message "*Specify thickness for rectangles <0'-0">:*" is displayed. Enter **10"** [ENTER].

5. At the command prompt, enter **-2',-2'** [ENTER].

6. In the command prompt area, the message "*Specify other corner point or [Dimensions]:*" is displayed. Enter **@34',34'** [ENTER].

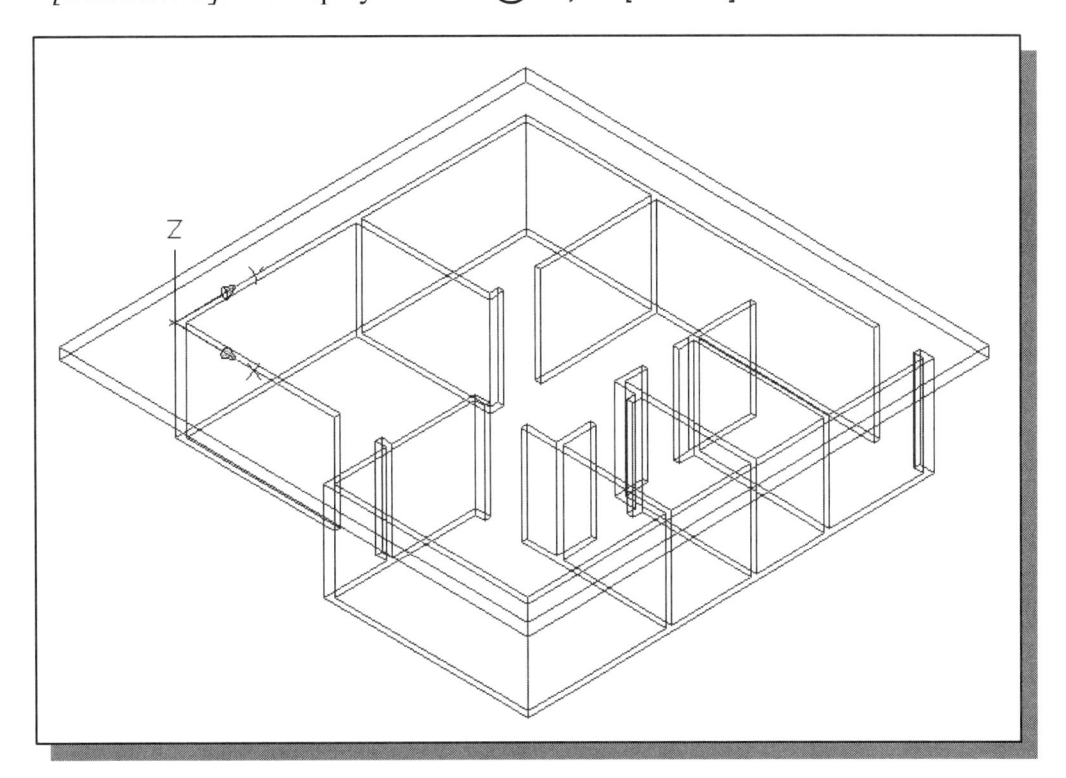

- Note that the coordinates entered are measured relative to the new UCS origin.

7. In the pull-down menus, select:

   [View] → [Hide]

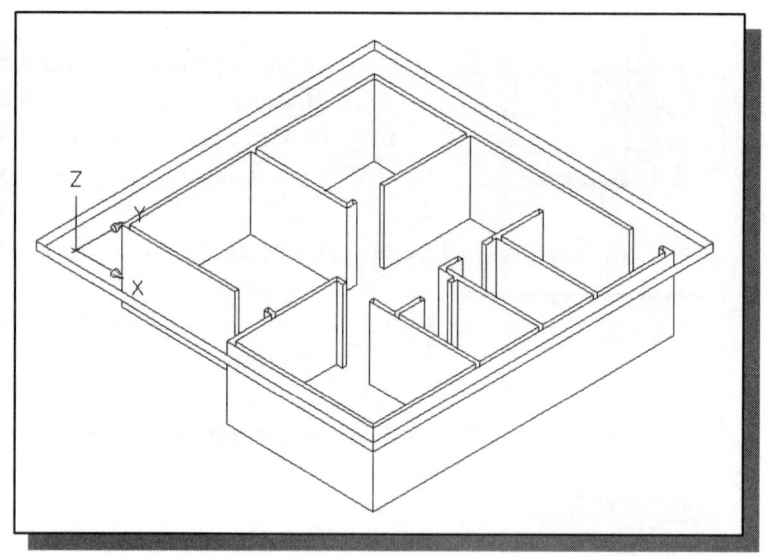

- The **Thickness** option extrudes the 2D design in the positive Z-direction relative to the current UCS, which is at the same Z-elevation of the plane of the 2D design. Note that the **Thickness** option only makes 2D objects look like 3D solids; only surfaces perpendicular to the plane of the 2D design are created. No top or bottom surfaces are created with the **Thickness** option.

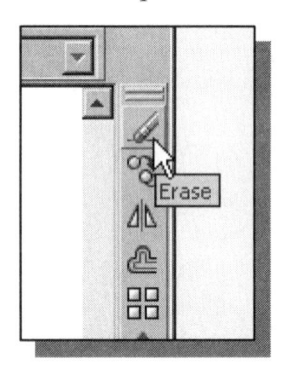

8. Pick **Erase** in the *Modify* toolbar. (The icon is the first icon in the *Modify* toolbar.) The message "*Select objects*" is displayed in the command prompt area and AutoCAD awaits us to select the objects to erase.

9. Pick any edge of the rectangle we just created.

10. **Right-mouse-click** inside the graphics window to accept the selection.

11. Reset the display to the wireframe image:
    [View] → [Shade] → [3D wireframe]

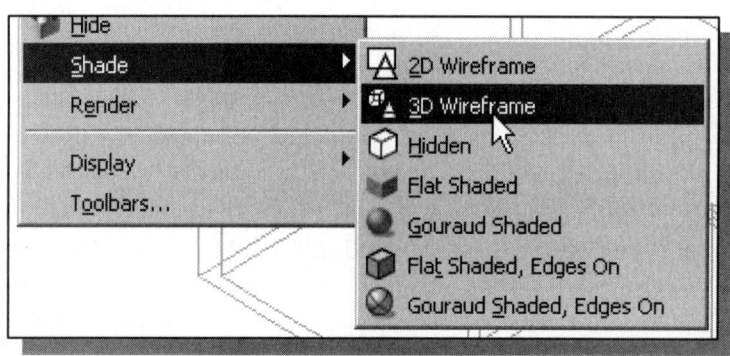

Rotating the UCS

- AutoCAD's **User Coordinate System (UCS)** can be repositioned and reoriented in 3D space. All of the UCS options are organized and can be found in the *Standard* toolbar.

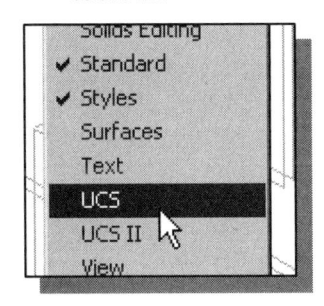

  1. Move the cursor on any icon in the *Standard* toolbar area and **right-mouse-click** once to display a list of toolbar menu groups.

  2. Select **UCS**, with the left-mouse-button, to display the *UCS* toolbar on the screen to assist the construction of the *Roof*.

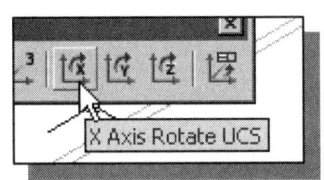

  3. Select the **X Axis Rotate UCS** option in the *UCS* toolbar as shown.

  4. In the command prompt area, the message "*Specify rotation angle about X axis<90>:*" is displayed.

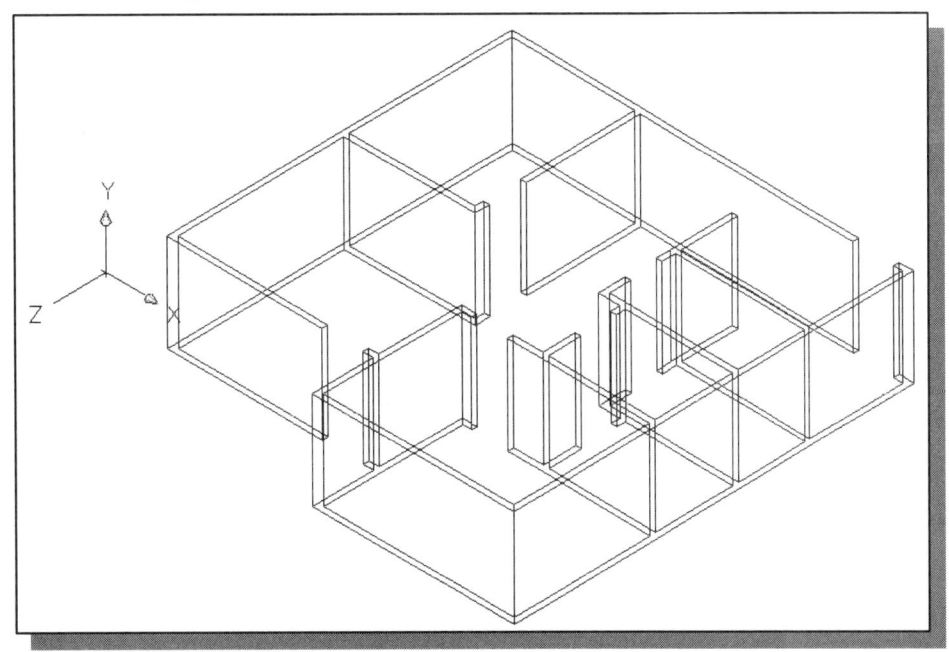

     Enter: **90 [ENTER]**.

- The UCS rotation is done using the standard engineering convention, where positive values are treated as rotating counter-clockwise. Also note that the *SNAP*, *GRID*, and *ORTHO* options all rotate in line with the UCS.

Sketching on the Rotated UCS

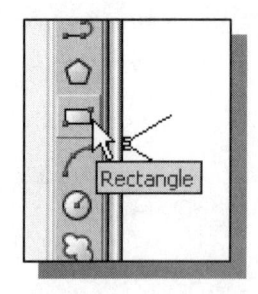

1. Select the **Rectangle** command icon in the *Draw* toolbar as shown.

2. In the *command prompt* area, the message "*Specify first corner point or [Chamfer/Elevation/Fillet/Thickness/Width]:*" is displayed. Move the cursor inside the graphics window and right-mouse-click to display the option menu.

3. In the *option menu*, select **Thickness** by clicking once with the left-mouse-button.

4. In the *command prompt* area, the message "*Specify thickness for rectangles <0'-10">:*" is displayed. Enter **-30'** **[ENTER]**.

5. At the *command prompt*, enter **-2',0'** **[ENTER]** as the first corner of the rectangle.

6. In the *command prompt* area, the message "*Specify other corner point or [Dimensions]:*" is displayed. Enter **@34',10"** **[ENTER]**.

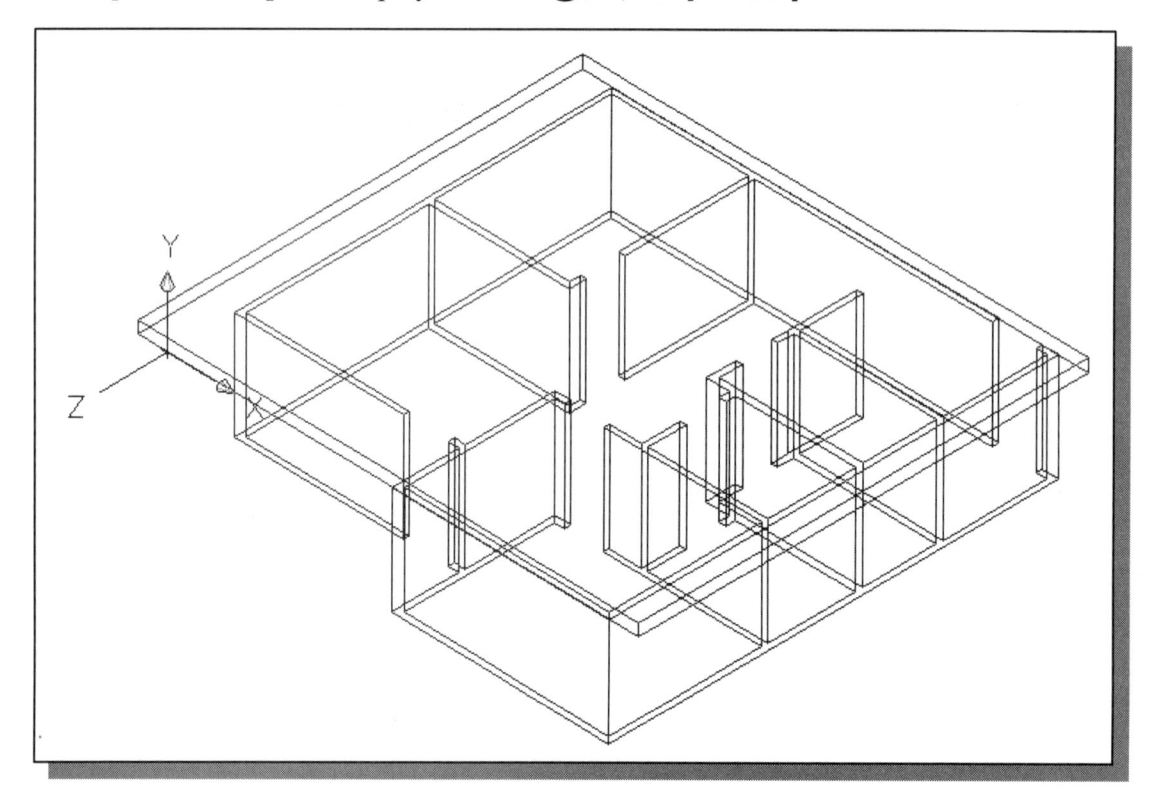

- The negative thickness extrudes the rectangle in the negative Z-direction of the current UCS.

Viewing the Design Using the Hidden Option

- Several options are available in AutoCAD to create more realistic images of 3D designs. The **Shade - Hidden** option can also be used to eliminate the hidden lines on the screen. The **Shade - Hidden** option performs a more complicated calculation of the display and usually produces a better image than the **Hide** option.

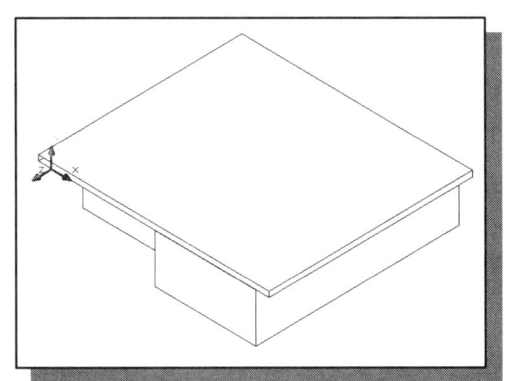

1. In the pull-down menus, select:

    [View] → **[Shade]** → **[Hidden]**

2. To display the wireframe image, select:

    [View] → **[Shade]** → **[3D Wireframe]**

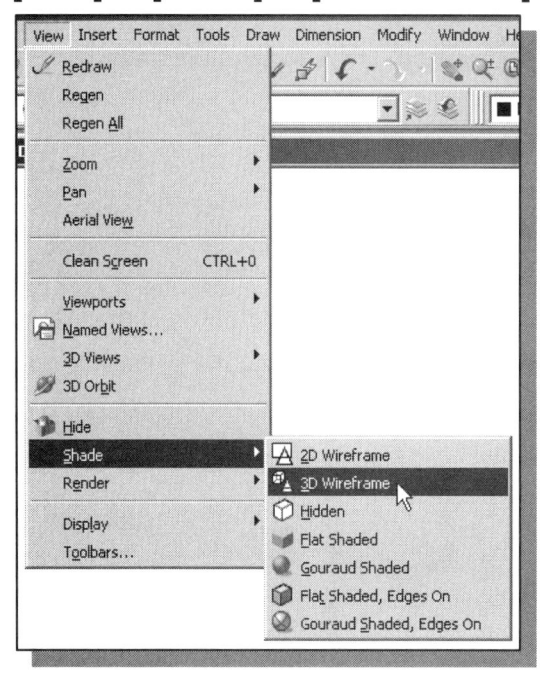

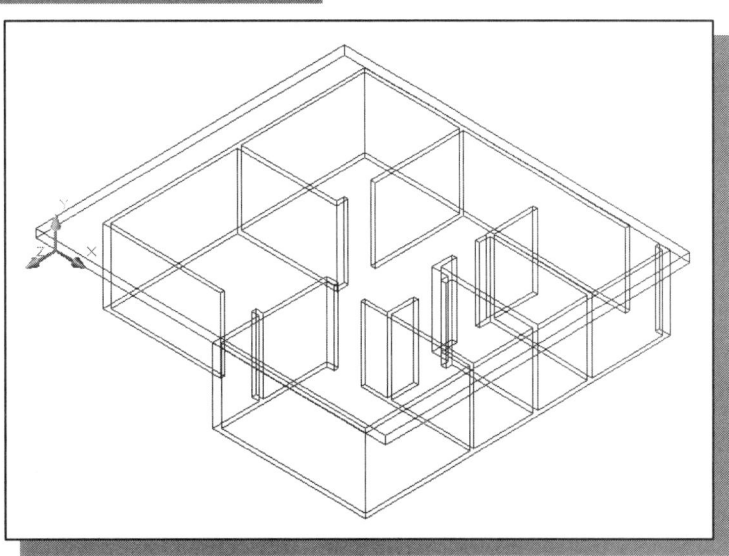

Questions:

1. What is the difference between *WCS* and *UCS*?

2. List and describe two available options to manipulate the UCS.

3. Describe some of the advantages of using the AutoCAD **Thickness** option over creating 2D or 3D models.

4. In the tutorial, which commands were used to eliminate the hidden lines on the screen?

5. What is the difference between a **Polyline** and a **Line**?

6. Identify the following commands:

   (a)

   (b)

   (c)

   (d)

Exercises:

1. Dimensions are in inches. Thickness : 0.5 & 1.0

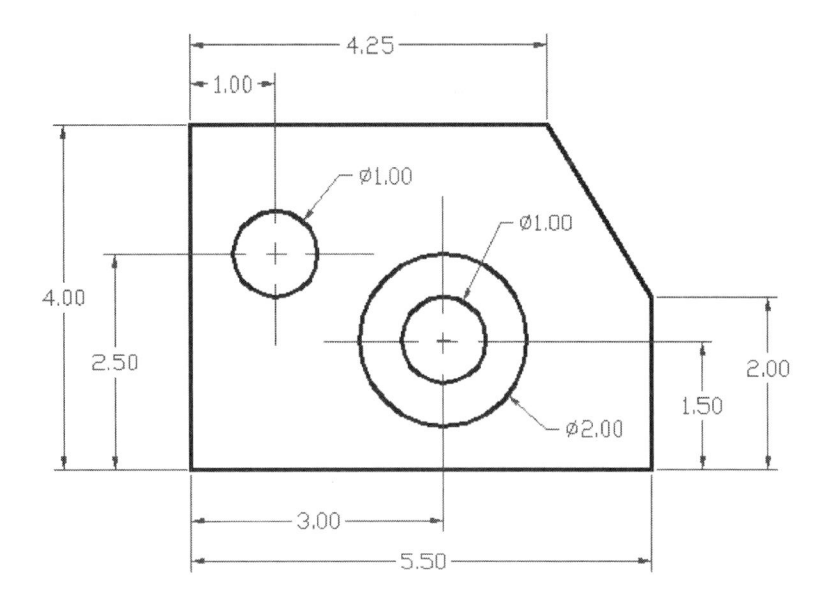

2. Wall thickness: 5 inch; height: 7 feet

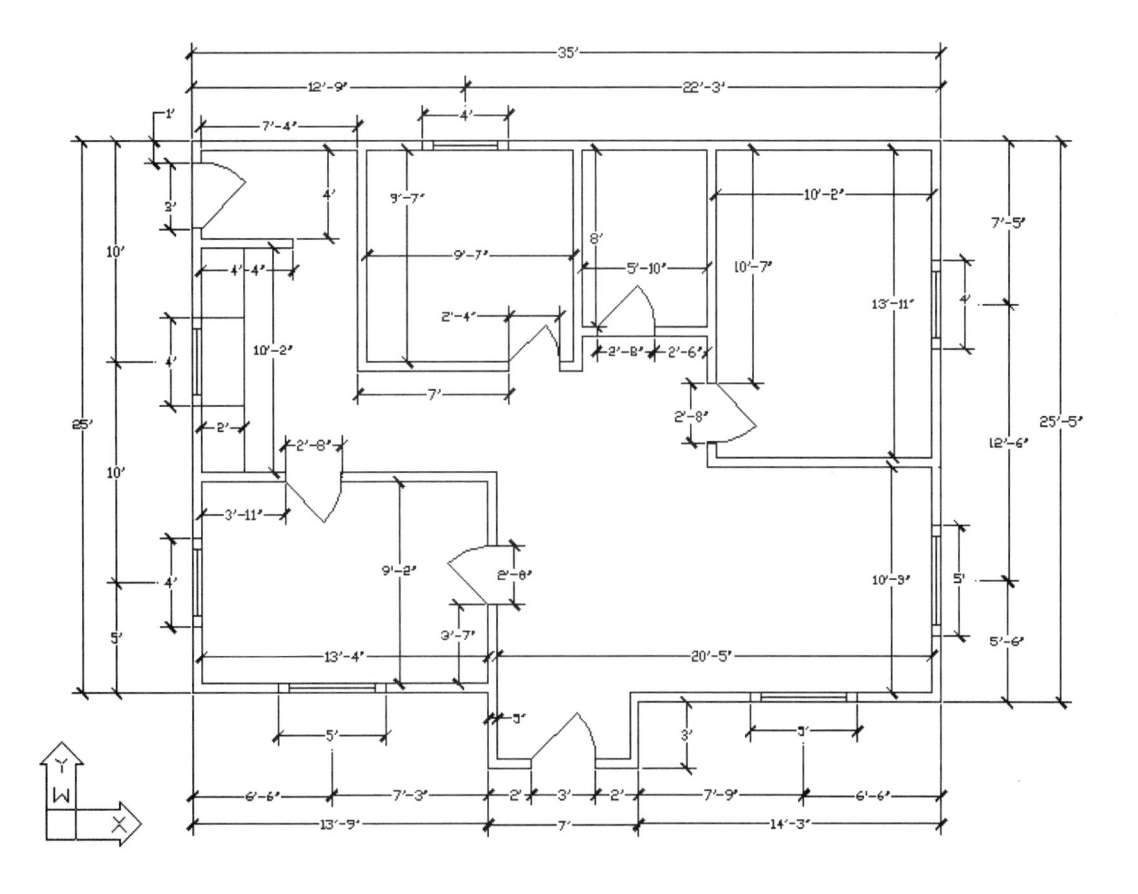

3. Dimensions are in inches. Thickness: 0.25

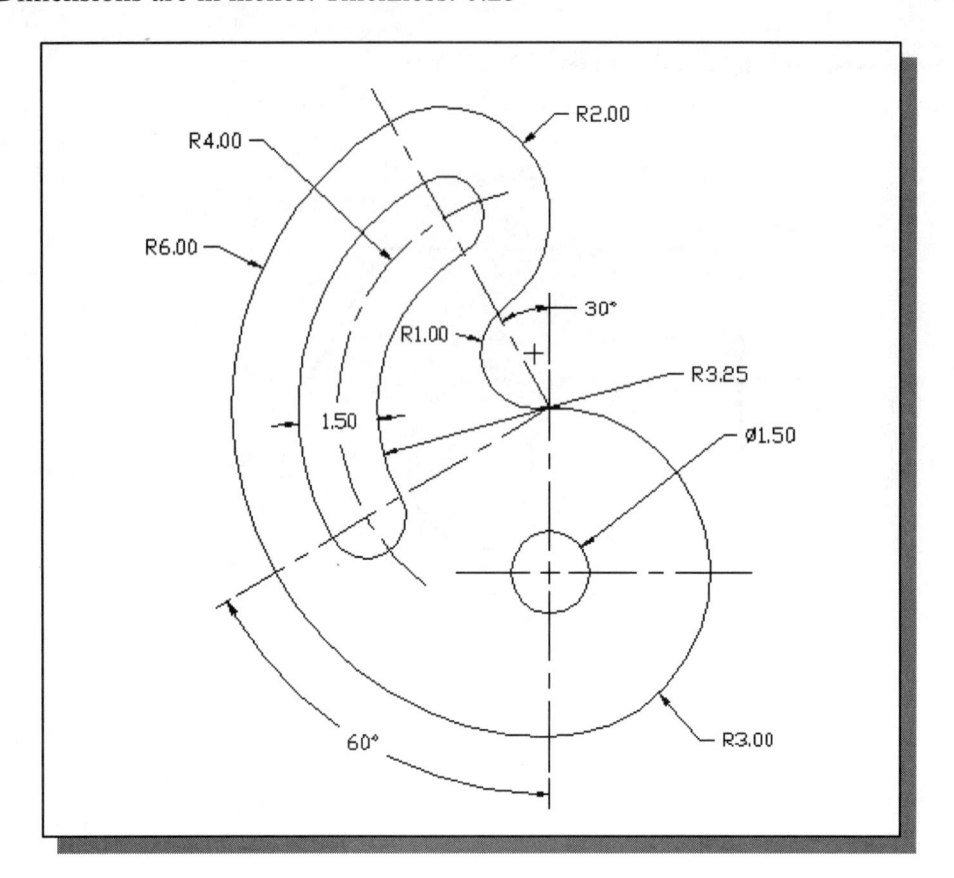

4. Dimensions are in inches. Thickness: 0.25

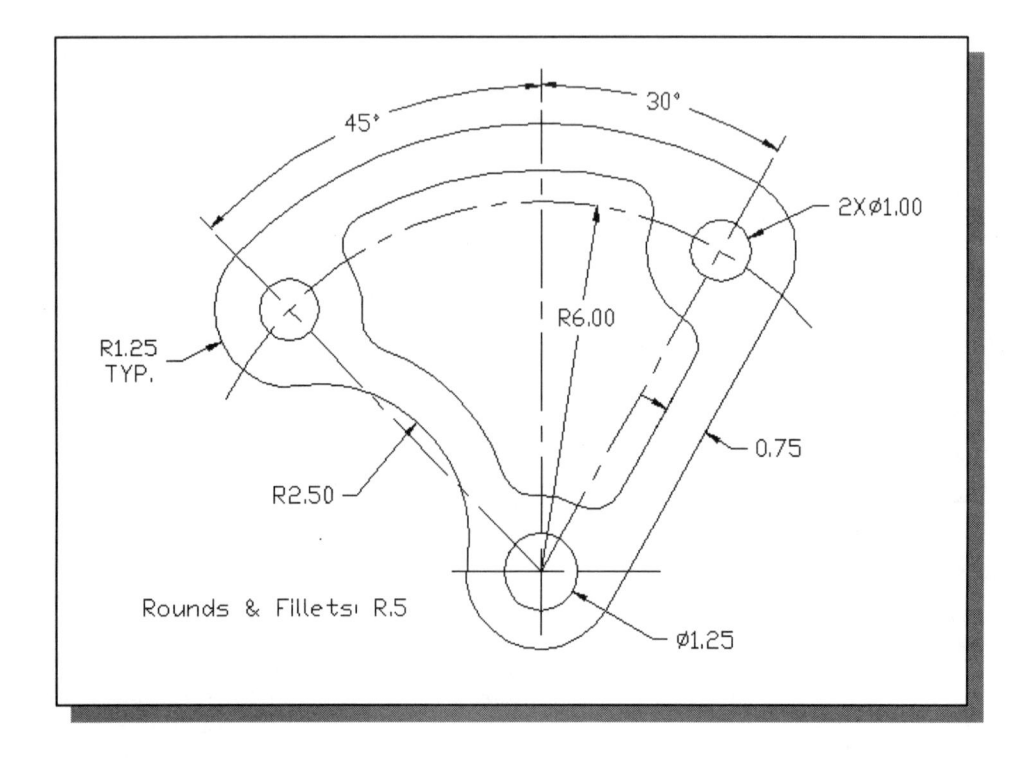

Chapter 3
3D Wireframe Modeling

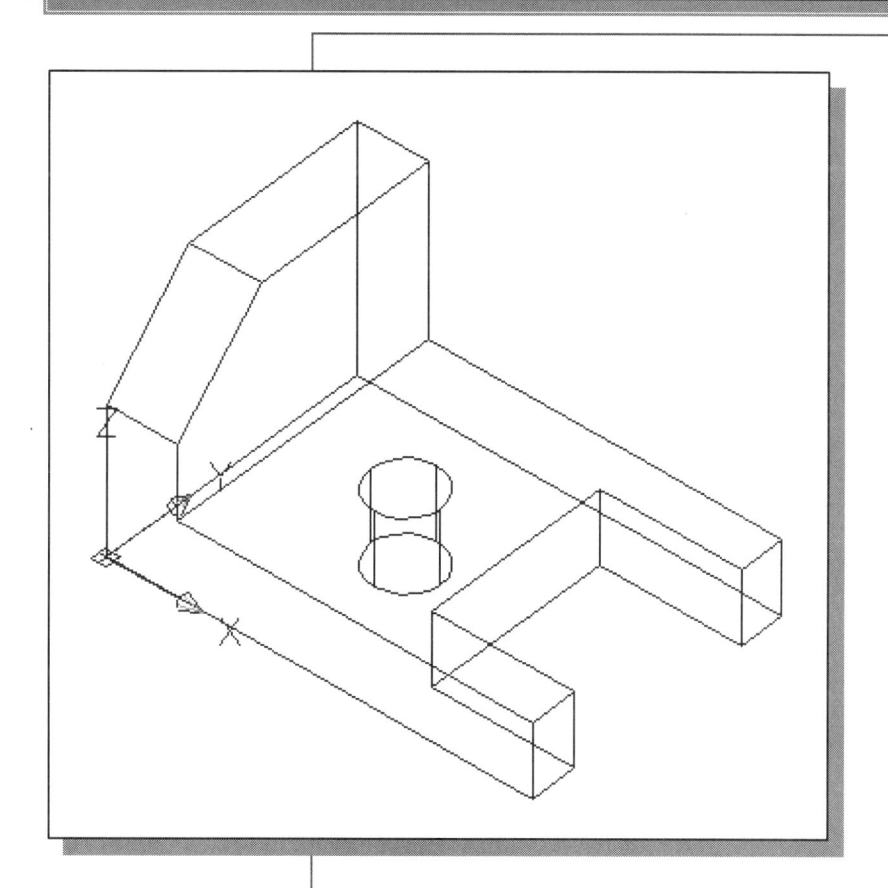

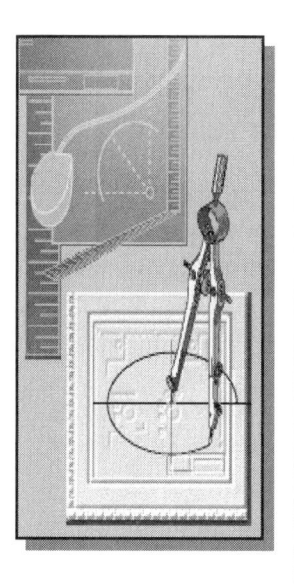

Learning Objectives

- **Using the Setup Wizard**
- **Create Wireframe Models**
- **Apply the Box Method in Creating Models**
- **Construct with the Copy Command**
- **Understand the Available 3D Coordinates Input Options**
- **Using the View Toolbar**
- **Setup and Using the TRIM options**

Introduction

The first true 3D computer model created on CAD systems in the late 1970s was the 3D wireframe model. Computer generated 3D wireframe models contain information about the locations of all the corners and edges in space coordinates. The 3D wireframe models can be viewed from any direction as needed and are in general reasonably good representations of 3D designs. But because surface definition is not part of a wireframe model, all wireframe images have the inherent problem of ambiguity. For example, in the figure displayed below, which corner is in front, corner A or corner B? The ambiguity problem becomes much more serious with complex designs that have many edges and corners.

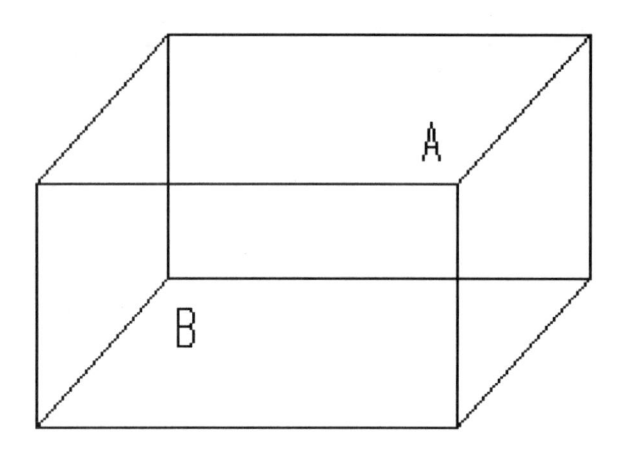

Wireframe Ambiguity: Which corner is in front, A or B?

The main advantage of using a 3D wireframe modeler to create 3D models is its simplicity. The computer hardware requirements for wireframe modelers are typically much lower than the requirements for surface and solid modelers. A 3D wireframe model, also known as a stick-figure model or a skeleton model, contains only information about the locations of all the corners and edges of the design in space coordinates. You should also realize that, in some cases, it could be quite difficult to locate some of the corner locations while creating a 3D wireframe model. Note that 3D wireframe modelers are usually used in conjunction with surfacing modelers, which we will discuss in the later chapters of this text, to eliminate the problem of ambiguity.

With most CAD systems, creating 3D wireframe models usually starts with constructing 2D entities in 3D space. Two of the most commonly used methods for creating 3D wireframe models are the **Box method** and the **2D Extrusion method**. As the name implies, the *Box method* involves the creation of a 3D box with the edges constructed from the overall height, width and depth dimensions of the design. The 3D wireframe model is typically completed by locating and connecting corners within the box.

The *2D Extrusion method* involves making copies of 2D geometries in specific directions. This method is similar to the 2½D extrusion approach illustrated in the previous chapter (Chapter 2) with several differences. First of all, we do not really extrude the wireframe entities; instead we simply make copies of wireframe entities in the desired directions. Secondly, constructed wireframe entities have true 3D space coordinates, while the *thickness* approach creates entities with no true 3D coordinates. And lastly, no surfaces are created in the 3D wireframe models.

In this chapter, we will illustrate the general procedure to construct a 3D wireframe model using both the box method and the 2D extrusion method. To illustrate the AutoCAD 3D construction environment, we will create the wireframe model using only the default UCS system, which is aligned to the world coordinate system. Repositioning and/or reorienting the User Coordinate System can be useful in creating 3D models. But it is also feasible to create 3D models referencing only a single coordinate system. One important note about creating wireframe models is that the construction techniques mostly concentrate on locating the space coordinates of the individual corners of the design. The ability to visualize designs in the form of 3D wireframe models is extremely helpful to designers and CAD operators. It is hoped that the experience of thinking and working on 3D wireframe models, as outlined in this chapter, will enhance one's 3D visualization ability.

The *Locator* Design

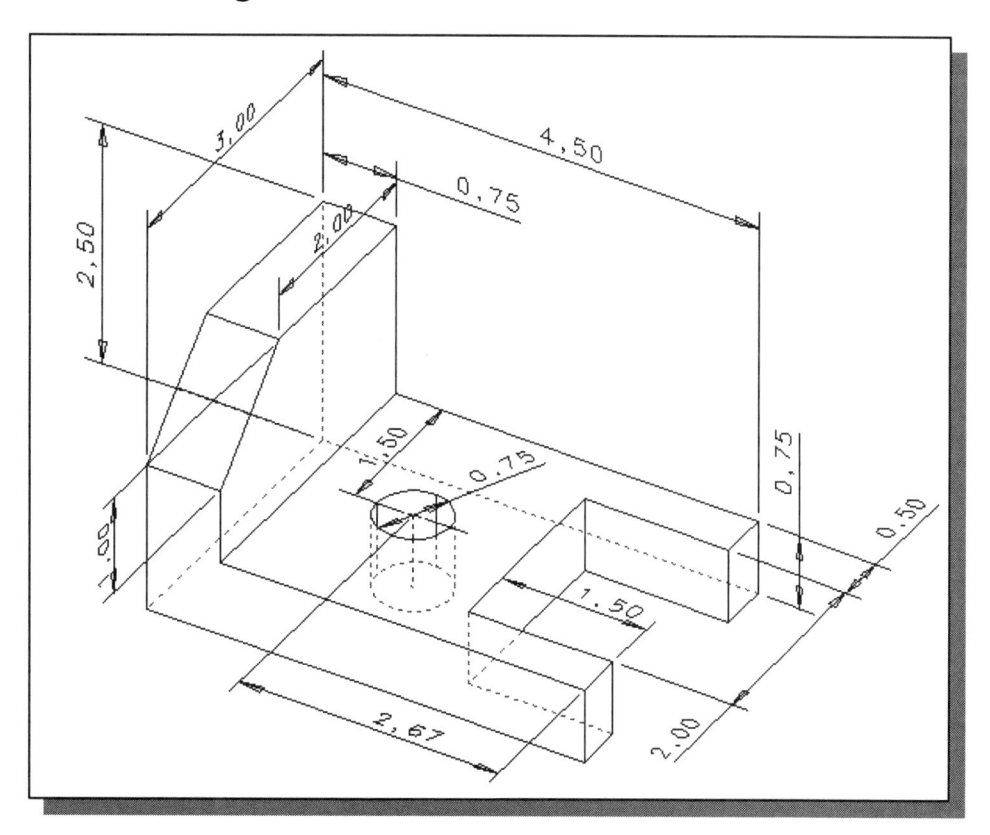

Starting Up AutoCAD® 2006

1. Select the **AutoCAD 2006** option on the *Program* menu or select the **AutoCAD 2006** icon on the *Desktop*. Once the program is loaded into the memory, the **AutoCAD® 2006** drawing screen will appear on the screen.

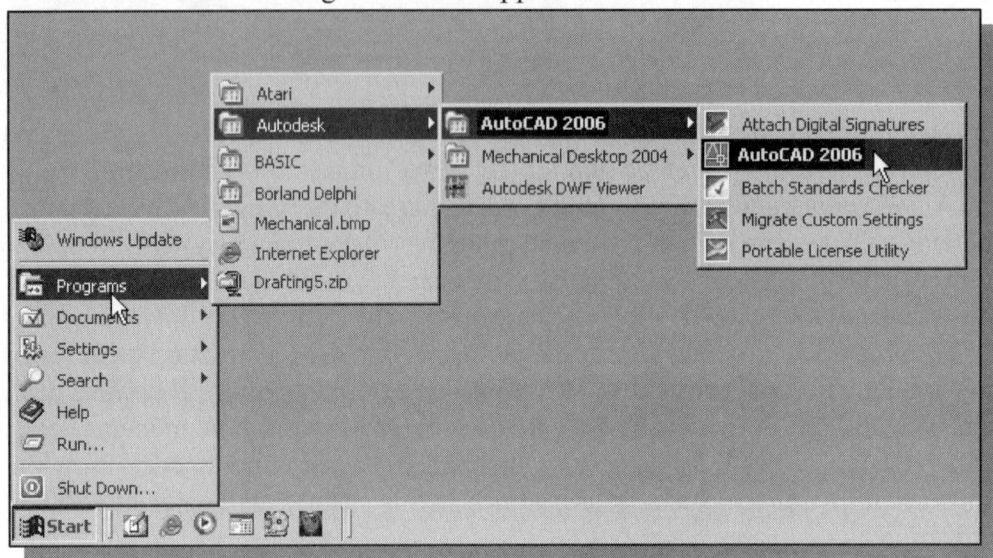

Using the Startup Options

❖ In **AutoCAD® 2006**, we can use the *setup wizards* to establish, step by step, several of the drawing settings. This option can be activated through the **Options** command.

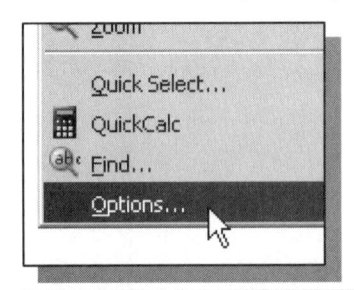

1. Inside the graphics area, **right-mouse-click** once to display the option list.

2. Select **Options**, with the left-mouse-button, to display the *Options* dialog box.

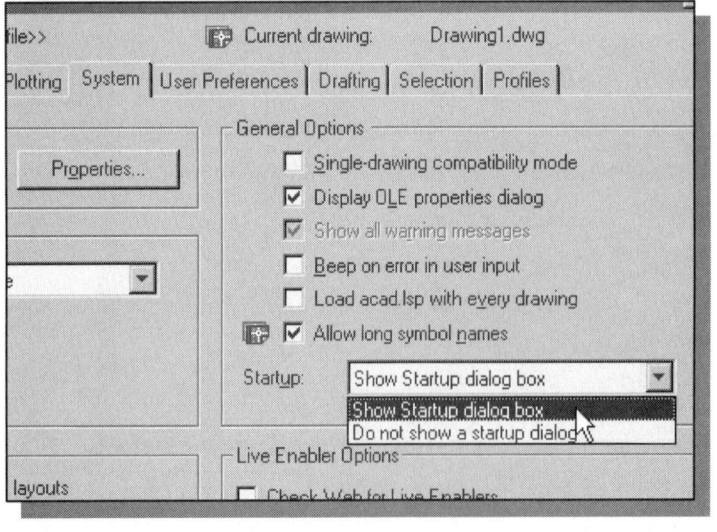

3. Switch to the **System** tab as shown.

4. In the *General Options* list, set the ***Startup*** option to **Show Startup dialog box** as shown.

5. Pick **OK** to exit the *Options* dialog box.

3D Wireframe Modeling　　3-5

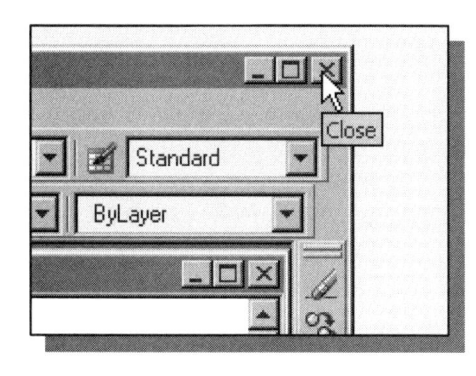

6. To show the effect of the *Startup* option, **exit** AutoCAD by clicking on the **Close** icon as shown.

7. Restart AutoCAD by selecting the **AutoCAD 2006** option through the *Start* menu.

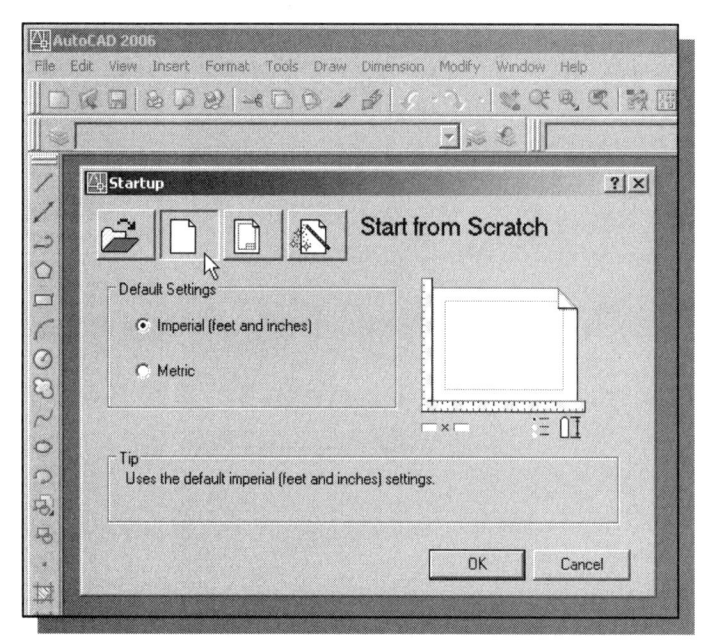

8. The *Startup* dialog box appears on the screen with different options to assist the creation of drawings. Move the cursor on top of the four icons and notice the four options available:
   (1) **Open a drawing**,
   (2) **Start from Scratch**,
   (3) **Use a Template** and
   (4) **Use a Setup Wizard**.

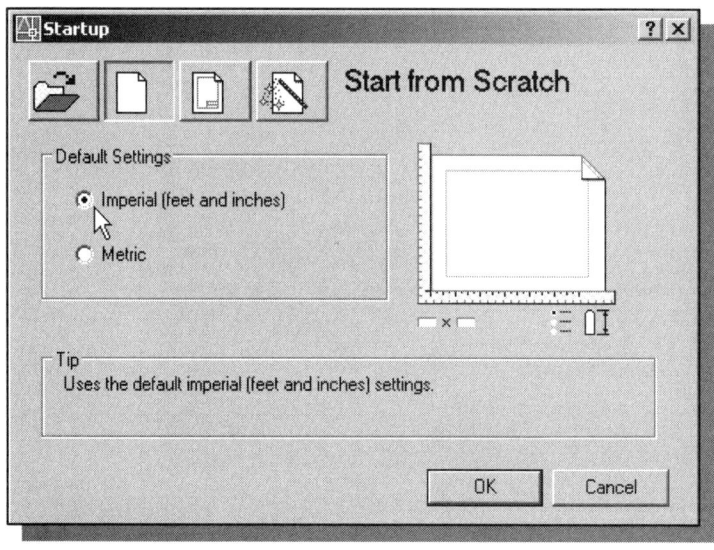

9. In the *Startup* dialog box, select the **Start from Scratch** option as shown in the figure.

10. Choose **Imperial** to use the English settings.

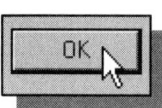

11. Click **OK** to accept the setting.

Creating the Rectangular Base of the Design

> We will first construct the wireframe geometry defining the rectangular base of the design.

1. In the *Status Bar* area, reset the options and turn **ON** the *GRID, POLAR, OSNAP, DYN, LWT* and *MODEL* options.

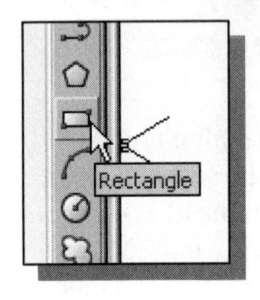

2. Select the **Rectangle** icon in the *Draw* toolbar. In the command prompt area, the message "*Specify first corner point or [Chamfer/Elevation/Fillet/Thickness/Width]:*" is displayed.

3. Place the first corner-point of the rectangle at the origin of the world coordinate system.
   Command: _line Specify first point: **0,0 [ENTER]**
       (Type **0,0** and press the **[ENTER]** key once.)

4. We will create a 4.5" × 3.0" rectangle by entering the absolute coordinates of the second corner.
   Specify other corner point or [Dimension]: **4.5,3 [ENTER]**

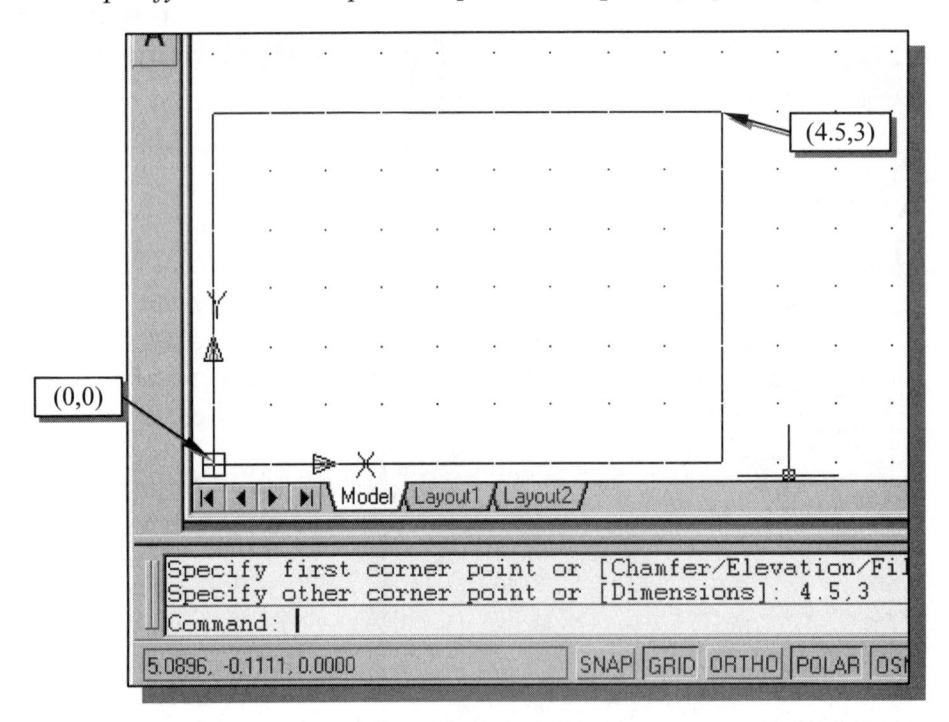

- The **Rectangle** command creates rectangles as *polyline* features, which means the four segments of a rectangle are created as a single object. In AutoCAD, rectangles are wireframe entities.

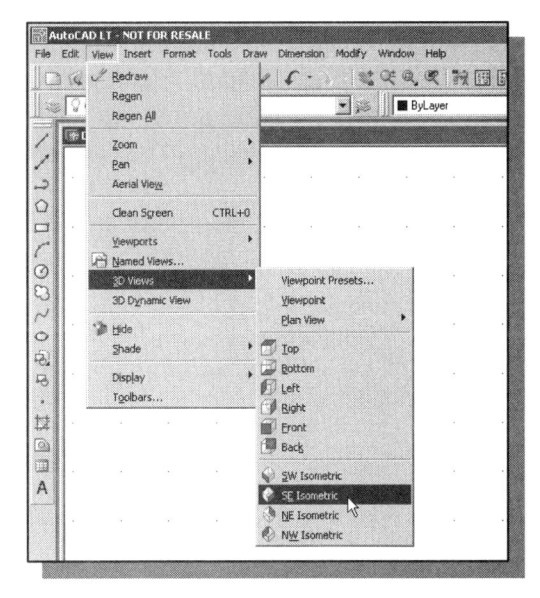

5. In the pull-down menus, select:

   [View] → [3D Views] → [SE Isometric]

   - Notice the orientation of the sketched 2D rectangle in relation to the displayed AutoCAD user coordinate system. By default, the 2D sketch-plane is aligned to the XY plane of the world coordinate system.

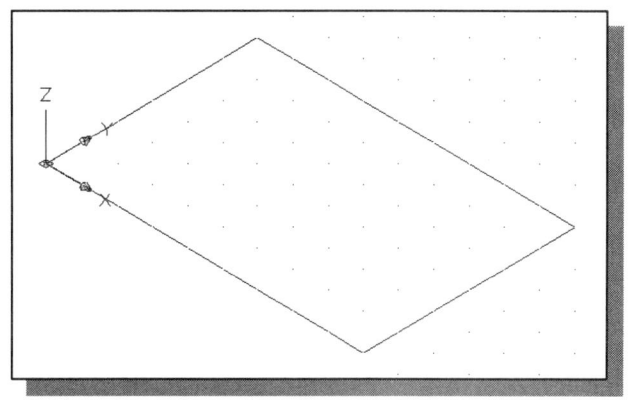

Create a 3D Box

- We will create a 3D box to define the 3D boundary of the design. We will do so by placing a copy of the base rectangle at the corresponding height elevation of the design. The dimensions of the 3D box are therefore based on the height, width and depth dimensions of the design.

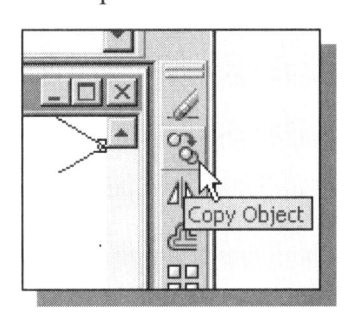

1. Click on the **Copy Object** icon in the *Modify* toolbar.

2. In the command prompt area, the message "*Select objects:*" is displayed. Pick any edge of the sketched rectangle.

3. Inside the graphics window, right-mouse-click once to accept the selection.

4. In the command prompt area, the message "*Specify base point or displacement, or [Multiple]:*" is displayed. Pick any corner of the sketched rectangle as a base point to create the copy.

5. In the command prompt area, the message "*Specify second point of displacement or <use first point as displacement>:*" is displayed.
    Enter: **@0,0,2.5 [ENTER]**
    (The three values are the X, Y and Z coordinates of the new location.)

3-8   AutoCAD® 2006 Tutorial: 3D Modeling

6. Select the **Zoom All** icon in the *Standard* toolbar to view the constructed geometry. If the **Zoom All** icon is not displayed, choose the icon by pressing down the left-mouse-button on the displayed icon to display the option list.

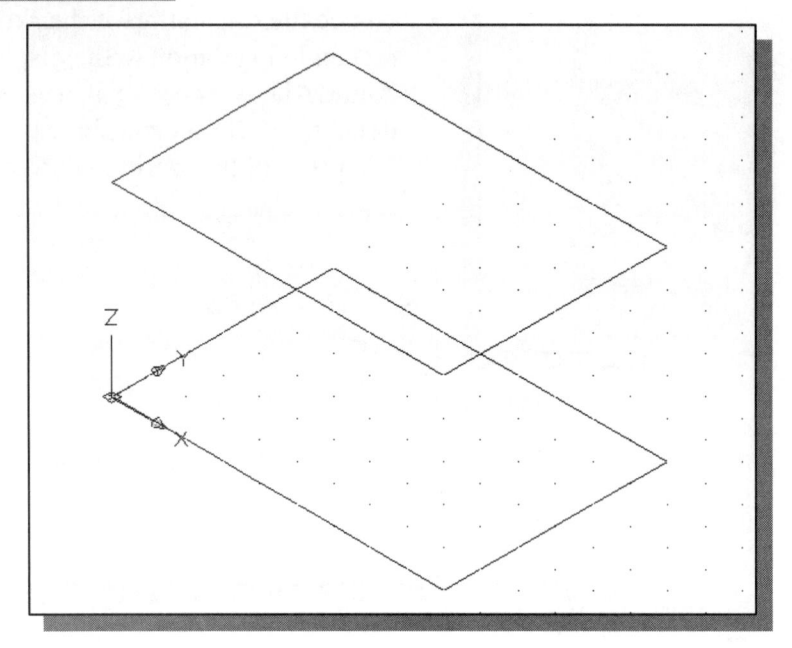

❖ The two rectangles represent the top and bottom of a 3D box defining the 3D boundary of the design. Note that the construction of the second rectangle was independent of the UCS, User Coordinate System; the UCS is still aligned to the world coordinate system.

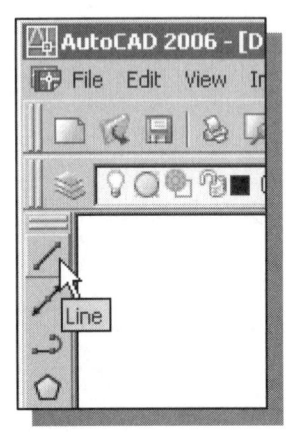

7. Select the **Line** icon in the *Draw* toolbar.

8. In the command prompt area, the message "*_line Specify first point:*" is displayed.

   Command: _line Specify first point: **0,0 [ENTER]**

9. In the command prompt area, the message "*Specify next point or [Undo]:*" is displayed.

   Command: _line Specify first point: **0,0,2.5 [ENTER]**

❖ Notice the Line command correctly identified the entered 3D coordinates of the second point. The default Z-coordinate, which is set by the AutoCAD UCS, is applied automatically whenever the Z-coordinates are omitted.

10. Inside the graphics window, right-mouse-click to activate the option menu and select **Enter** with the left-mouse-button to end the Line command.

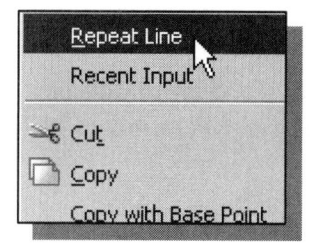

11. Inside the graphics window, **right-mouse-click** to bring up the popup option menu.

12. Pick **Repeat Line** with the left-mouse-button in the popup menu to repeat the last command.

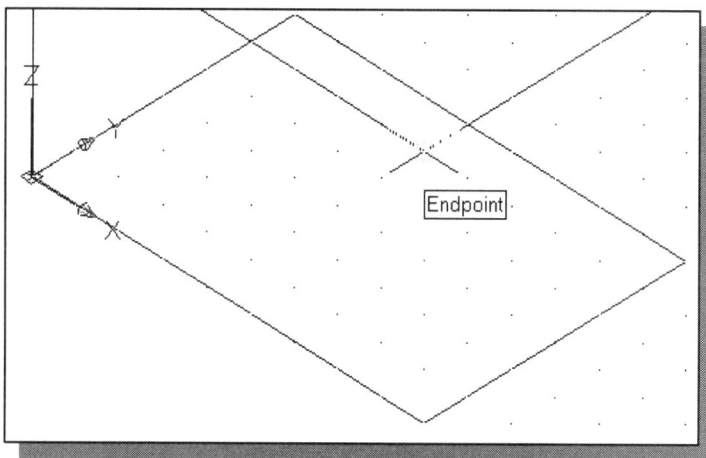

13. Move the cursor on top of the top front corner as shown. Note that the AutoCAD's *OSNAP* and *OTRACK* features identify geometric features, such as endpoints, automatically.

14. Left-click once to select the endpoint as shown.

15. Create a line connecting to the endpoint directly below the previously selected point.

16. On your own, complete the 3D box by creating the two lines connecting the back corners of the 3D box as shown.

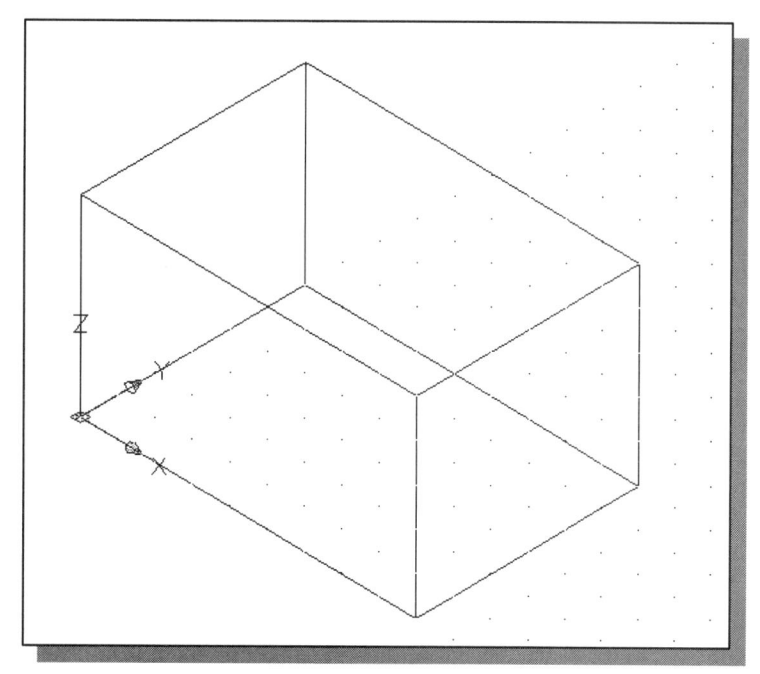

Object Snap Toolbar

1. Move the cursor to the *Standard* toolbar area and **right-mouse-click** once on any icon in the toolbar area to display a list of toolbar menu groups.

   ❖ AutoCAD provides many toolbars for access to frequently used commands, settings, and modes. The *Standard*, *Object Properties*, *Draw*, and *Modify* toolbars are displayed by default. The *check marks* in the list identify the toolbars that are currently displayed on the screen.

2. Select **Object Snap**, with the left-mouse-button, to display the *Object Snap* toolbar on the screen.

   ❖ *Object Snap* is an extremely powerful construction tool available on most CAD systems. During an entity's creation operations, we can snap the cursor to points on objects such as endpoints, midpoints, centers, and intersections. For example, we can turn on *Object Snap* and quickly draw a line to the center of a circle, the midpoint of a line segment, or the intersection of two lines.

3. On your own, move the cursor over the icons in the *Object Snap* toolbar and read the description of each icon in the *Status Bar* area.

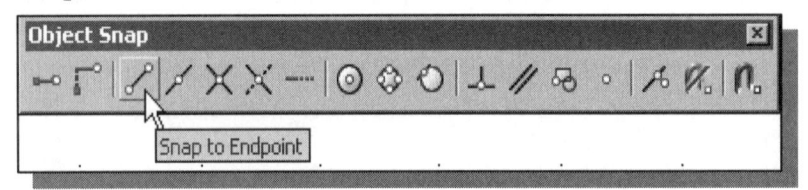

Using the Snap Options to Locate the Top Corners

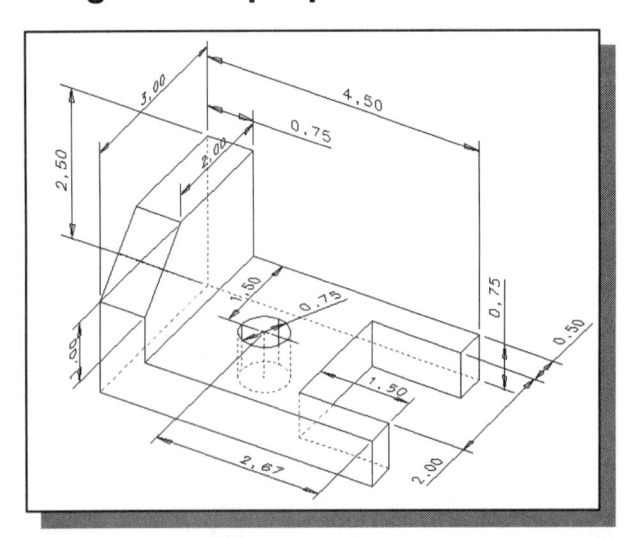

❖ We will use the *Object Snap* options to identify the locations of the top corners of the model.

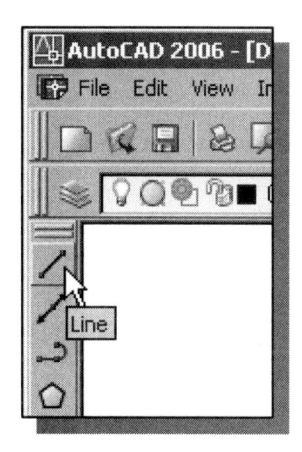

1. Select the **Line** icon in the *Draw* toolbar.

2. In the command prompt area, the message "*_line Specify first point:*" is displayed. Select **Snap From** in the *Object Snap* toolbar.

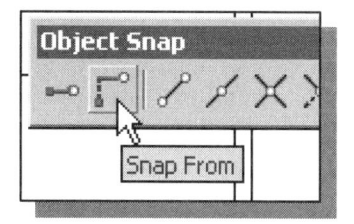

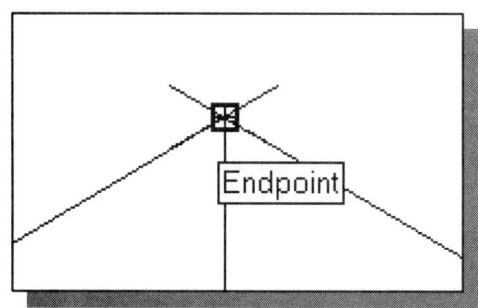

3. Select the **top back corner** as the reference point as shown.

4. In the command prompt area, the message "*_line Specify first point: from Base point:<Offset>:*" is displayed.

   Command: **@0.75,0,0 [ENTER]**

❖ By using the relative coordinate input method, we can locate the position of any point in 3D space. Note that the entered coordinates are measured relative to the current UCS.

5. In the command prompt area, the message "*Specify next point or [Undo]:*" is displayed.
   Command: Specify next point or [Undo]: **@0,-2,0 [ENTER]**

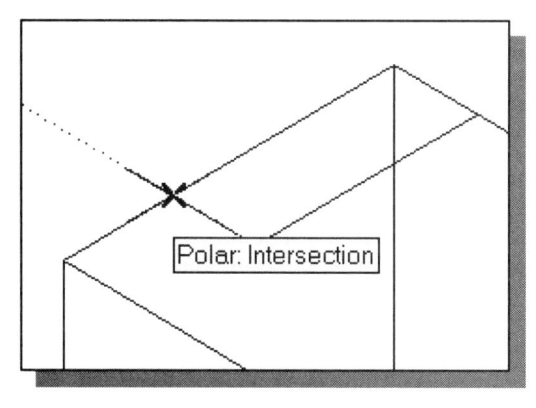

6. Move the cursor toward the left to create a perpendicular line. Select a location that is on the back line as shown, notice the displayed *Object Snap/Tracking* tips: Polar: Intersection.

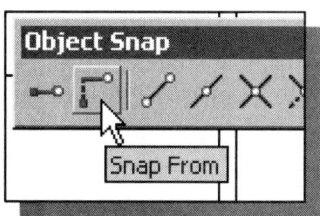

7. In the command prompt area, the message "*Specify next point or [Undo]:*" is displayed. Select **Snap From** in the *Object Snap* toolbar.

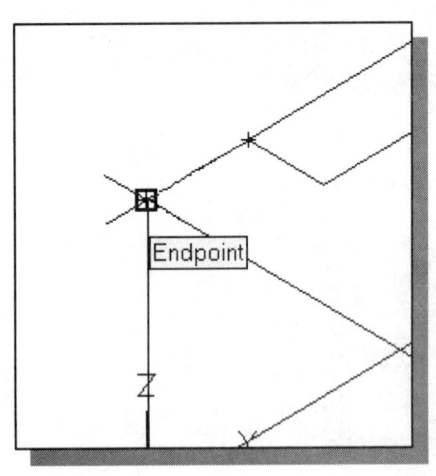

8. Select the **top front corner** as the reference point as shown.

9. In the command prompt area, the message "*Specify next point or [Close/Undo]:_from Base point <Offset>:*" is displayed.

   Command: **@0,0,-1 [ENTER]**

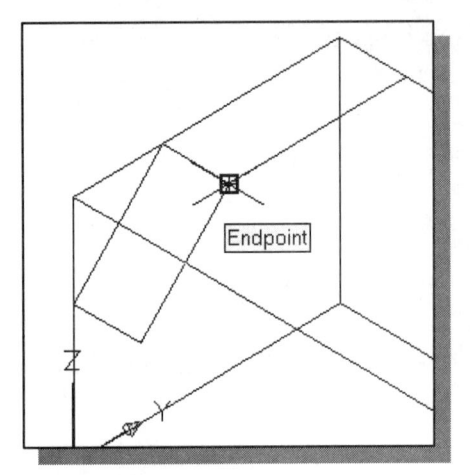

10. In the command prompt area, the message "*Specify next point or [Undo]:*" is displayed.

    Command: **@0.75,0,0 [ENTER]**

11. Move the cursor to the top corner as shown in the figure.

❖ Using the *Object Snap* options and the *relative coordinate input method* allow us to quickly locate points in 3D space.

12. Inside the graphics window, right-mouse-click to activate the option menu and select **Enter** with the left-mouse-button to end the Line command.

Using the *Copy* Option to Create Additional Edges

❖ The Copy option can also be used to create additional edges of the wireframe model.

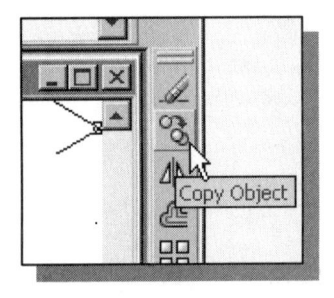

1. Click on the **Copy Object** icon in the *Modify* toolbar.

2. In the command prompt area, the message "*Select objects:*" is displayed. Pick any edge of the bottom rectangle.

3. Inside the graphics window, **right-mouse-click** once to accept the selection.

4. In the command prompt area, the message "*Specify base point or displacement, or [Multiple]:*" is displayed. Pick any corner of the base rectangle to be used as a base point to create the copy.

5. In the command prompt area, the message "*Specify second point of displacement or <use first point as displacement>:*" is displayed.
   Enter: **@0,0,0.75 [ENTER]**

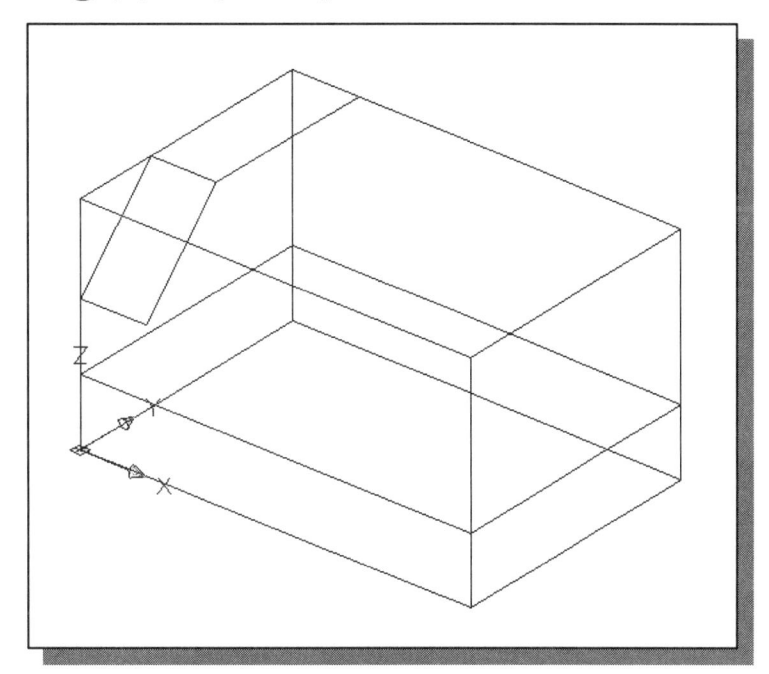

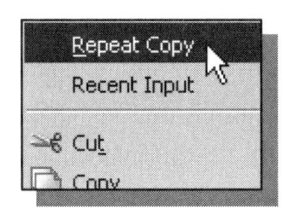

6. Inside the graphics window, right-mouse-click to bring up the popup option menu.

7. Pick **Repeat Copy Object** with the left-mouse-button in the popup menu to repeat the last command.

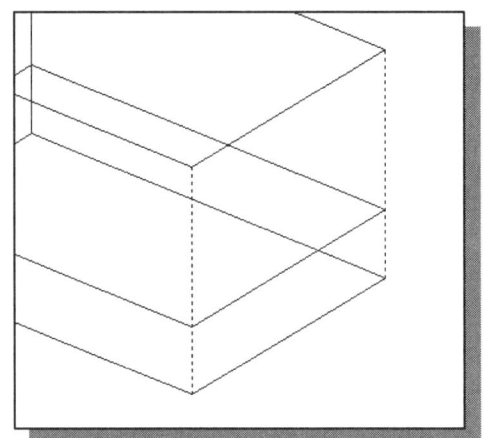

8. Pick the two **vertical lines** on the right side of the 3D box as shown.

9. Inside the graphics window, right-mouse-click once to accept the selection.

10. In the command prompt area, the message "*Specify base point or displacement, or [Multiple]:*" is displayed. Pick the **top back corner** of the wireframe as a base point to create the copy.

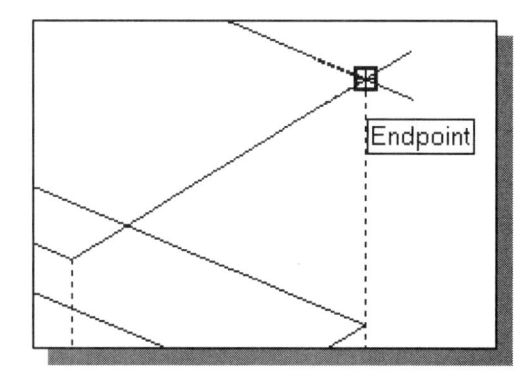

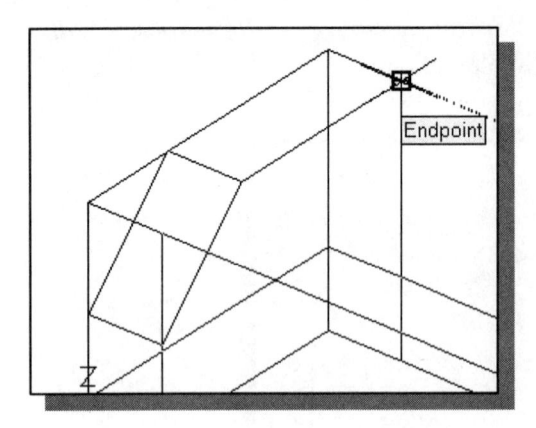

11. In the command prompt area, the message "*Specify second point of displacement or <use first point as displacement>:*" is displayed. Pick the **top back corner** of the wireframe model as shown.

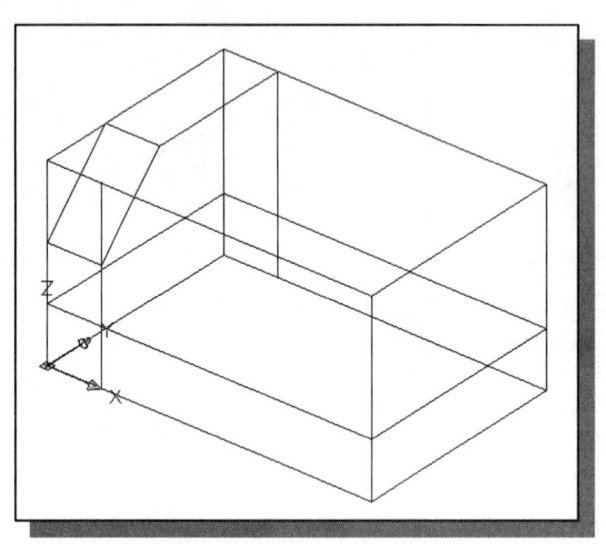

❖ The copy option is an effective way to create additional edges of wireframe models, especially when multiple objects are involved. With wireframe models, the emphasis is placed on the corners and edges of the model.

Using the Trim Command

❖ The Trim command can be used to shorten objects so that they end precisely at selected boundaries.

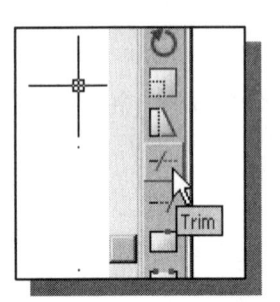

1. Select the **Trim** command icon in the *Modify* toolbar. In the command prompt area, the message "*Select boundary edges... Select objects:*" is displayed.

    ❖ First we will select the objects that define the boundary edges to which we want to trim the object.

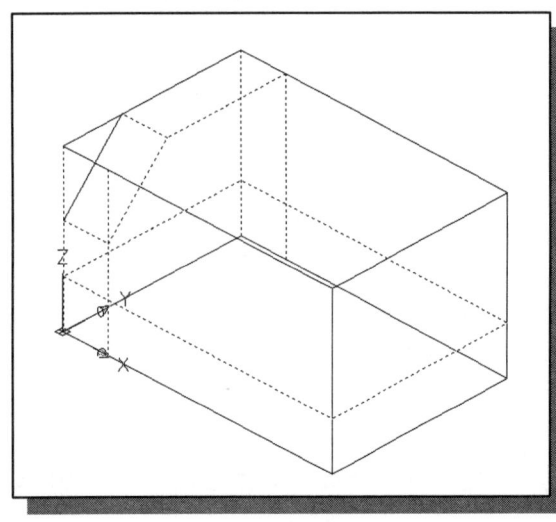

2. Pick the highlighted edges as shown in the figure; these edges are the *boundary edges*.

3. Inside the graphics window, **right-mouse-click** to accept the selection of boundary edges and proceed with the Trim command.

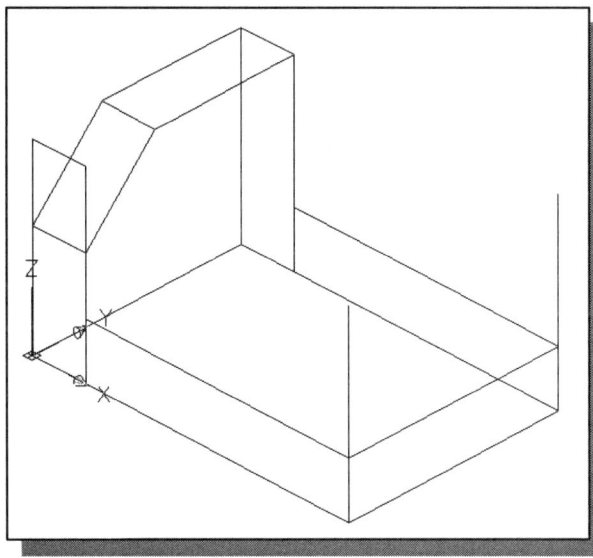

3D Wireframe Modeling   3-15

4. The message "*Select object to trim or shift-select object to extend or [Project/Edge/Undo]:*" is displayed in the command prompt area. Pick the portions of the entities to be trimmed so that the model appears as shown.

❖ Note that entities perpendicular to the UCS XY plane cannot be trimmed with the default AutoCAD trim settings.

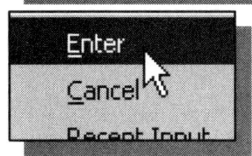

5. Inside the graphics window, right-mouse-click to bring up the option menu and select **Enter** to end the Trim command.

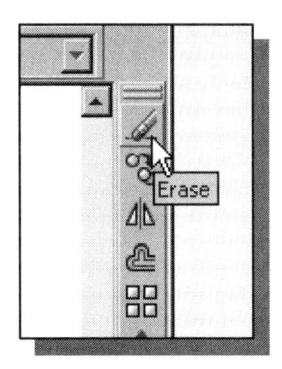

6. Pick **Erase** in the *Modify* toolbar. (The icon is the first icon in the *Modify* toolbar.) The message "*Select objects*" is displayed in the command prompt area and AutoCAD awaits us to select the objects to erase.

7. Pick the **top edge** on the front face as shown.

8. **Right-mouse-click** inside the graphics window to accept the selection.

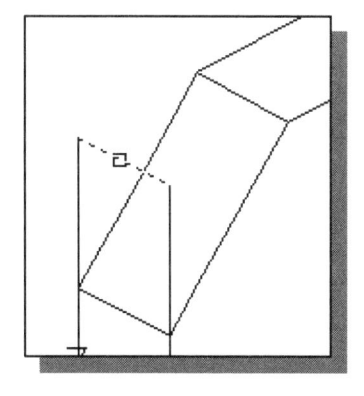

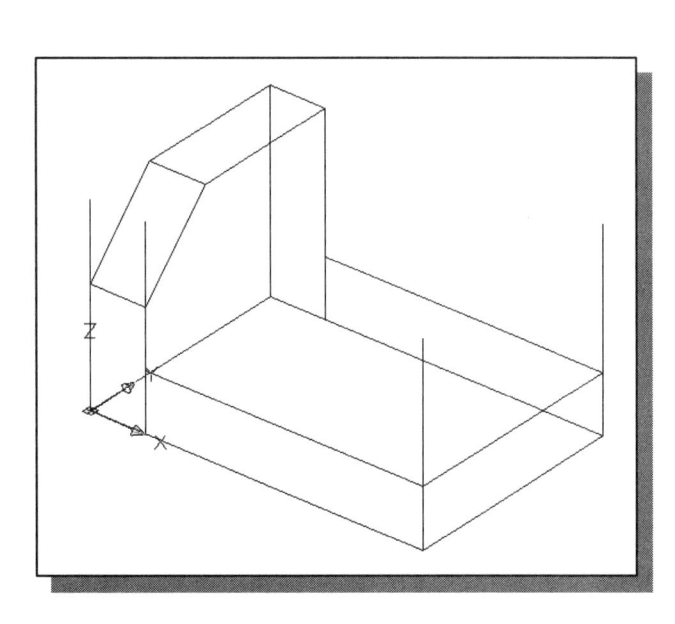

Using the Trim-Project-View Option

❖ The **Trim-Project-View** option allows us to trim entities that are perpendicular to the UCS XY plane.

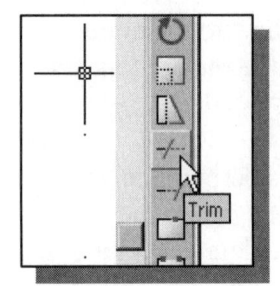

1. Select the **Trim** command icon in the *Modify* toolbar. In the command prompt area, the message *"Select boundary edges... Select objects:"* is displayed.

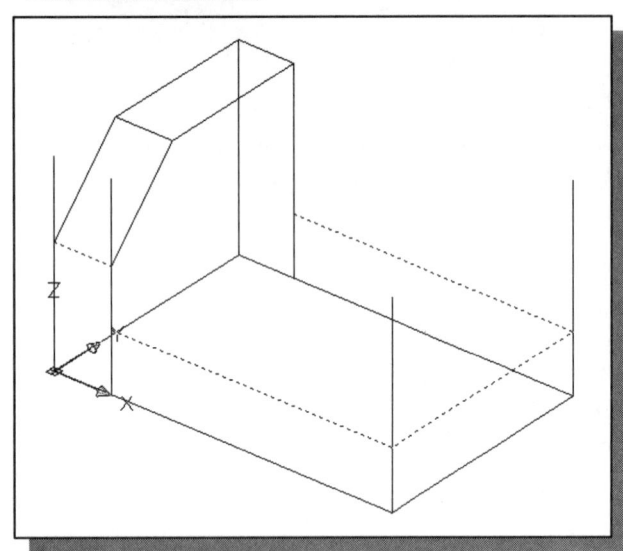

2. Pick the highlighted edges as shown in the figure; these edges are the *boundary edges*.

3. Inside the graphics window, right-mouse-click to accept the selected boundary edges.

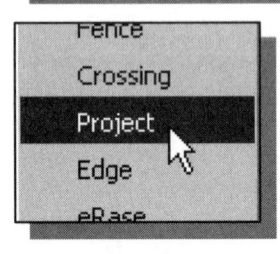

4. The message *"Select object to trim or shift-select object to extend or [Project/Edge/Undo]:"* is displayed in the command prompt area. Inside the graphics window, right-mouse-click to bring up the option menu and select **Project** as shown.

❖ Note that the default **Trim-Projection** is set to the *UCS XY* plane. Any entities that are perpendicular to the projection plane cannot be trimmed.

5. The message *"Enter a projection option [None/Ucs/View] <UCS>:"* is displayed in the command prompt area. Inside the graphics window, right-mouse-click to bring up the option menu and select **View** as shown.

❖ Changing the projection setting to **View** allows us to modify basically any viewable entities.

3D Wireframe Modeling   3-17

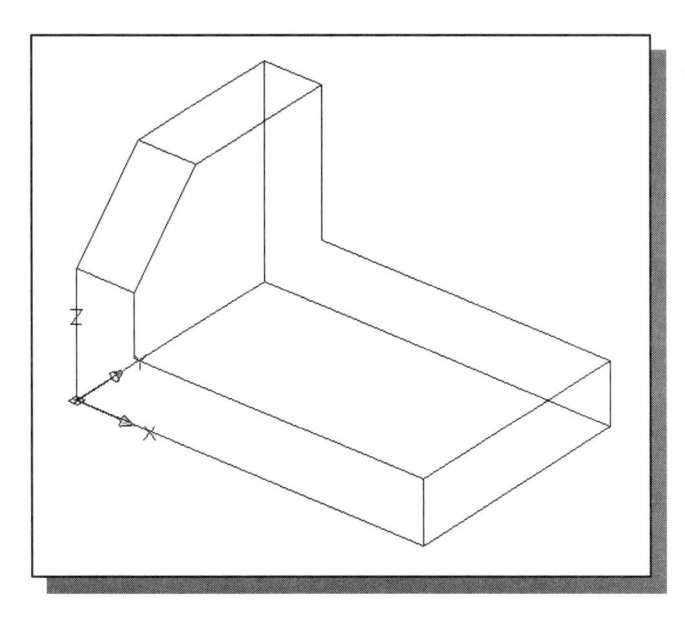

6. The message "*Select object to trim or shift-select object to extend or [Project/Edge/Undo]:*" is displayed in the command prompt area. Pick the portions of the entities to be trimmed so that the model appeared as shown.

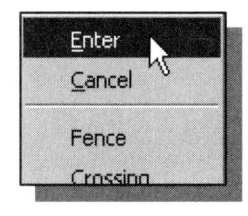

7. Inside the graphics window, right-mouse-click to bring up the option menu and select **Enter** to end the Trim command.

8. On your own, use the Line command to complete the inside corner of the wireframe model as shown.

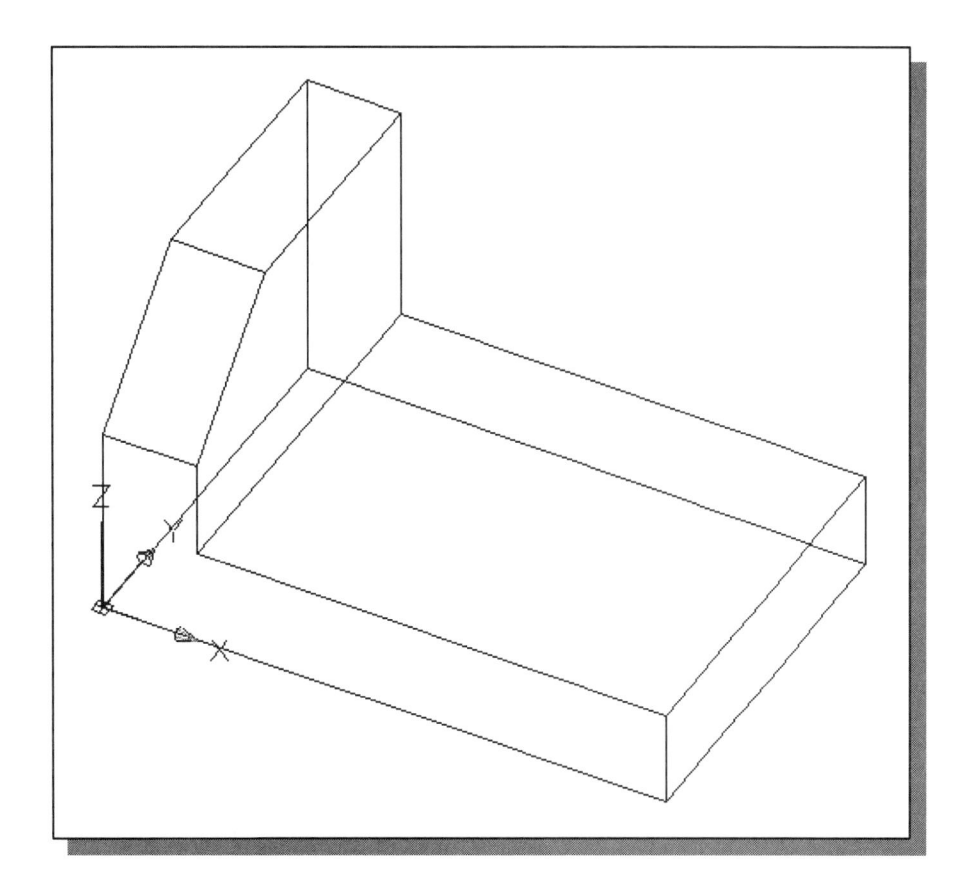

Using the View Toolbar

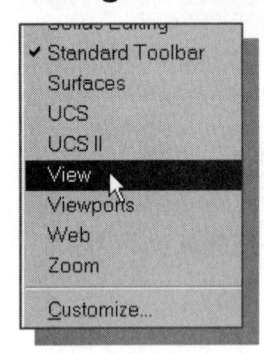

1. Move the cursor on top of any icon and **right-click** on any icon of the *Standard* toolbar to display a list of toolbar menu groups.

2. Select **View**, with the left-mouse-button, to display the *View* toolbar on the screen.

➢ The *View* toolbar contains two sections of icons that allow us to quickly switch to standard 2D and 3D views.

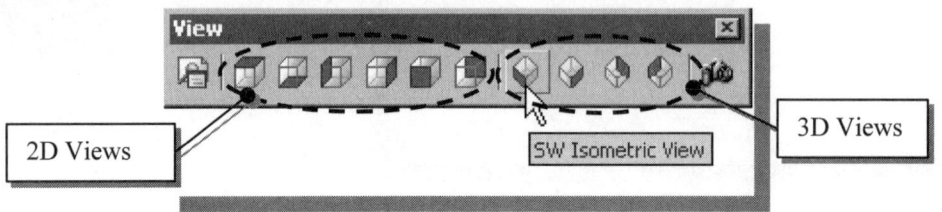

3. On your own, changing to the different viewing positions by selecting the icons in the *View* toolbar.

Dynamic Rotation – 3D Orbit

1. Select **3D Orbit** in the *View* pull-down menu.
   [View] → [3D Orbit]

❖ The 3D-Orbit view displays an ***arcball***, which is a circle, divided into four quadrants by smaller circles. 3D-Orbit enables us to manipulate the view of 3D objects by clicking and dragging with the left-mouse-button.

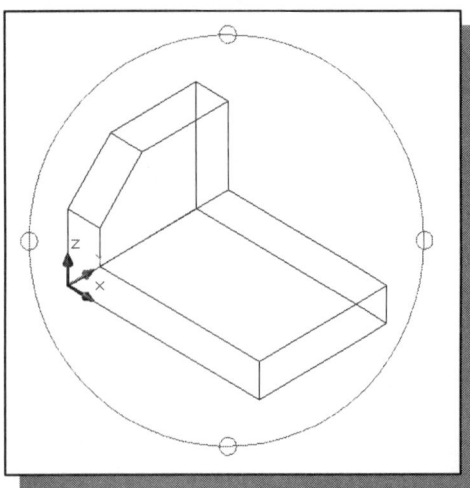

2. Inside the *arcball*, press down the left-mouse-button and drag it up and down to rotate about the screen X-axis. Dragging the mouse left and right will rotate about the screen Y-axis.

3. Move the cursor to different locations on the screen, outside the *arcball* or on one of the four small circles, and experiment with the real-time dynamic rotation feature of the **3D-Orbit** command.

Using the *Offset* Command to Create Parallel Edges

❖ The AutoCAD **Offset** option can also be used to create edges that are at specific distances to existing 3D wireframe edges. Prior to creating the parallel edges, we will first convert one of the polylines to individual line segments.

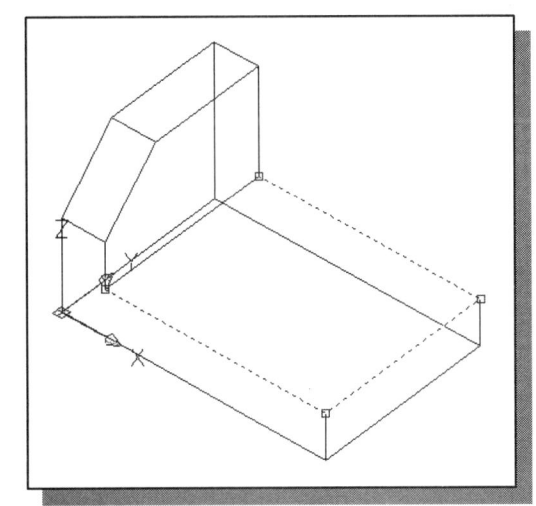

1. Pick the top right edge of the wireframe model.

   - **AutoCAD® 2006** provides a flexible graphical user interface that allows users to select graphical entities BEFORE the command is selected (*Pre-selection*), or AFTER the command is selected (*Post-selection*). The polyline we have pre-selected consists of three line-segments.

2. Select the **Explode** command icon in the *Modify* toolbar as shown.

3. Pick **Offset** in the *Modify* toolbar as shown.

4. In the command prompt area, the message "*Specify offset distance or [Through] <Through>:*" is displayed.
   Enter: **0.5 [ENTER]**

5. Select the front line segment of the polyline we just exploded.

6. In the command prompt area, the message "*Specify point on side to offset:*" is displayed. Pick a location that is above the line to create the first parallel line.

7. Select the back line segment of the polyline we exploded and create another parallel line as shown.

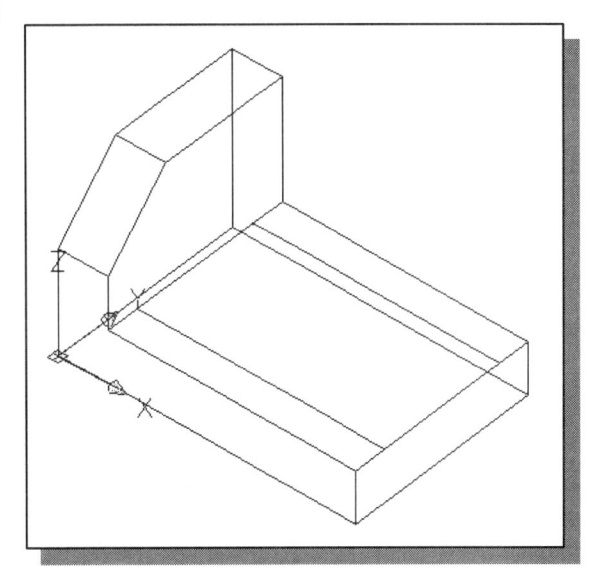

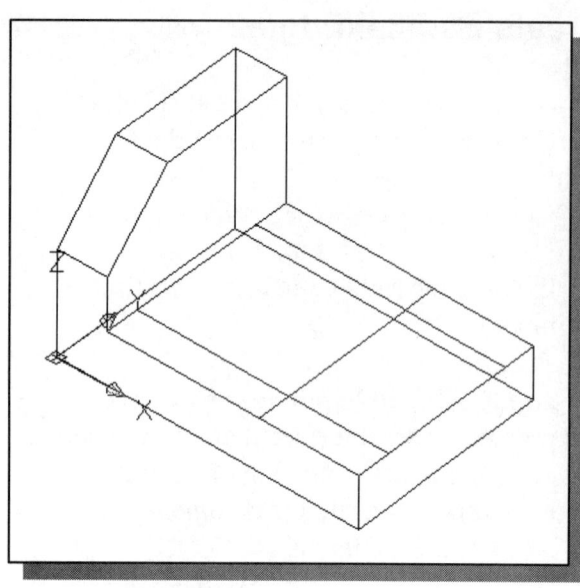

8. On your own, repeat the above steps and create another parallel line (distance: **1.5**) as shown.

   - In creating parallel edges, the **Offset** command can be viewed as an alternative to the **Copy** command. In 3D wireframe modeling, the focus is on identifying the edges and corners of the model. Most of the 2D construction tools are also applicable in 3D environment.

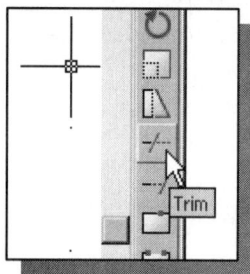

9. Select the **Trim** command icon in the *Modify* toolbar.

10. In the command prompt area, the message "*Select boundary edges... Select objects:*" is displayed. Pick the highlighted edges as shown in the figure; these edges are the *boundary edges*.

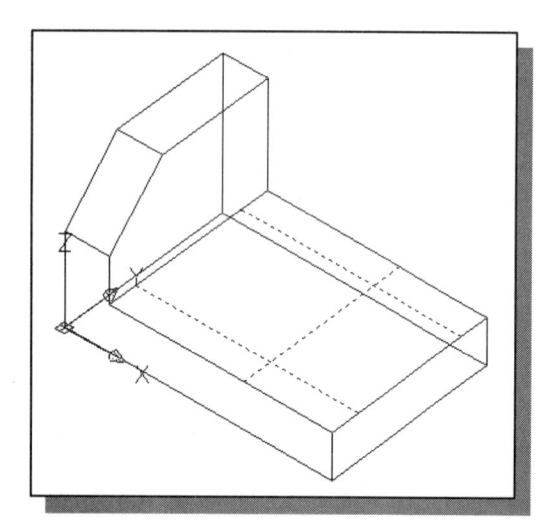

11. Inside the graphics window, right-mouse-click to accept the selection of boundary edges and proceed with the **Trim** command.

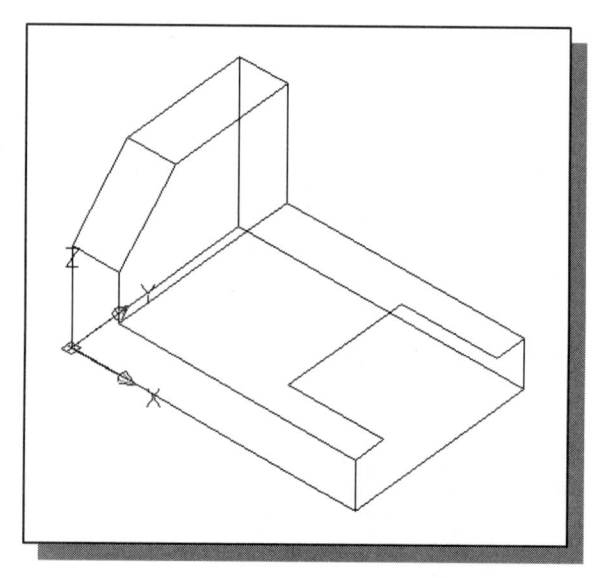

12. On your own, trim the line segments so that the wireframe model appears as shown.

Creating a Circle above the UCS Sketch Plane

1. Select the **Circle** command icon in the *Draw* toolbar.

- By default, the XY plane of the UCS defines the sketching plane for constructing 2D geometric entities.

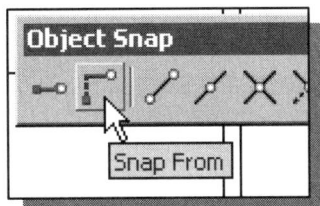

2. In the command prompt area, the message "*_circle Specify center point for circle or[3P/2P/Ttr (tan tan radius)]:*" is displayed. Select **Snap From** in the *Object Snap* toolbar.

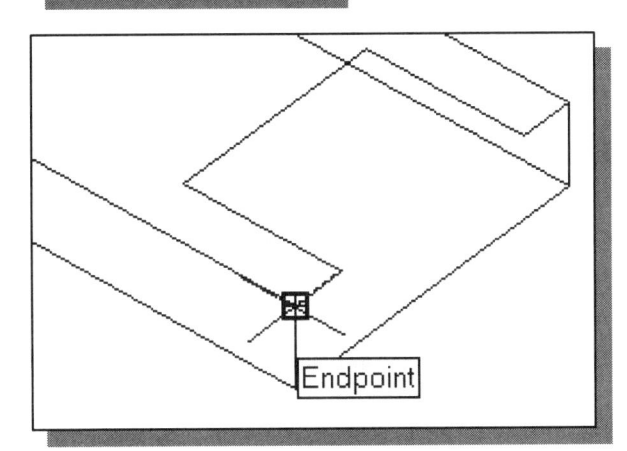

3. Select the **top right corner** as the reference point as shown.

4. In the command prompt area, the message "*Specify next point or [Close/Undo]:_from Base point <Offset>:*" is displayed.

   Command: **@-2.67,1.5,0 [ENTER]**

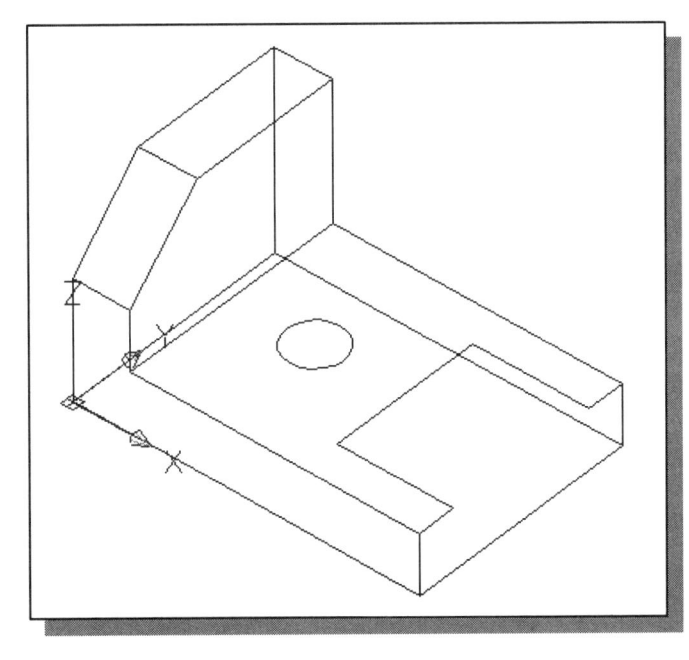

5. In the command prompt area, the message "*Specify radius of circle or [Diameter]:*" is displayed.

   Command: **0.375 [ENTER]**

- The circle is created above the sketching plane with the **Snap From** option.

Completing the Wireframe Model

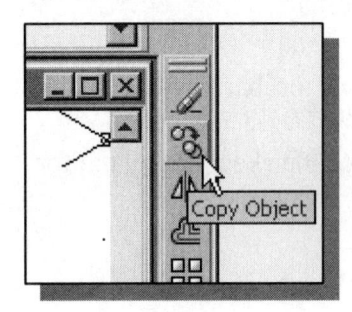

1. Click on the **Copy Object** icon in the *Modify* toolbar.

2. In the command prompt area, the message "*Select objects:*" is displayed. Pick the edges and the circle as shown in the figure.

3. Inside the graphics window, right-mouse-click once to accept the selection.

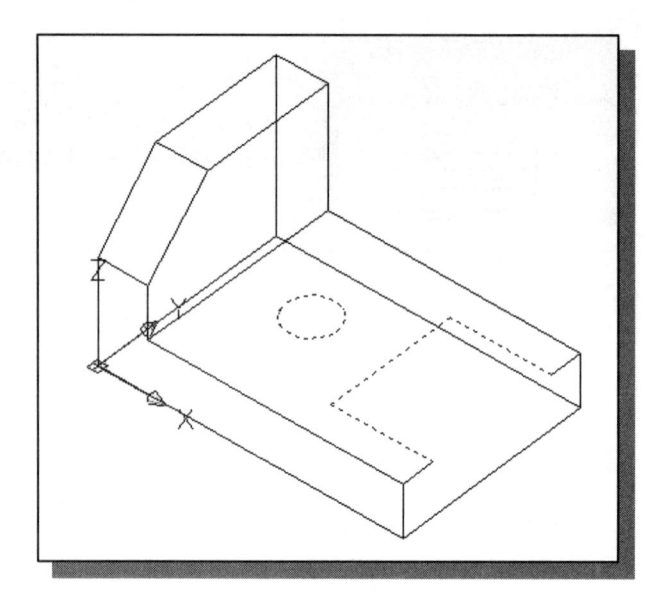

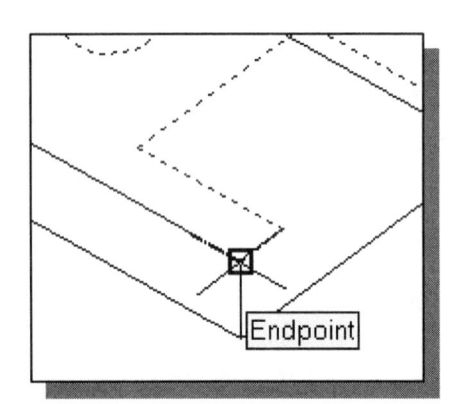

4. In the command prompt area, the message "*Specify base point or displacement, or [Multiple]:*" is displayed. Pick the front right corner as shown.

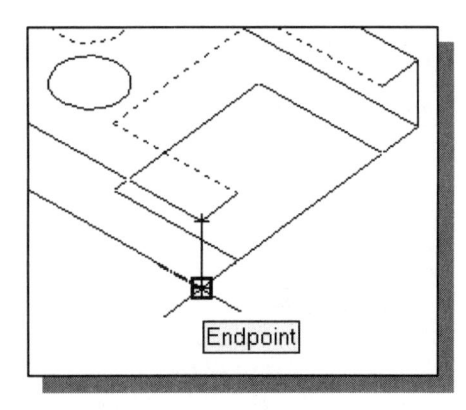

5. In the command prompt area, the message "*Specify second point of displacement or <use first point as displacement>:*" is displayed. Pick the bottom-right corner as shown.

3D Wireframe Modeling   3-23

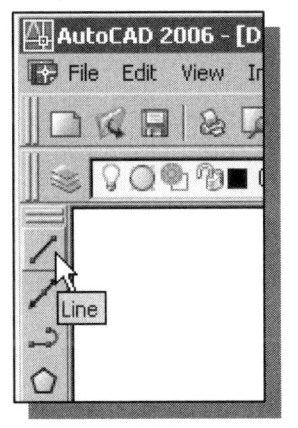

6. Select the **Line** icon in the *Draw* toolbar.

7. On your own, create the lines connecting the corners of the created edges as shown in the below figure.

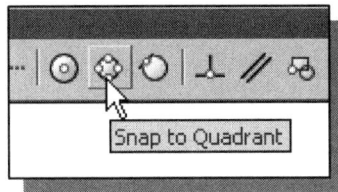

8. Use the **Snap to Quadrant** option to create edges in between the two circles.

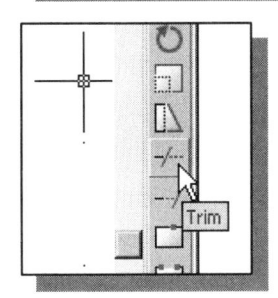

9. Select the **Trim** command icon in the *Modify* toolbar.

10. On your own, trim the center portion of the bottom right edge and complete the wireframe model as shown.

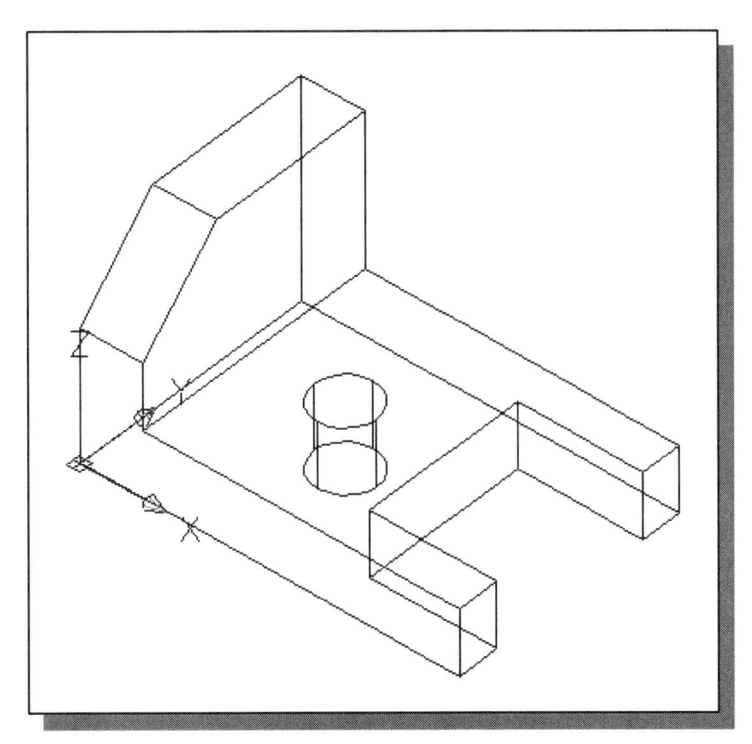

Questions:

1. Describe some of the control options available with the **3D Orbit** command.

2. List and describe two different methods to create 3D edges from existing 3D edges in **AutoCAD® 2006**.

3. How many of the UCS options were used to create the 3D model in this chapter and how many were used to create the model in the previous chapter?

4. When and why would you use the **Trim-Project-View** option?

5. Identify the following commands:

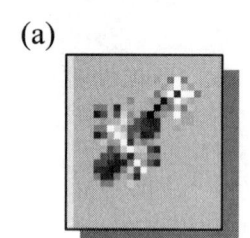

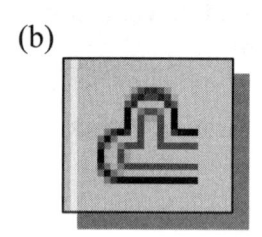

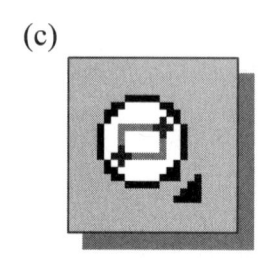

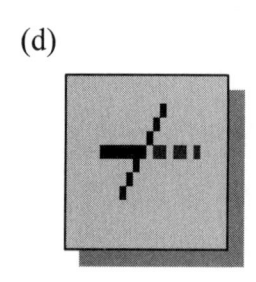

   (a)

   (b)

   (c)

   (d)

Exercises: All dimensions are in inches.

1.

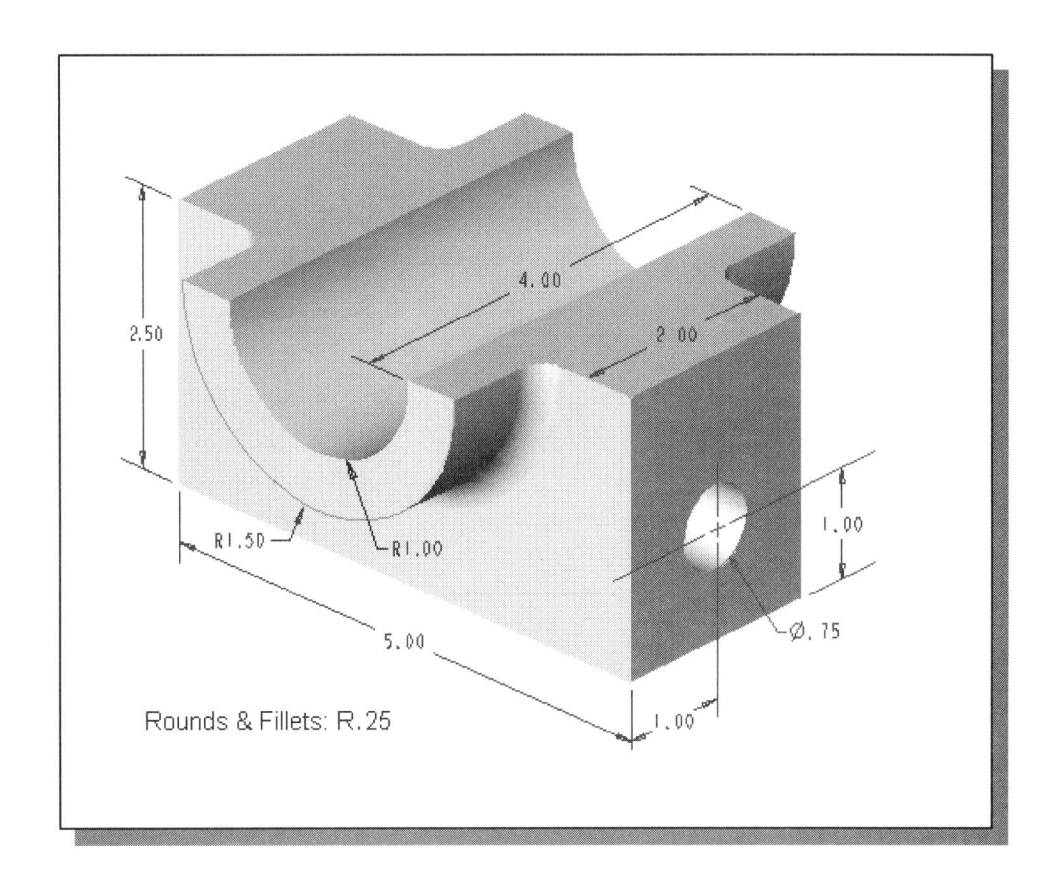

2.

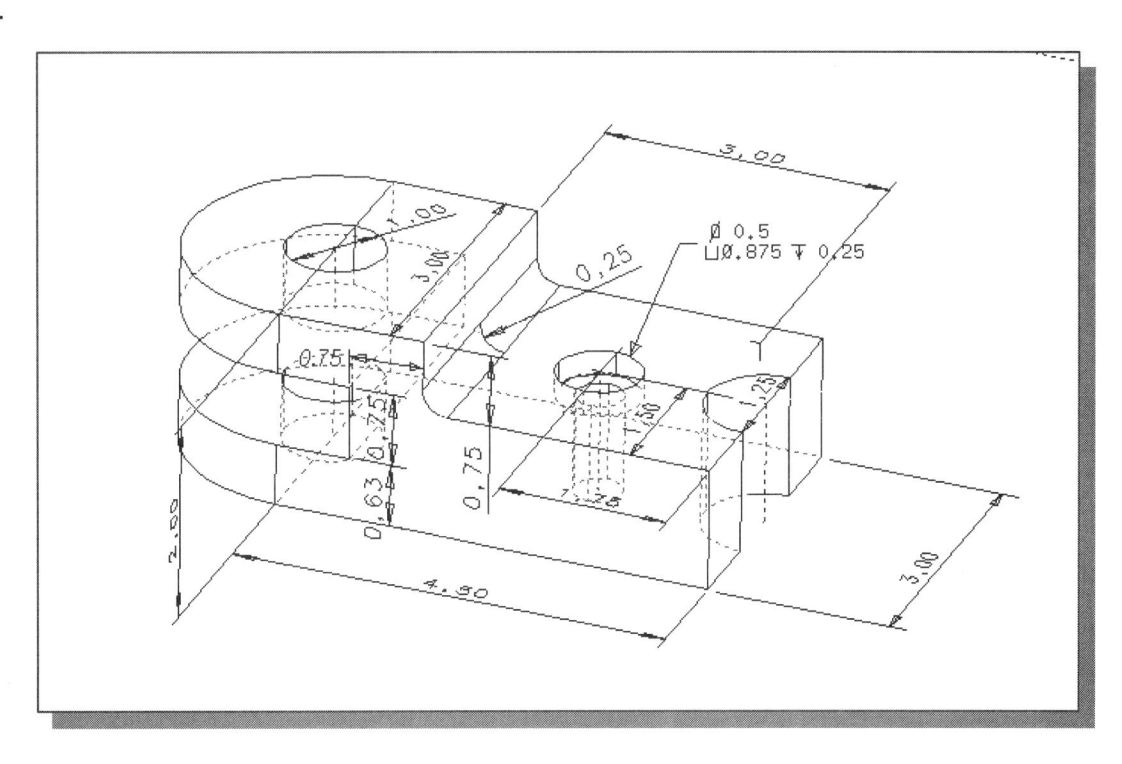

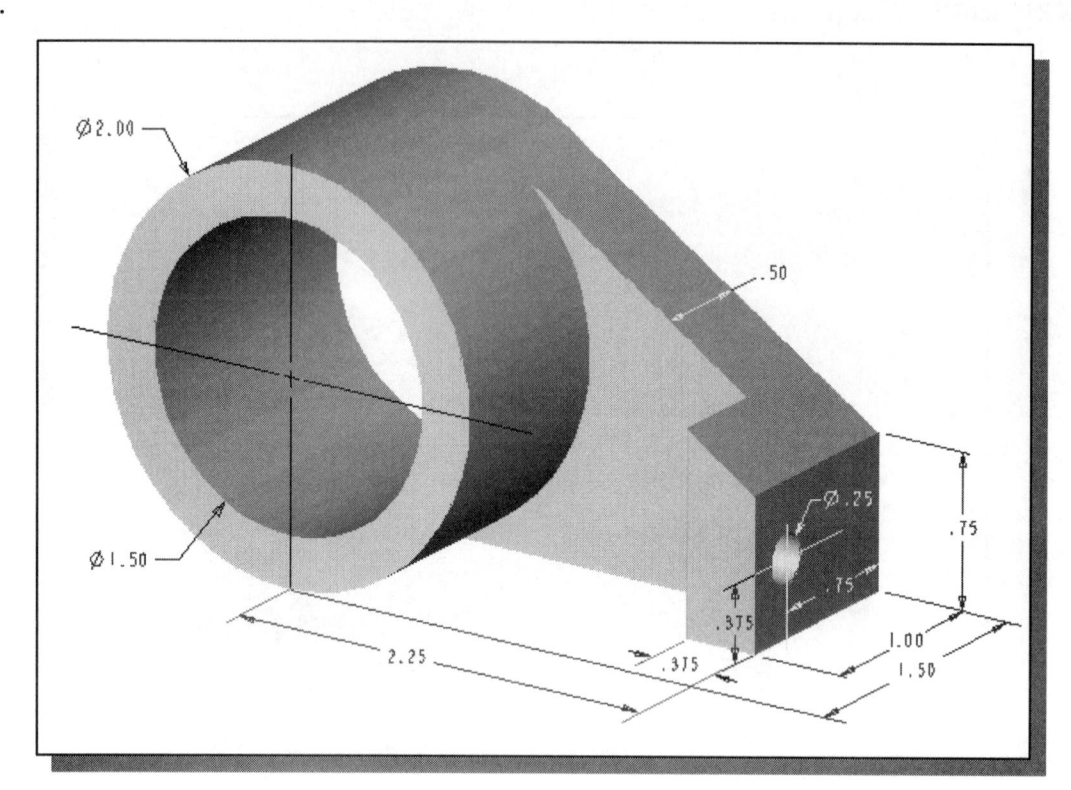

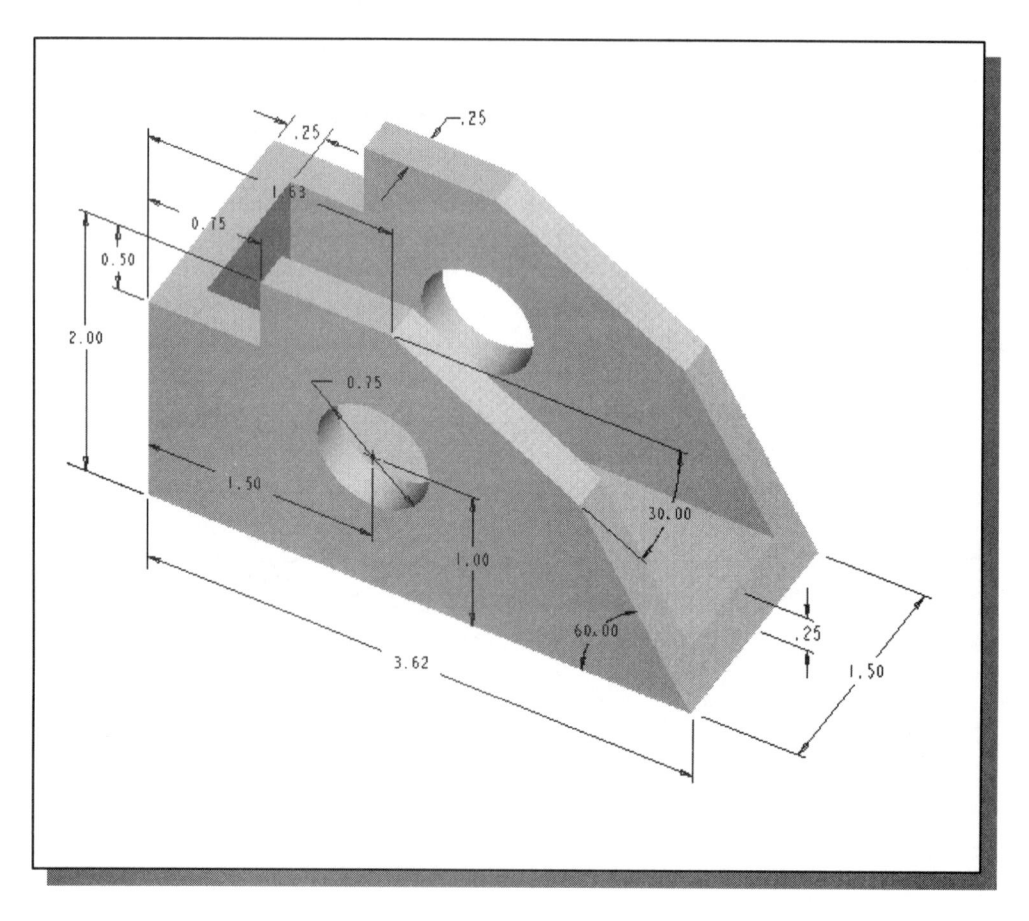

3.

4.

Chapter 4
UCS, Viewports and Wireframe Modeling

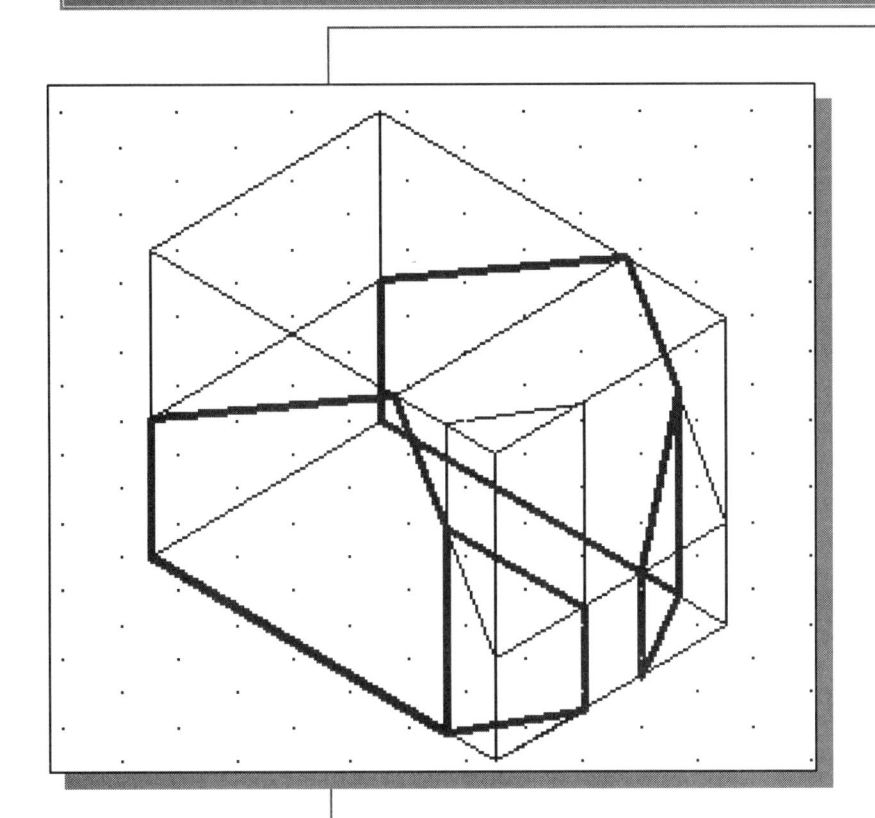

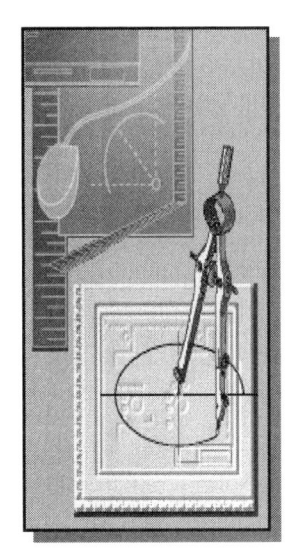

Learning Objectives

- **Understand the Use of UCS on More Complex Designs**
- **Display and Orient 3D Views**
- **Apply and Expand the Basic Wireframe Construction Techniques**
- **Using Additional Tools to Assist the Creation of 3D Wireframe Models**
- **Use the Named Views Option**

Introduction

In the previous chapter, we examined some of the basic techniques to create 3D wireframe models. Again, a wireframe model is a skeletal description of a 3D object. There are no surfaces in a wireframe model; it consists only of points, lines, and curves that describe the edges of the object.

In this chapter, we will further discuss the techniques and wireframe modeling tools available in AutoCAD. Instead of creating the corners and edges of wireframe models by entering the precise 3D coordinates, the use of construction lines and referencing techniques are illustrated. The construction of edges in 3D space can be simplified by using the AutoCAD User Coordinate System. We will also use the multiple viewports option, which is available in most CAD systems, to aid the construction of 3D wireframe models.

It is important to point out that most of the construction tools used in creating 2D designs, such as *Object Snap* and *Polar Tracking*, as well as the **Copy**, **Mirror**, and **Array** commands, are also applicable to the constructions of 3D wireframe models.

The *V-Block* Design

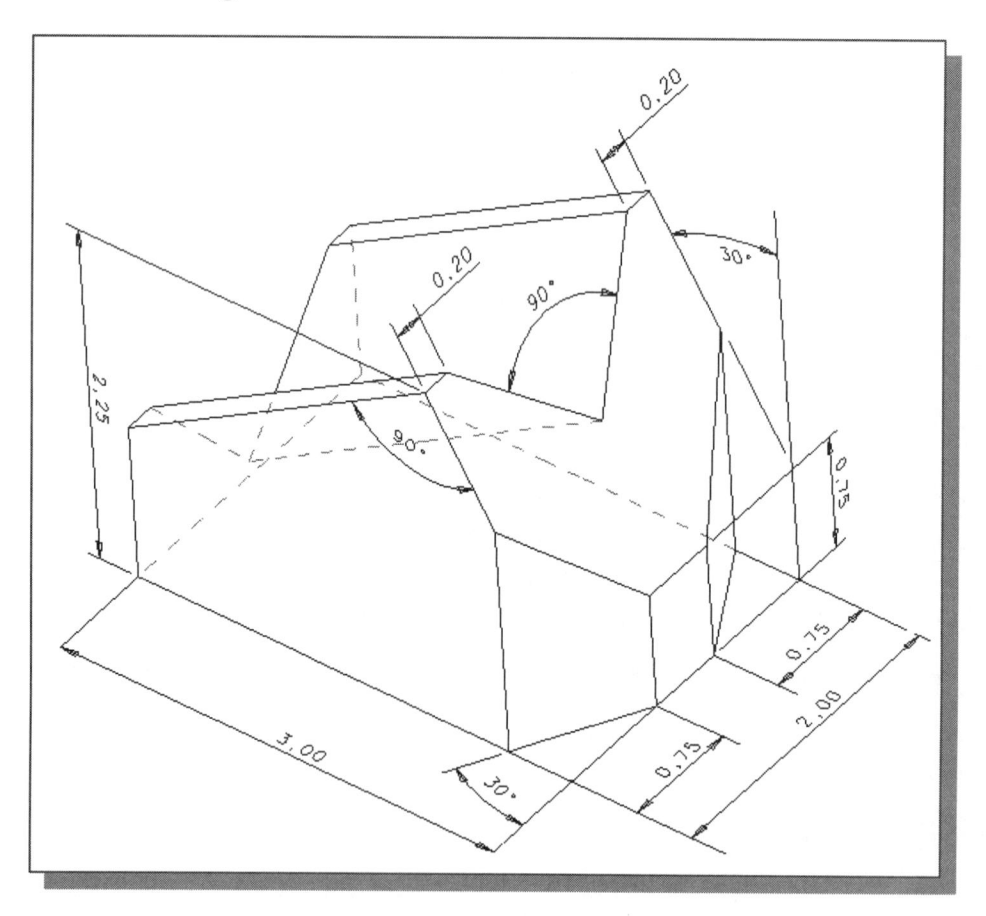

Starting Up AutoCAD® 2006

1. Select the **AutoCAD 2006** option on the *Program* menu or select the **AutoCAD 2006** icon on the *Desktop*.

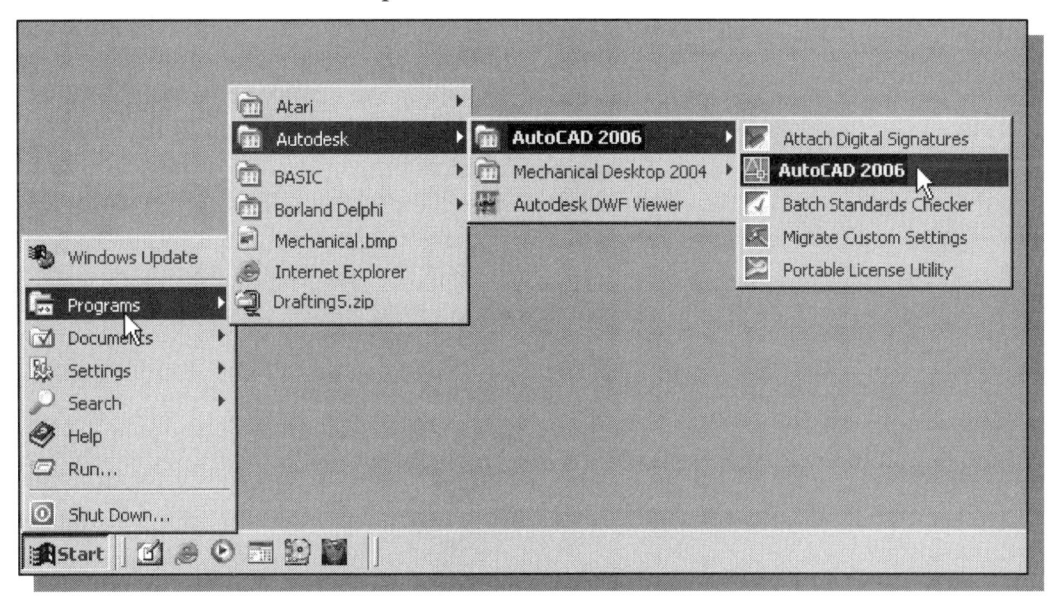

2. In the startup window, select **Start from Scratch**, as shown in the figure below.

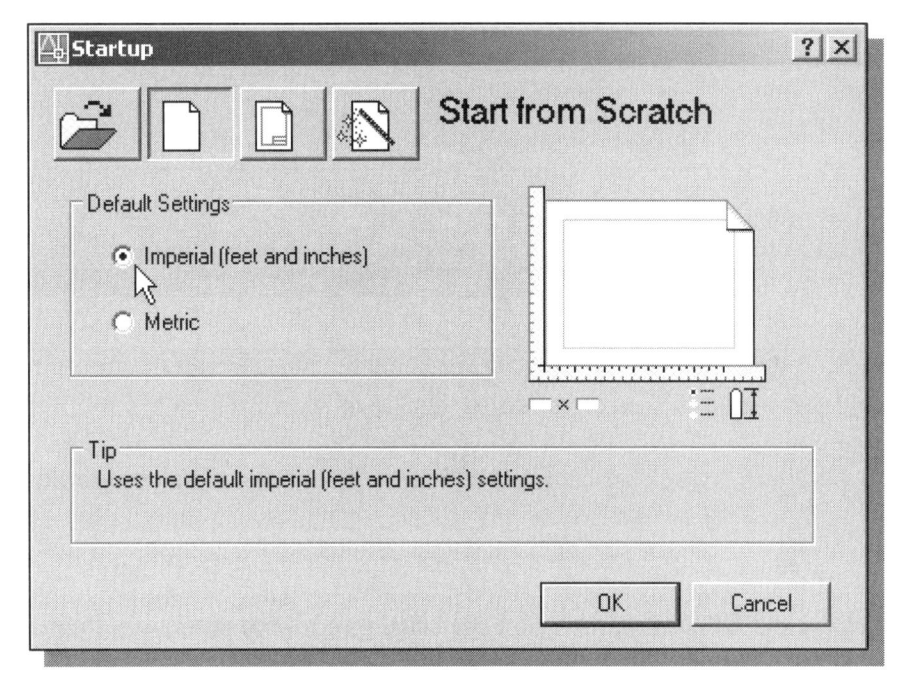

3. In the *Default Settings* section, pick **Imperial (feet and inches)** as the drawing units.

4. Pick **OK** in the startup dialog box to accept the selected settings.

Layers Setup

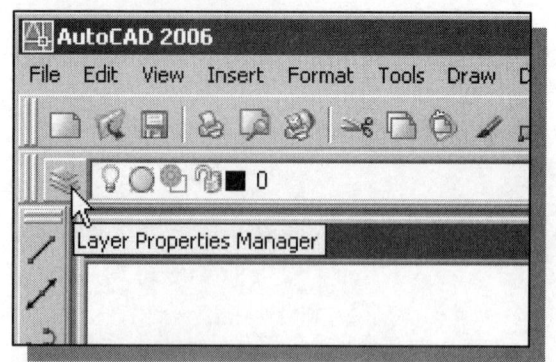

1. Pick **Layer Properties Manager** in the *Object Properties* toolbar. The *Layer Properties Manager* dialog box appears. AutoCAD creates a default layer, *Layer 0*, which we cannot rename or delete. *Layer 0* has special properties used by the system.

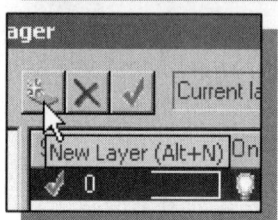

2. Click on the **New** icon to create new layers.

3. Create **two new layers** with the following settings:

Layer	Color	LineType	LineWeight
Construction	White	Continuous	Default
Object	Cyan	Continuous	0.30mm

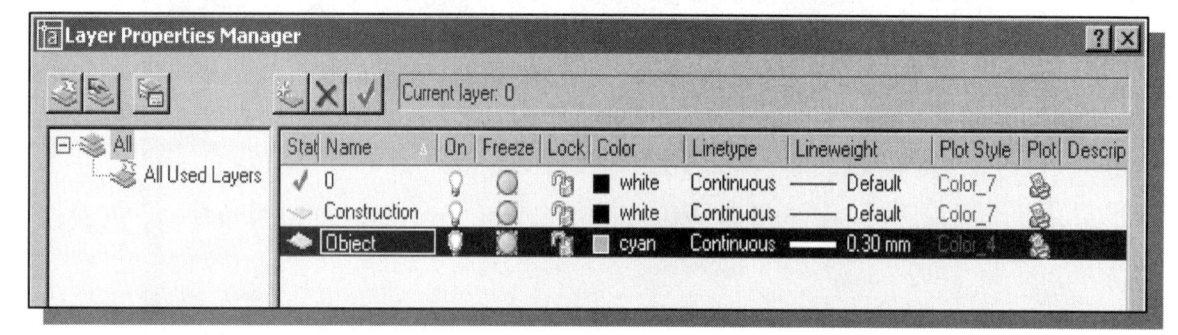

4. Highlight the layer *Construction* in the list of layers.

5. Click on the **Current** icon to set layer *Construction* as the *Current Layer*.

6. Click on the **OK** button to accept the settings and exit the *Layer Properties Manager* dialog box.

7. In the *Status Bar* area, reset the options and turn **ON** the *GRID, POLAR, OSNAP, OTRACK, DYN, LWT* and *MODEL* options.

Creating the Rectangular Base of the Design

➤ We will first use construction geometry to define the outer edges of the design.

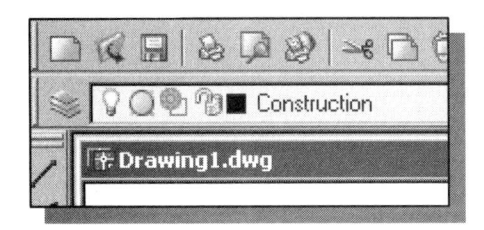

1. In the *Layer Control* box, confirm that layer *Construction_Lines* is set as the *Current Layer*.

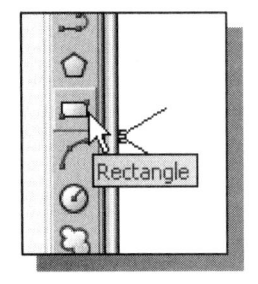

2. Select the **Rectangle** icon in the *Draw* toolbar. In the command prompt area, the message "*Specify first corner point or [Chamfer/Elevation/Fillet/Thickness/Width]:*" is displayed.

3. Place the first corner point of the rectangle near the lower left corner of the screen. Do not be overly concerned about the actual coordinates of the location; the drawing space is as big as you can imagine.

4. We will create a 3" × 2" rectangle by entering the coordinates of the second corner using the relative coordinates input method.

   Specify other corner point or [Dimension]: **@3,2 [ENTER]**

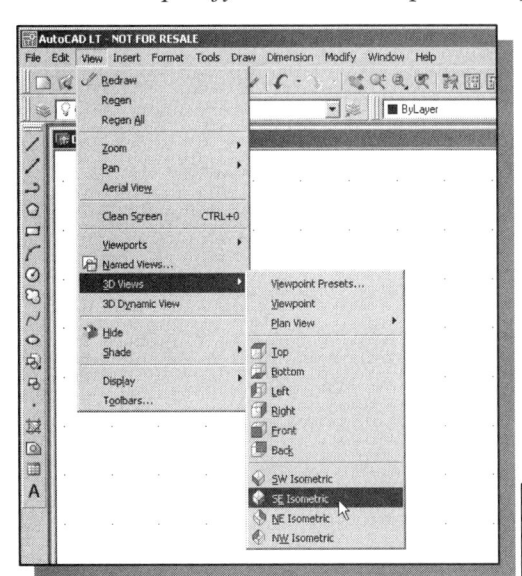

5. In the pull-down menus, select:

   [View] → [3D Views] → [SE Isometric]

   • Notice the orientation of the sketched 2D rectangle in relation to the displayed AutoCAD user coordinate system. By default, the 2D sketch-plane is aligned to the XY plane of the world coordinate system.

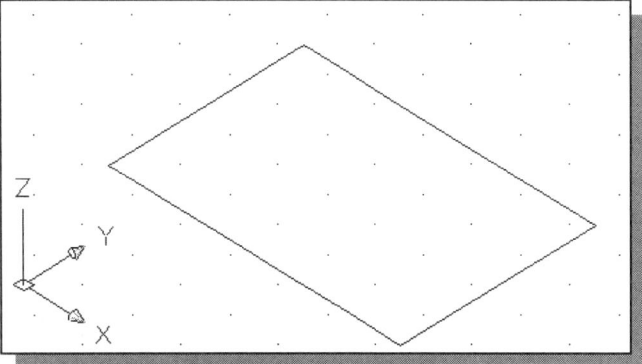

Creating a Wireframe Box

- Next, we will use the **Copy** command to define the 3D boundary of the design.

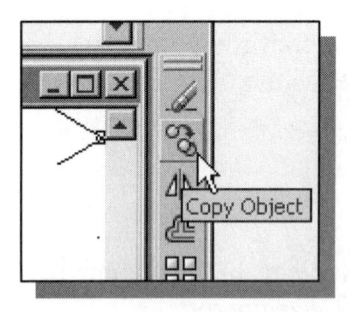

    1. Click on the **Copy Object** icon in the *Modify* toolbar.

    2. In the command prompt area, the message "*Select objects:*" is displayed. Pick any edge of the sketched rectangle.

    3. Inside the graphics window, right-mouse-click once to accept the selection.

    4. In the command prompt area, the message "*Specify base point or displacement, or [Multiple]:*" is displayed. Pick any corner of the sketched rectangle as a base point to create the copy.

    5. In the command prompt area, the message "*Specify second point of displacement or <use first point as displacement>:*" is displayed.
       Enter: **@0,0,2.25 [ENTER]**

    ➢ Note that the three values represent the distance measured in the X, Y and Z directions relative to the current UCS.

    6. Use the **Zoom All** command to adjust the display.

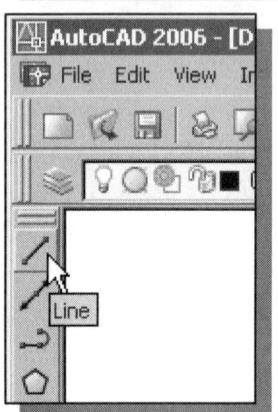

    7. On your own, use the **Line** command to create the four lines connecting the four-corners of the two rectangles as shown.

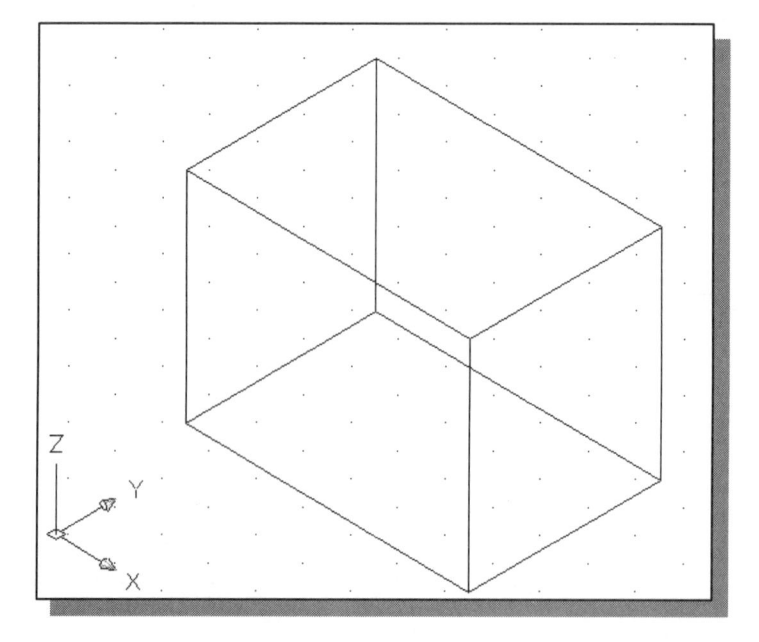

Using the View and the UCS Toolbars

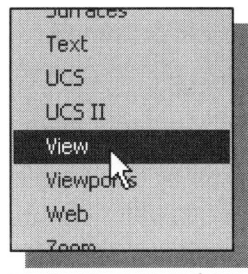

1. Move the cursor on top of any icon and **right-click** on any icon of the *Standard* toolbar to display a list of toolbar menu groups.

2. Select **View**, with the left-mouse-button, to display the *View* toolbar on the screen.

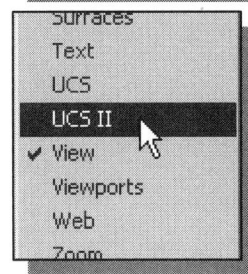

3. Move the cursor to the *Standard* toolbar area and **right-click** on any icon of the *Standard* toolbar to display a list of toolbar menu groups.

4. Select **UCS II**, with the left-mouse-button, to display the *UCSII* toolbar on the screen.

➢ The options available in *UCS II* toolbar allow us quickly align the UCS to the preset standard 2D views.

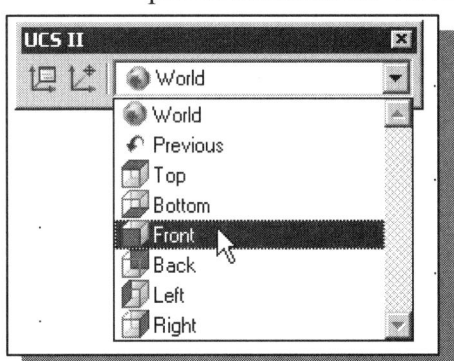

5. Note that the current *UCS* is aligned to the *World Coordinate System*. Select **Front** in the *UCS II* toolbar as shown.

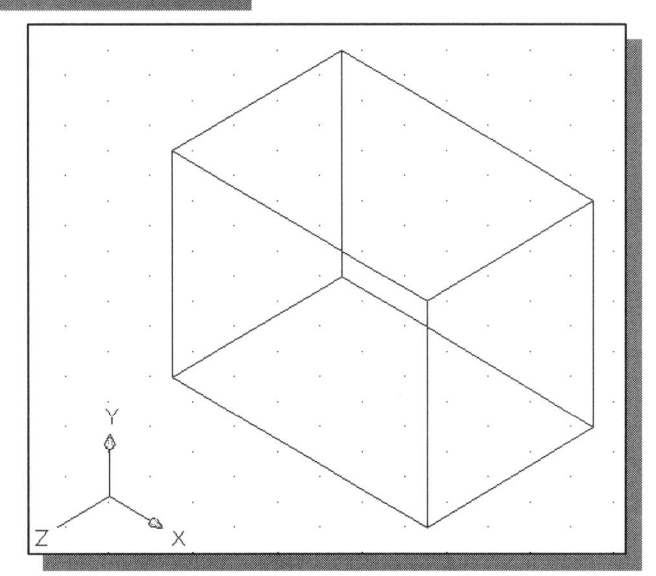

➢ Note that the UCS has been adjusted and it is now aligned parallel to the front view of the wireframe box as shown in the figure.

Creating Construction Lines in the Front View

1. Select the **Line** command icon in the *Draw* toolbar. In the command prompt area, the message "*_line Specify first point:*" is displayed.

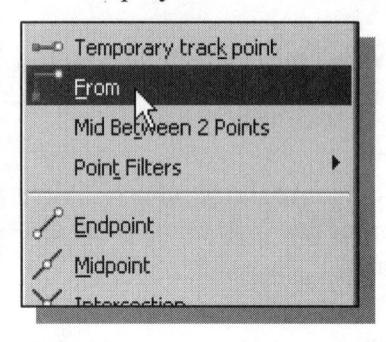

2. Inside the graphics window, hold down the **[SHIFT]** key and **right-mouse-click** once to bring up the *Object Snap* shortcut menu.

3. Select the **From** option in popup window.

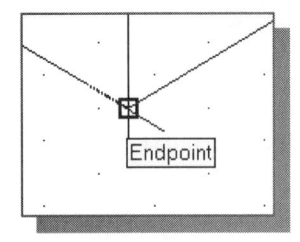

4. Pick the lower right corner of the bottom horizontal line in the front view as the base point.

5. In the command prompt area, the message " *Specify next point or [Undo]:*" is displayed. Enter: **@0,0.75 [ENTER]**

6. We will set up the *Polar Tracking* option to assist the creation of the next corner. **Right-mouse-click** once on the *POLAR* option located at the bottom of the graphics window.

7. In the option menu, select **Settings** by clicking once with the left-mouse-button.

8. In the *Drafting Settings* dialog box, set the *Polar Increment Angle* setting to **30** degrees.

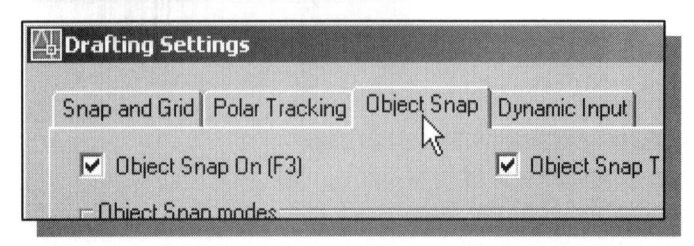

9. Click on the **Object Snap** tab to switch to the *Object Snap* option settings.

UCS, Viewports and Wireframe Modeling   4-9

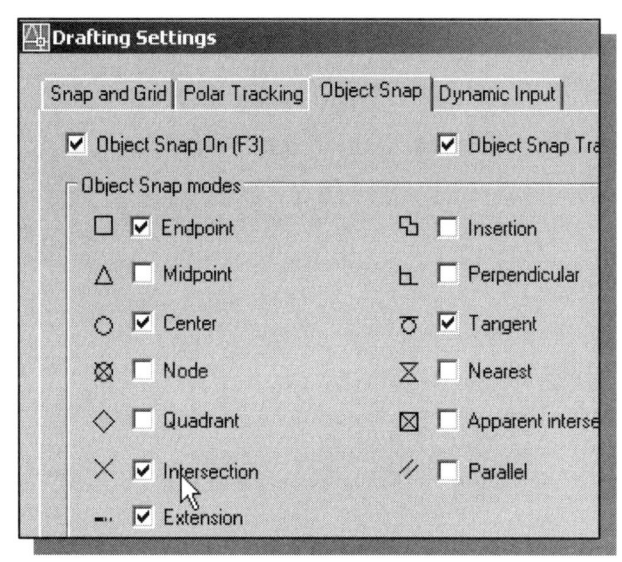

10. Confirm that the *Intersection* and *Extension* options are switched *ON* as shown.

11. Click on the **OK** button to accept the settings.

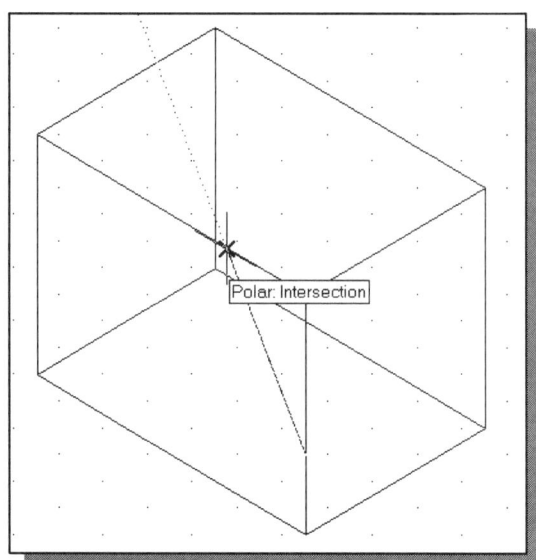

12. In the command prompt area, the message "*Specify next point or [Undo]:*" is displayed. Using the *Polar Tracking* option (30 degrees increment), create a line as shown in the figure.

13. Move the cursor near the left vertical edge of the front view and create the second line perpendicular to the last line we created.

14. Inside the graphics window, **right-mouse-click** to activate the option menu and select **Enter** with the left-mouse-button to end the Line command.

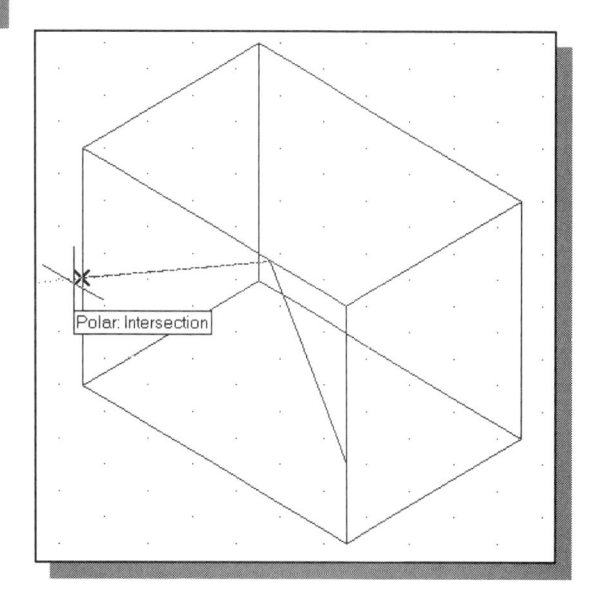

Copying in the Negative Z Direction

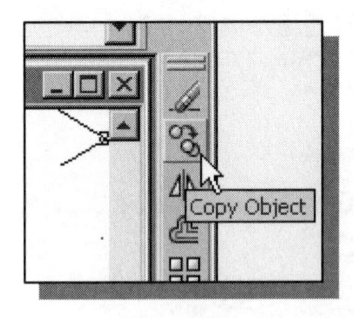

1. Click on the **Copy Object** icon in the *Modify* toolbar.

2. In the command prompt area, the message "*Select objects:*" is displayed. Pick the two edges we just created.

3. Inside the graphics window, right-mouse-click once to accept the selection.

4. In the command prompt area, the message "*Specify base point or displacement, or [Multiple]:*" is displayed. Pick any corner of the wireframe mode to use as a base point to create the copy.

5. In the command prompt area, the message "*Specify second point of displacement or <use first point as displacement>:*" is displayed.

    Enter: **@0,0,-2.0 [ENTER]**

➤ Note that the negative value represents the distance measured in the negative Z direction relative to the current UCS.

6. On your own, create three lines to connect the corners of the two sets of lines we just created, as shown below.

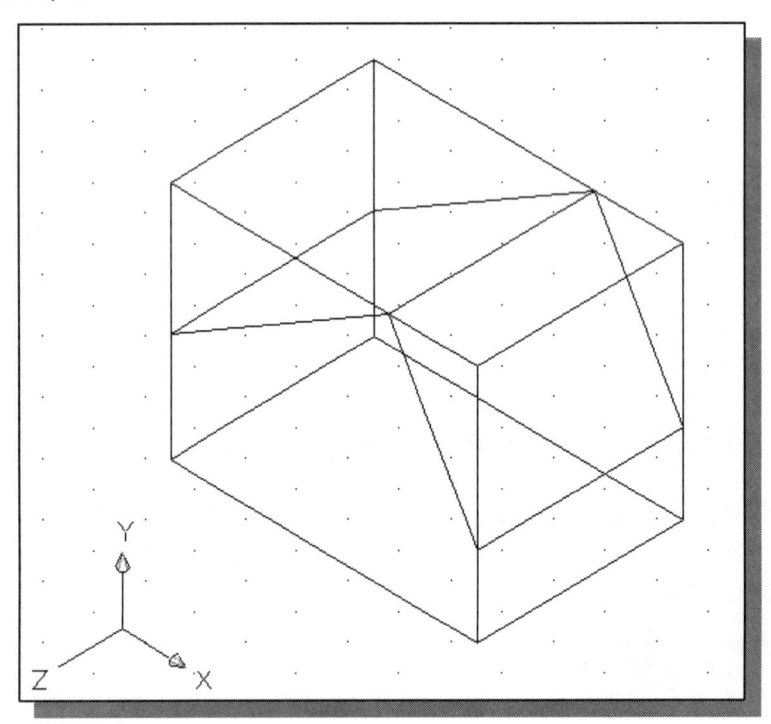

Creating an Inclined Line at the Base of the Model

1. Select **Top** in the *UCS II* toolbar as shown.

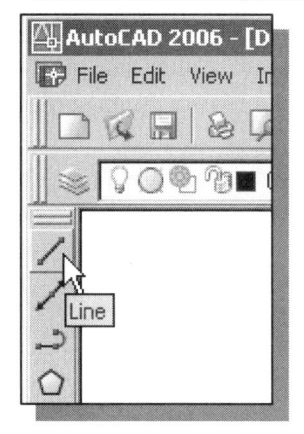

2. Select the **Line** command icon in the *Draw* toolbar. In the command prompt area, the message "*_line Specify first point:*" is displayed.

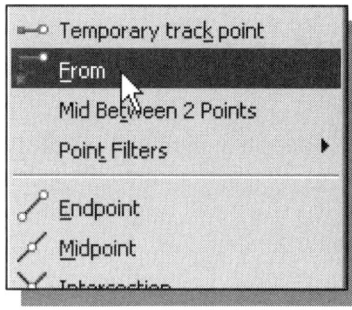

3. Inside the graphics window, hold down the [**SHIFT**] key and **right-mouse-click** once to bring up the *Object Snap* shortcut menu.

4. Select the **From** option in popup window.

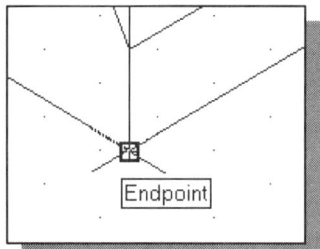

5. Pick the lower right corner of the bottom horizontal line as the base point as shown in the figure.

6. We will place the first point above the selected corner. Enter: **@0,0.75 [ENTER]**

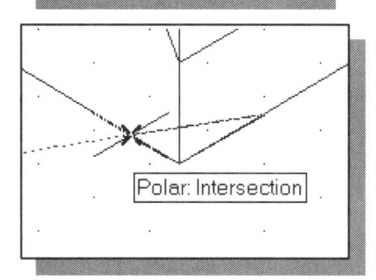

7. In the command prompt area, the message "*Specify next point or [Undo]:*" is displayed. Using the *Polar Tracking* option (30 degrees increment), create a line as shown in the figure.

8. Inside the graphics window, right-mouse-click to activate the option menu and select **Enter** with the left-mouse-button to end the Line command.

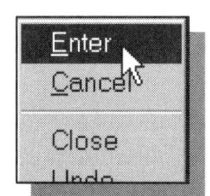

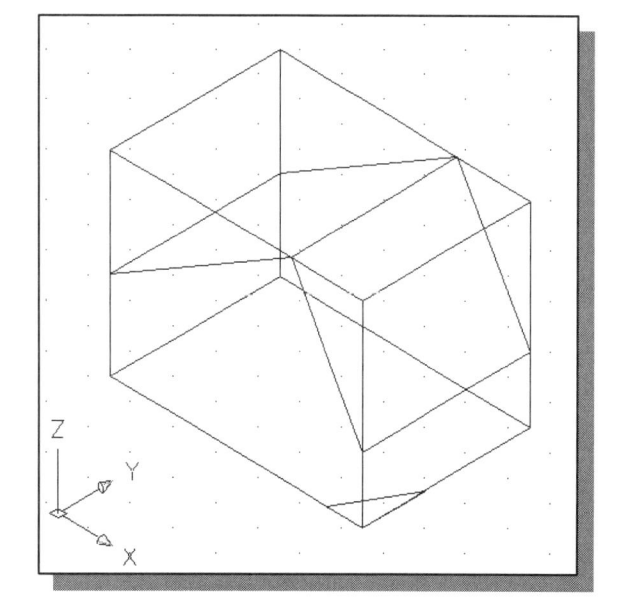

➤ On your own, use the **Copy** and **Line** commands to create additional edges so that the wireframe model appeared as shown below.

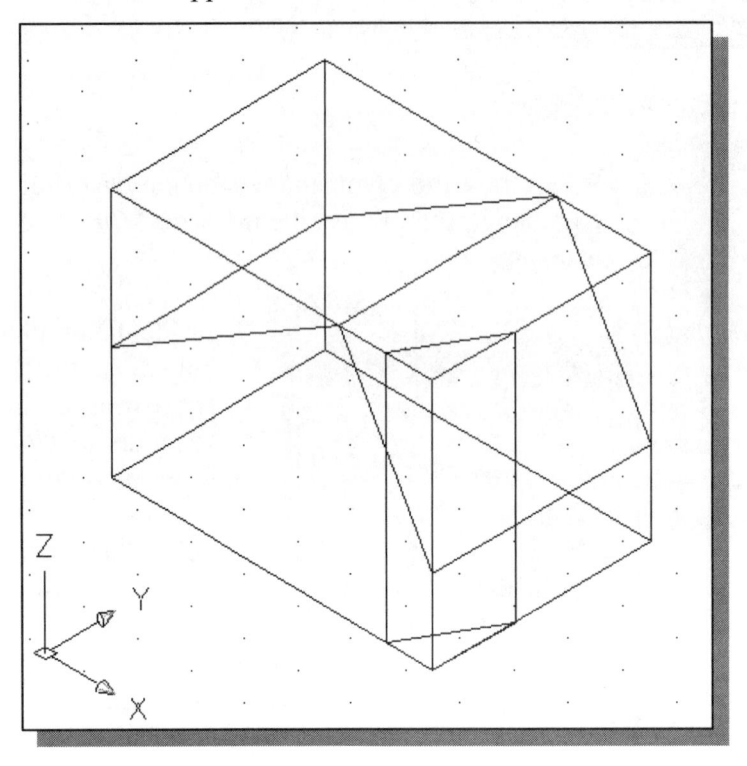

Creating Object Lines

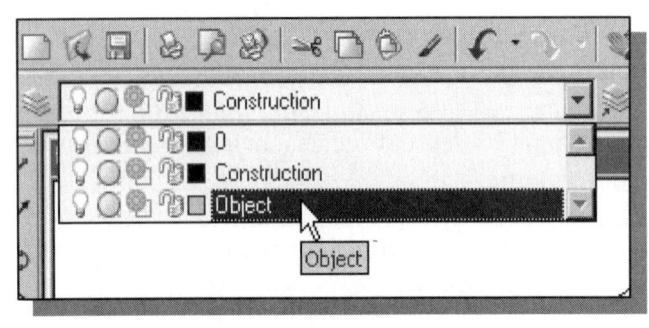

1. On the *Object Properties* toolbar, choose the **Layer Control** box with the left-mouse-button.

2. Move the cursor over the name of layer **Object**; the tool tip "*Object*" appears.

3. **Left-mouse-click once** on the layer name and now layer *Object* is set as the *Current Layer*.

4. Select the **Line** command icon in the *Draw* toolbar. In the command prompt area, the message "*_line Specify first point:*" is displayed.

5. Pick the lower left corner of the bottom of the wireframe model as the starting point and create the four line segments as shown.

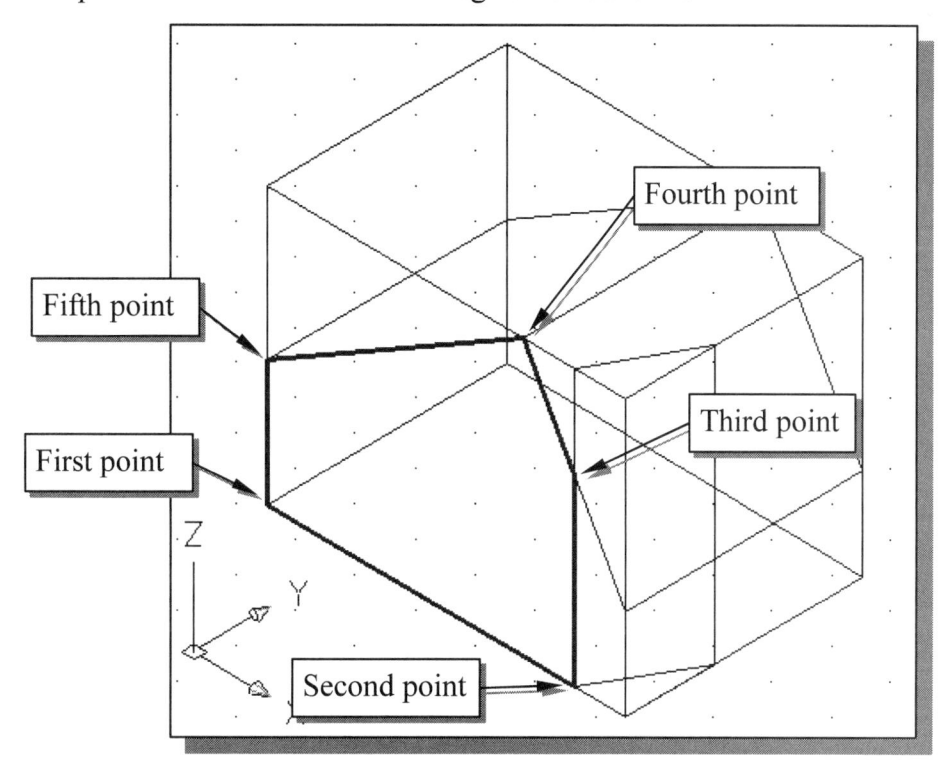

- Note that even though the UCS is aligned to the base plane of the model, the *OSNAP* option enables us to select points on the front plane of the model.

➢ On you own, create three additional lines representing the edges of the inclined face of the design as shown.

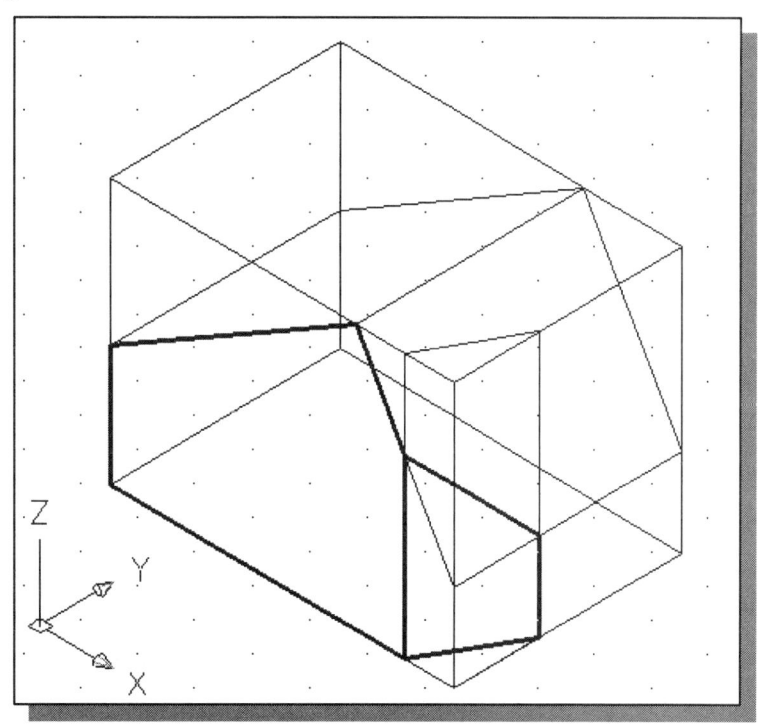

Multiple Viewports

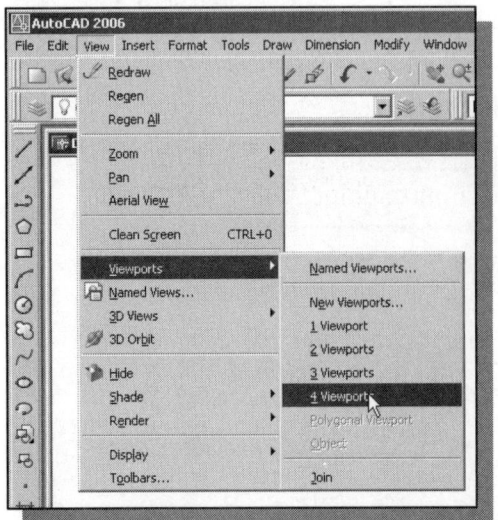

1. In the pull-down menus, select:

   [View] → [Viewports] → [4 viewports]

   ➢ AutoCAD allows us to display different VIEWS of the model simultaneously. Only one viewport is activated at one time. The viewport with the darker border is the active viewport. Move the cursor inside each viewport and notice the different cursor-display inside the active viewport.

2. Activate the top **left viewport** by left-clicking once inside the viewport.

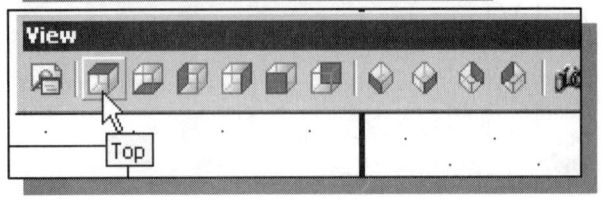

3. Select **Top View** in the *View* toolbar to change the display of the viewport.

4. On your own, repeat the above steps and use the **Realtime Zoom** command to set and adjust the displays of the viewports as shown below.

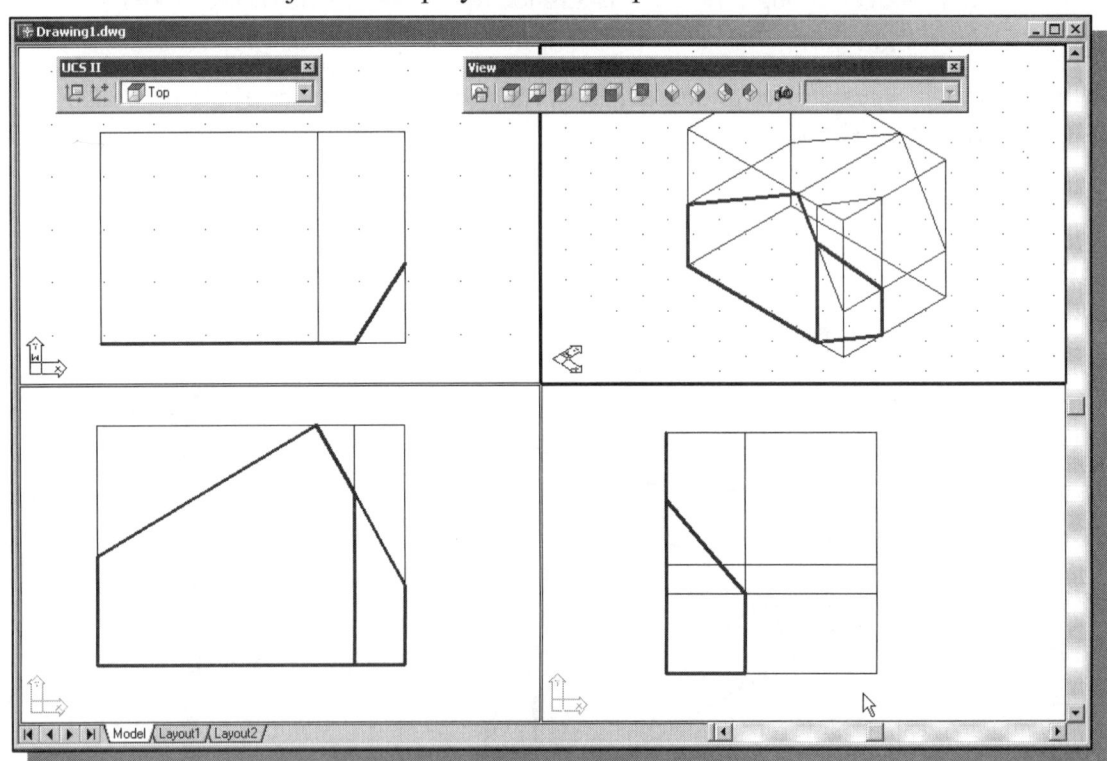

➢ Note that the displayed UCS icons at the lower left corner of each viewport correspond to the orientation of the wireframe model in each viewport.

UCS, Viewports and Wireframe Modeling   4-15

Using the Mirror Command

1. Activate the top left viewport by left-clicking once inside the viewport.

2. *Pre-select* all the objects in the active viewport by using a selection window.

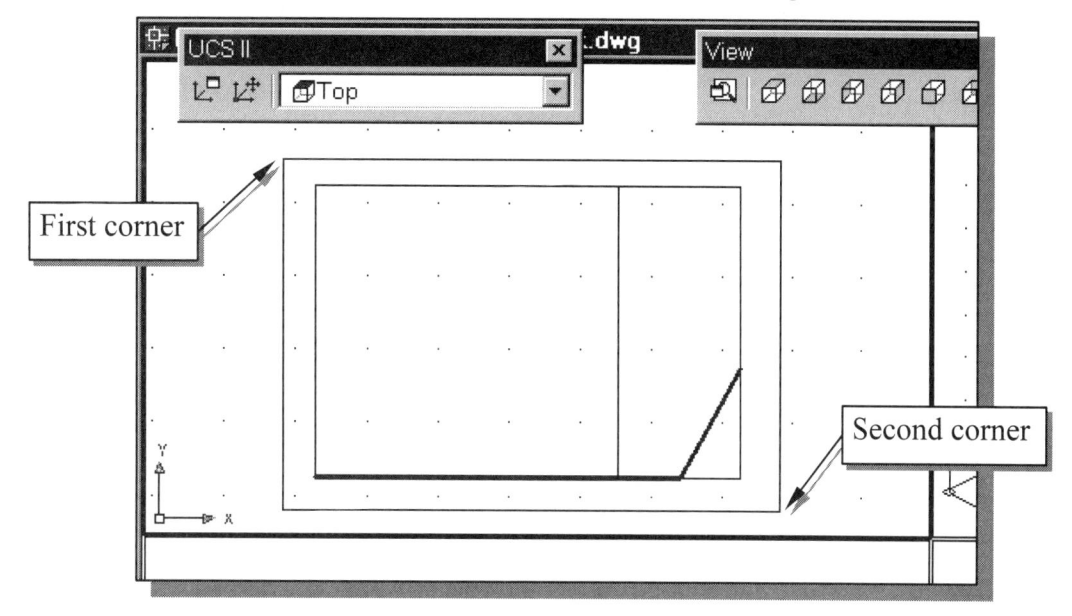

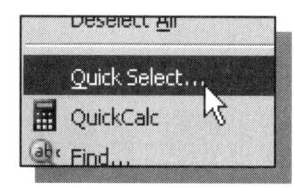

3. Inside the *active viewport*, **right-mouse-click** to bring up the popup option menu and select the **Quick Select** option.

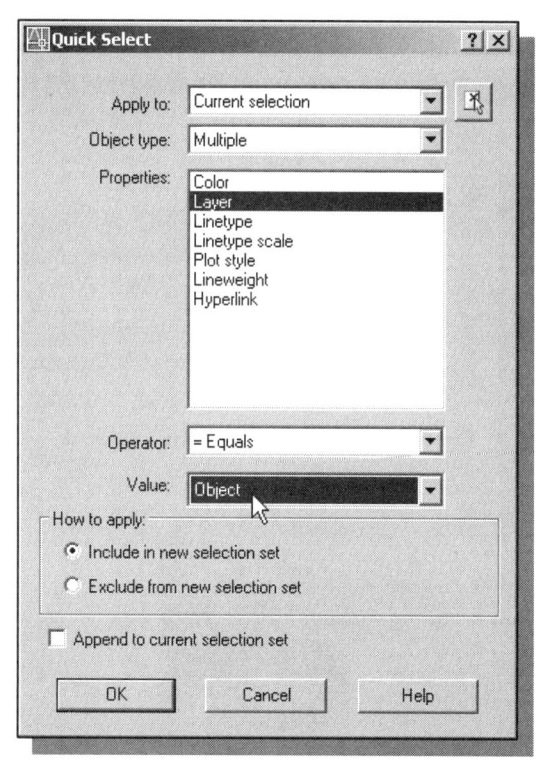

4. In the *Quick Select* dialog box, select **Layer** from the *Properties* list.

5. Set the *Value* box to **Object_Lines**.

6. In the *How to apply* section, confirm the *Include in new selection set* option is selected.

7. Switch *OFF* the *Append to current selection set*.

8. Click on the **OK** button to accept the settings.

- AutoCAD will now **filter out** objects that are not on layer *Object*.

9. Click on the **Mirror** icon in the *Modify* toolbar.

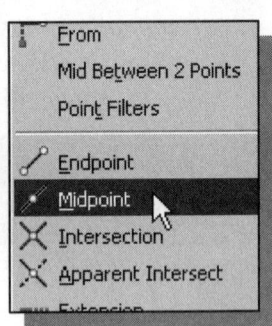

10. Inside the graphics window, hold down the **[SHIFT]** key and **right-mouse-click** once to bring up the *Object Snap* shortcut menu.

11. Select the **Midpoint** option in popup window.

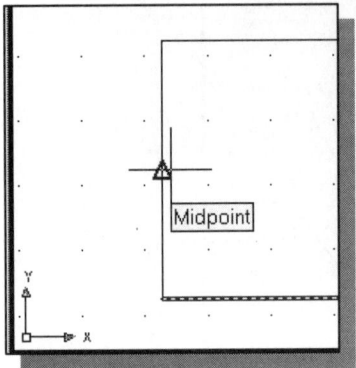

12. Inside the active viewport, click on the left vertical line of the wireframe model. With the *Midpoint* snap option, the midpoint of the line is selected

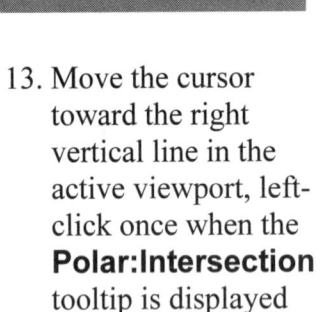

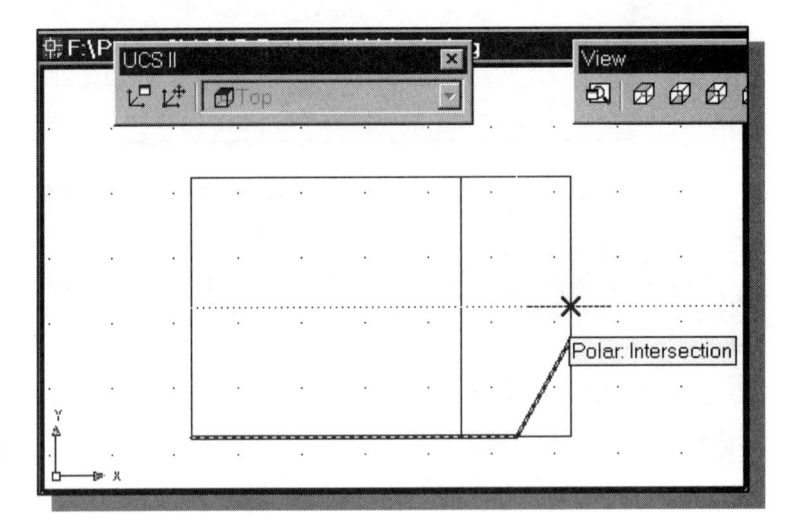

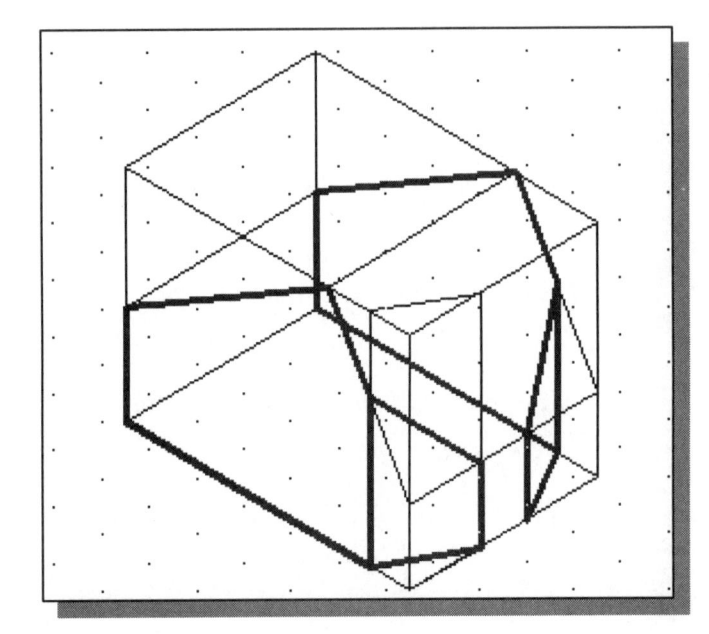

13. Move the cursor toward the right vertical line in the active viewport, left-click once when the **Polar:Intersection** tooltip is displayed as shown.

14. In the command prompt area, the message "*Delete Source Objects? [Yes/No] <N>:*" is displayed. Inside the graphics window, right-mouse-click once to bring up the option menu and select **No** to keep both sets of objects.

UCS, Viewports and Wireframe Modeling   4-17

Turn *OFF* the Construction Lines

1. On the *Object Properties* toolbar, choose the **Layer Control** box with the left-mouse-button.

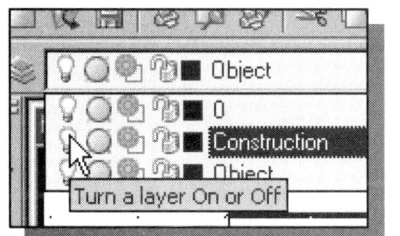

2. Move the cursor over the light-bulb icon for layer *Construction*, **left-mouse-click once**, and notice the icon color is changed to a gray tone color, representing the layer (layer *Construction*) is turned *OFF*.

3. In the pull-down menus, select:

   [View] → [Viewports] → [1 viewport]

   - Viewports can also be used as a way of checking the integrity of the constructed 3D models. Feel free to switch to multiple viewports in the following sections of this tutorial.

4. On your own, create the additional object lines connecting the two sets of objects as shown below.

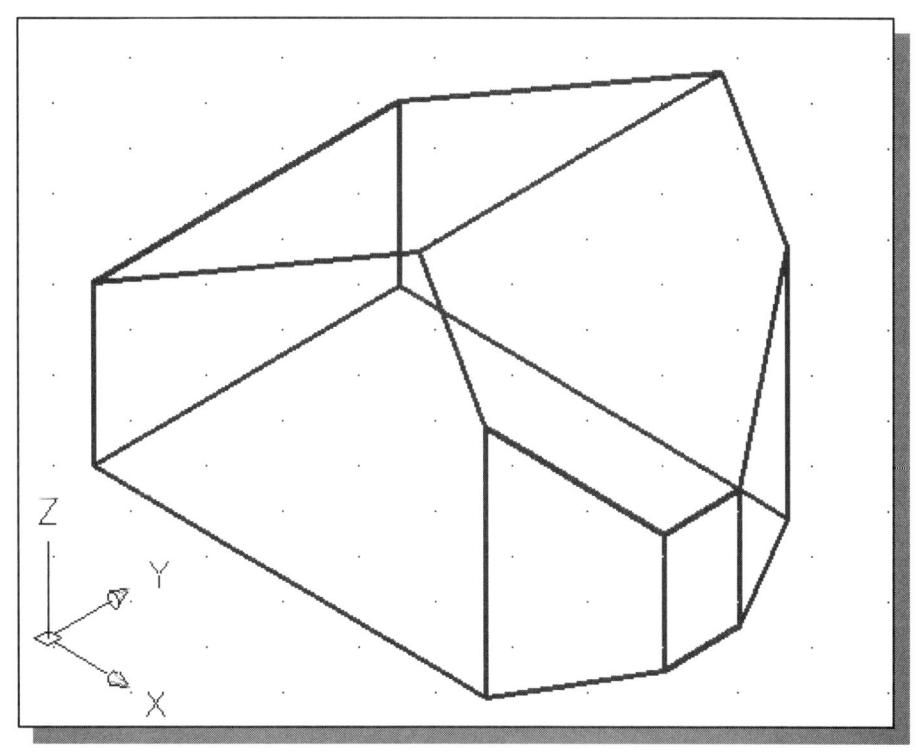

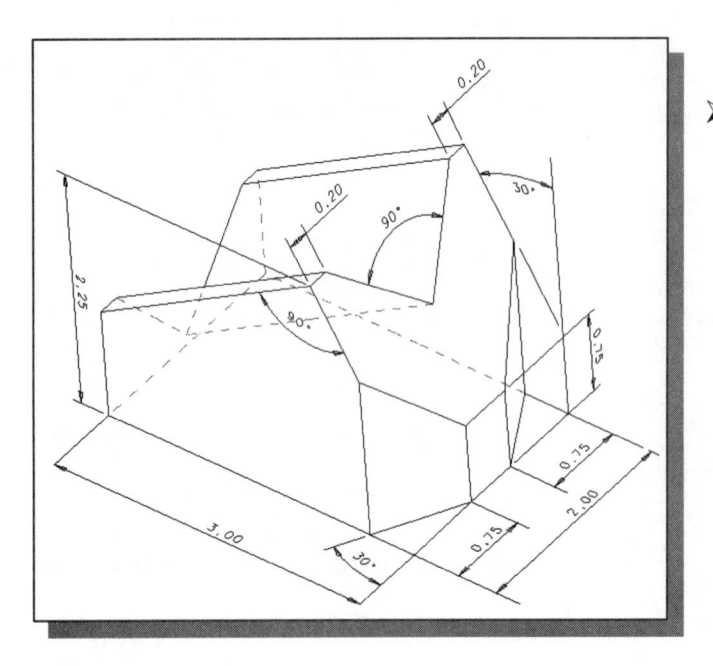

> The V-cut feature of the design is the last feature, and perhaps the most difficult one, to create in the CAD system. Based on your knowledge of modeling with AutoCAD, how would you create this feature? What are the more difficult aspects of creating this feature?

Creating a New UCS

> Besides using the preset UCS planes, AutoCAD also allows us to change our viewpoint and create new UCS planes. We will create an UCS plane aligning to the inclined plane of the design.

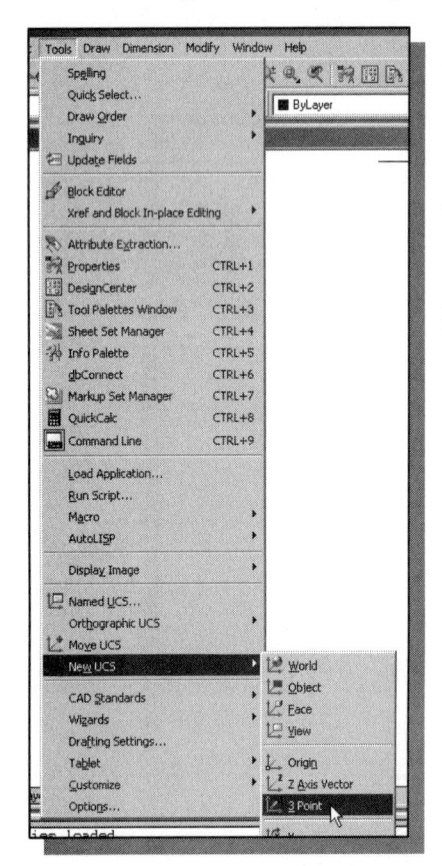

1. In the pull-down menus, select:

   [Tools] → [New UCS] → [3 point]

   - Note that other options are also available to aid the creation of new UCS planes.

2. In the command prompt area, the message "*Specify New Origin point <0,0,0>:*" is displayed. Pick the lower-left corner of the inclined plane as shown.

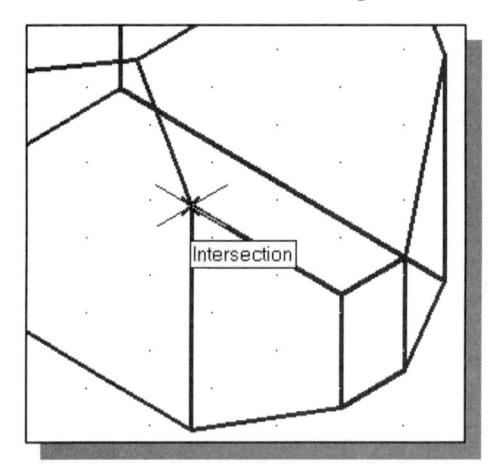

UCS, Viewports and Wireframe Modeling   4-19

3. In the command prompt area, the message "*Specify point on positive portion of X-axis:*" is displayed. Pick the lower right corner of the inclined plane as shown.

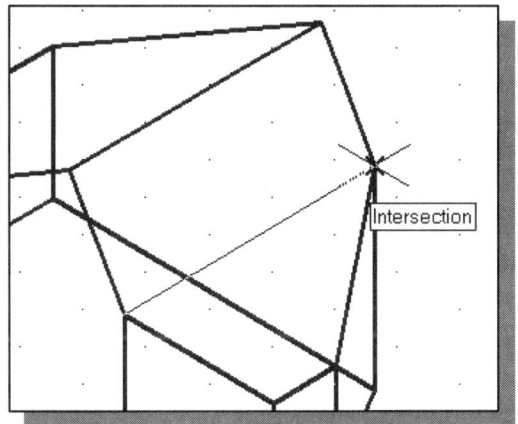

4. In the command prompt area, the message "*Specify point on positive portion of X-axis:*" is displayed. Pick the top left corner of the inclined plane as shown.

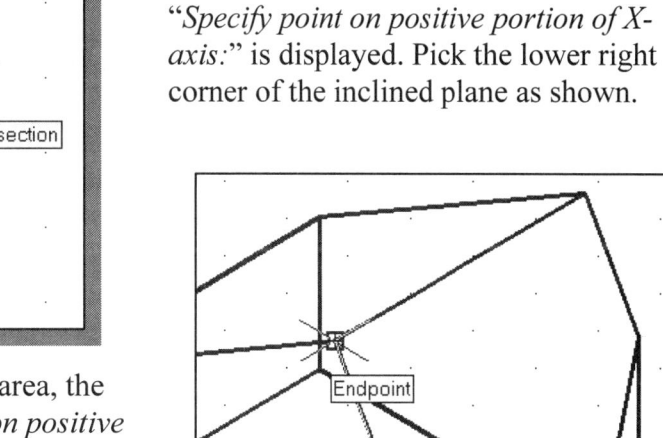

5. In the pull-down menus, select

   [View] → [Display] → [UCS Icon] → [Properties]

6. In the *UCS Icon* window, switch the *UCS icon style* to **2D** as shown.

7. Click on the **OK** button to accept the settings.

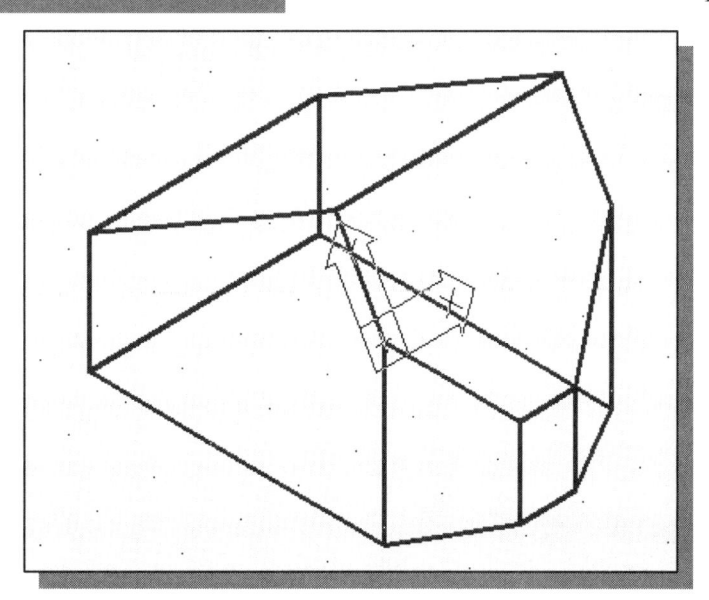

Creating a New Named View

* Besides using the preset views, AutoCAD also allows us to change our viewpoint and create new views, which can be saved and restored by name for convenient access. We will create a new view aligned to the new UCS plane we just created.

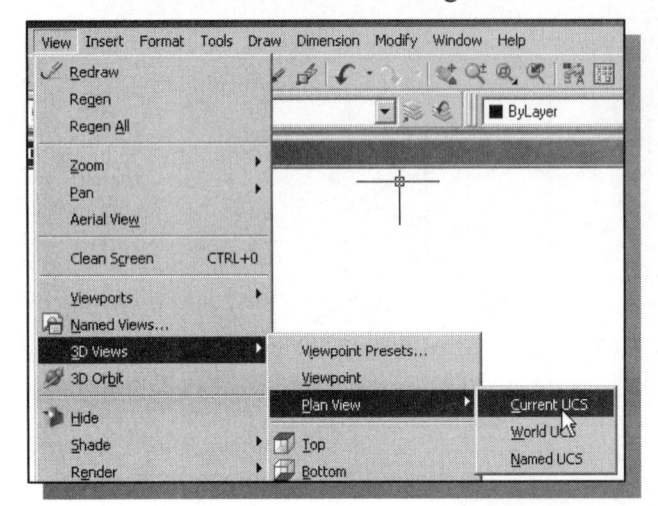

1. In the pull-down menus, select

   [View] → [3D Views] → [Plan View] → [Current UCS]

* The graphics window is now adjusted to display the current UCS plane.

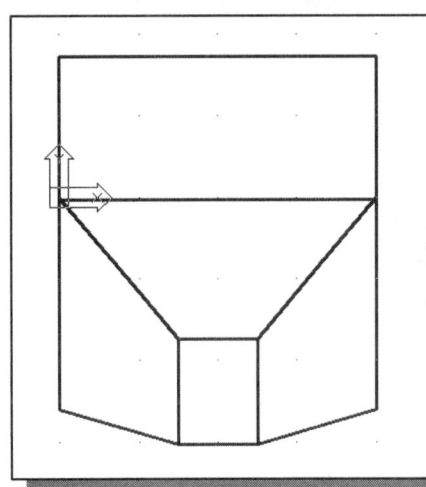

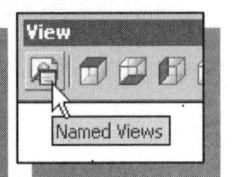

2. Click on the **Named Views** icon in the *View* toolbar.

3. In the *View* dialog box, click on the **New** button to create a new named view.

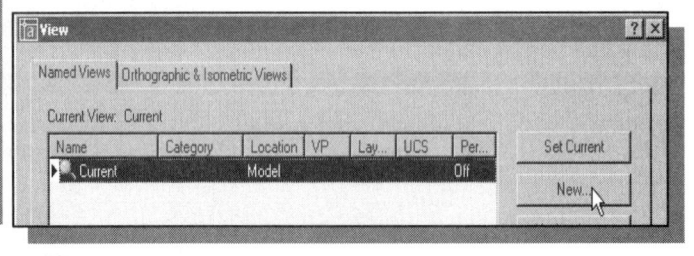

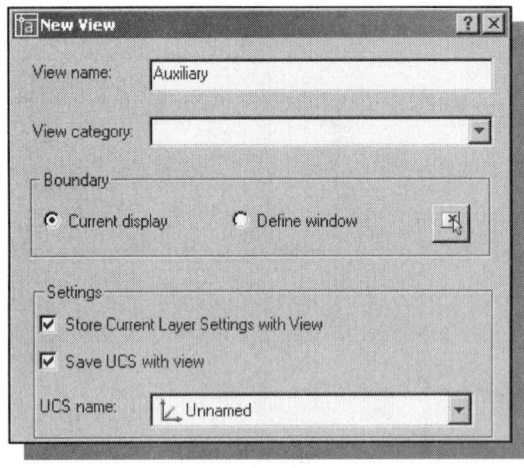

4. In the *New View* dialog box, enter **Auxiliary** as the new *View name* and confirm the *Current display* option and the *Save UCS with view* option are switched *ON* as shown.

5. Click on the **OK** button to accept the settings and create the named view.

6. In the *View* dialog box, the new named view (*Auxiliary*) is added to the *Named Views* list. We can also switch to this view by using the **Set Current** option.

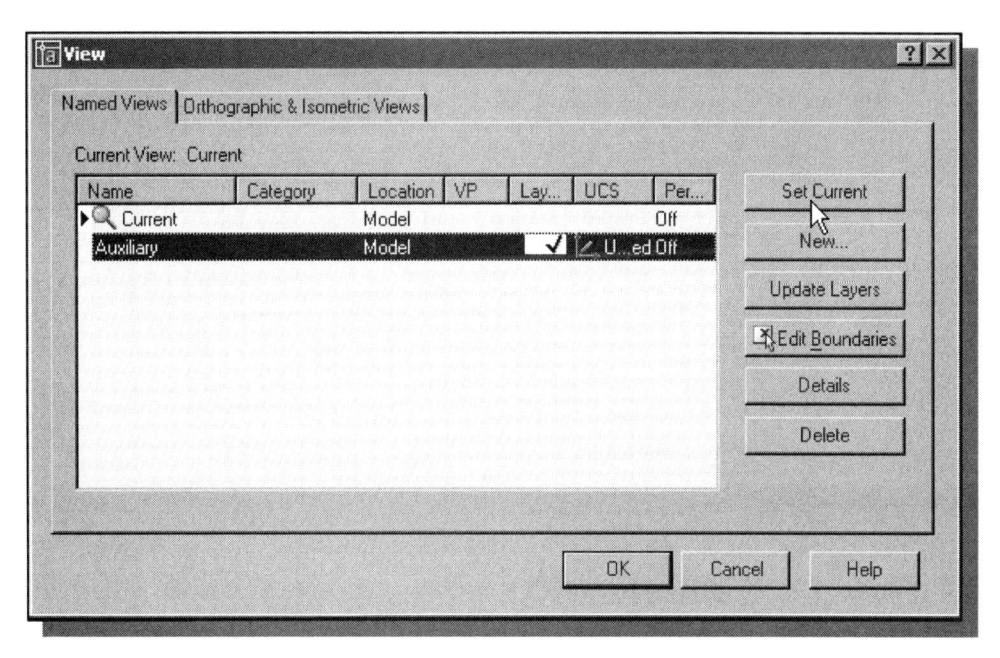

7. Click on the **OK** button to accept the settings.

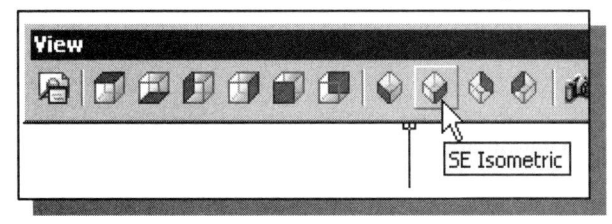

8. In the *View* toolbar, select **SE Isometric View**.

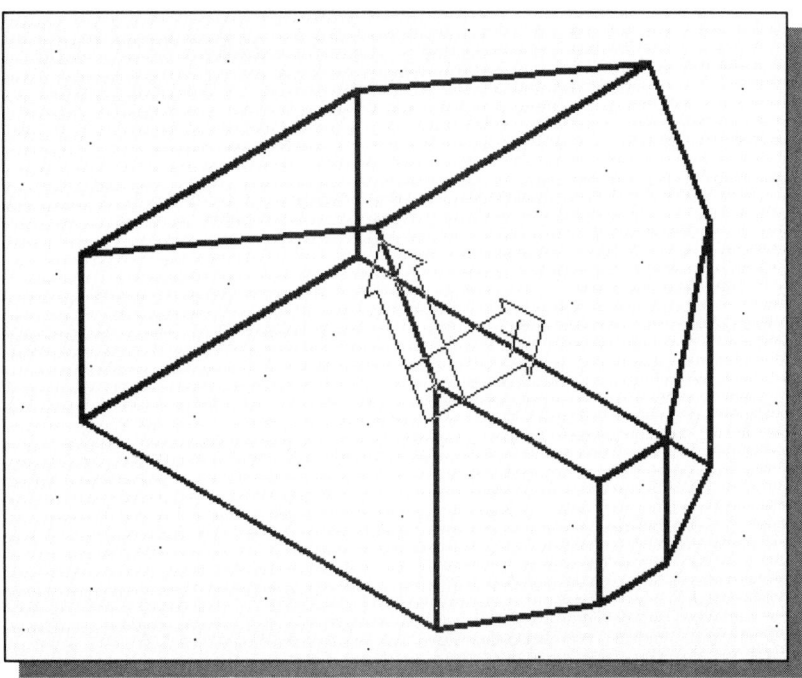

Creating the V-Cut Feature on the Inclined Plane

1. On your own, create two lines that are **0.2"** away from the top corners of the model, which are also rotated **45 degrees** relative to the top edge of the model.

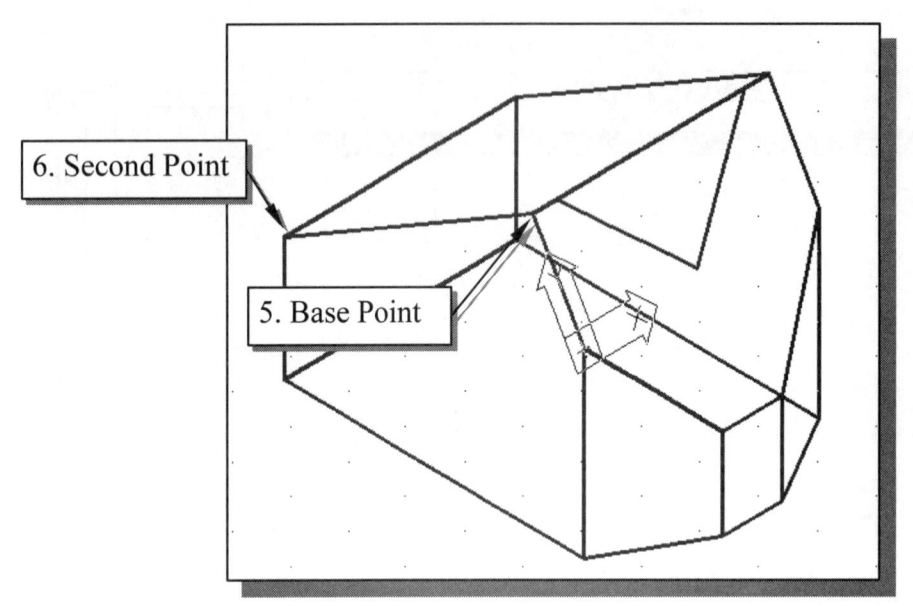

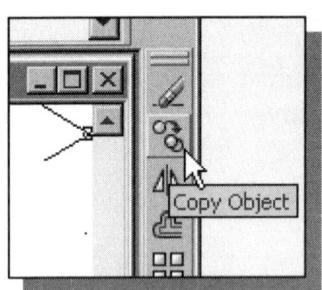

2. Click on the **Copy Object** icon in the *Modify* toolbar.

3. In the command prompt area, the message "*Select objects:*" is displayed. Pick the two inclined lines we just created.

4. Inside the graphics window, **right-mouse-click** once to accept the selections.

5. In the command prompt area, the message "*Specify base point or displacement, or [Multiple]:*" is displayed. Pick the top left corner as the base point to create the copy.

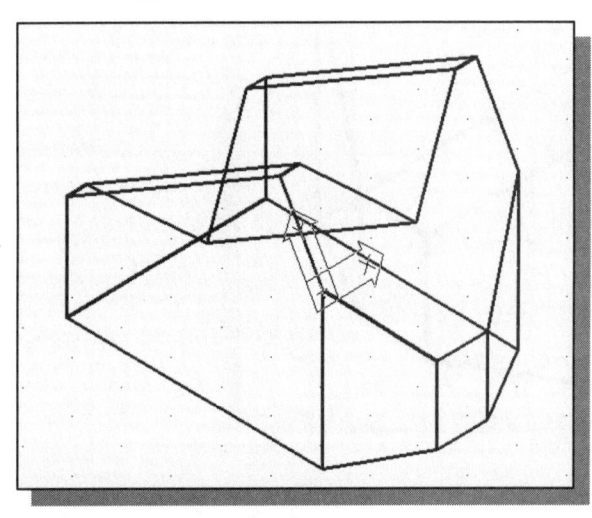

6. In the command prompt area, the message *"Specify second point of displacement or <use first point as displacement>:"* is displayed. Pick the corresponding corner on the backside of the model as shown in the above figure.

7. On your own, create addition lines to define the edges of the V-cut and use the **Trim** command to complete the model as shown.

UCS, Viewports and Wireframe Modeling 4-23

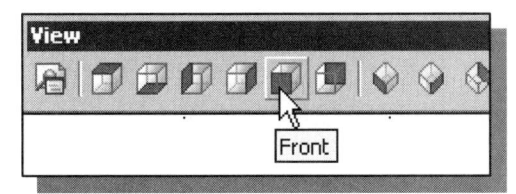

8. Click on the **Front View** icon in the *View* toolbar.

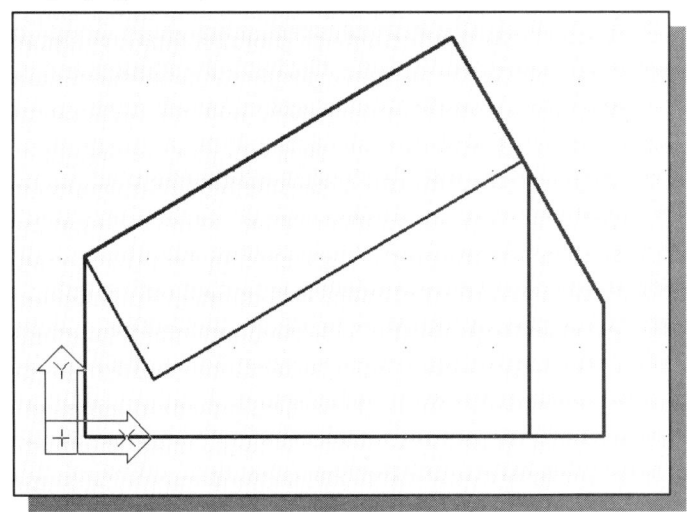

> Note that the V-cut we created, using the **Copy** command, does not represent a cut that passes through the entire block.

Extend the Cut and GRIP Editing

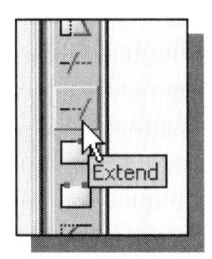

1. Select the **Extend** command icon in the *Modify* toolbar.

2. In the command prompt area, the message "*Select boundary edges... Select objects:*" is displayed. Pick the left vertical line as the *boundary edge*.

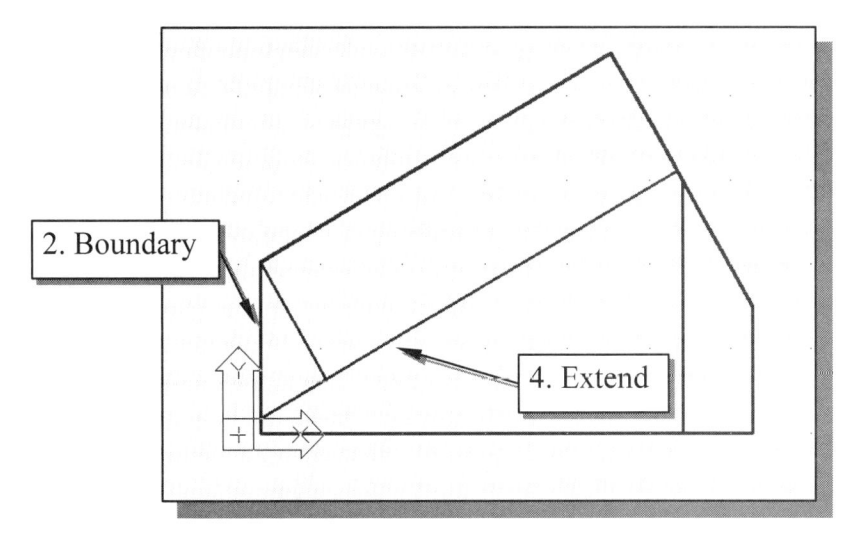

3. Inside the graphics window, **right-mouse-click** to accept the selection.

4. The message "*Select object to Extend or [Project/Edge/Undo]:*" is displayed in the command prompt area. Pick the bottom edge of the V-cut to extend.

5. Inside the graphics window, **right-mouse-click** to activate the option menu and select **Enter** with the left-mouse-button to end the Extend command.

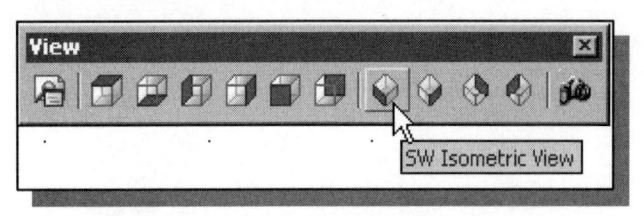

6. Click on the **SW Isometric View** icon to reset the display to the preset isometric view.

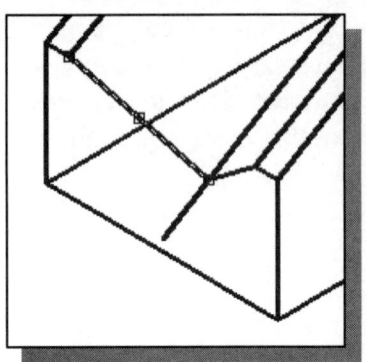

7. ***Pre-select*** the top edge of the *V-block* as shown.

8. Left-click once on the lower right grip point on the selected line.

- Selecting the grip points located at the end of the line or arc allows us to STRETCH the selected line.

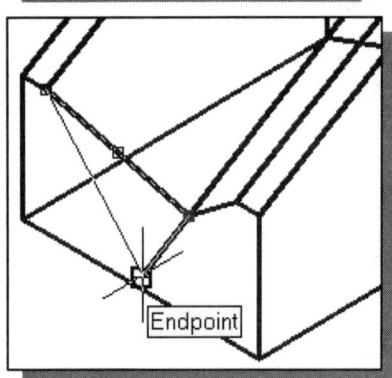

9. Move the cursor toward and select the lower corner of the V-cut as shown.

➢ On your own, repeat the above steps and use the grip editing options and complete the wireframe model as shown below.

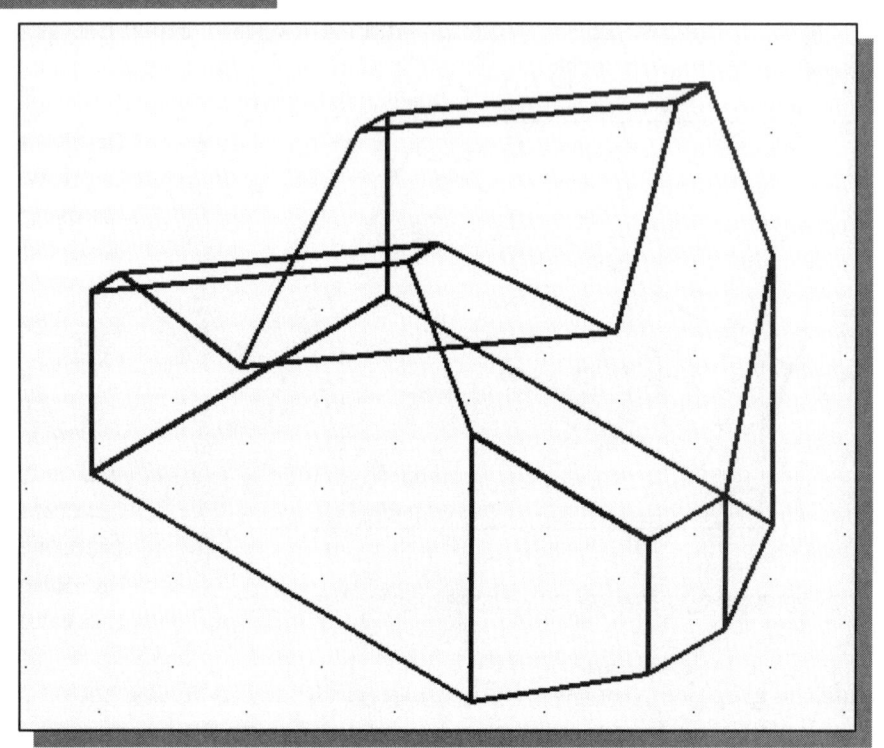

Questions:

1. List and describe three different types of 3D computer geometric modeling software available today?

2. What is a *Named View*?

3. What is the difference between **Stretch** and **Extend**?

4. List and describe the methods available in AutoCAD to create a new UCS plane.

5. What is the difference between **Copy** and **Mirror** in AutoCAD?

6. Identify and describe the following commands:

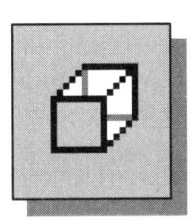

   a)

   b)

   c)

   d)

Exercises: (Unless otherwise specified, dimensions are in inches.)

1.

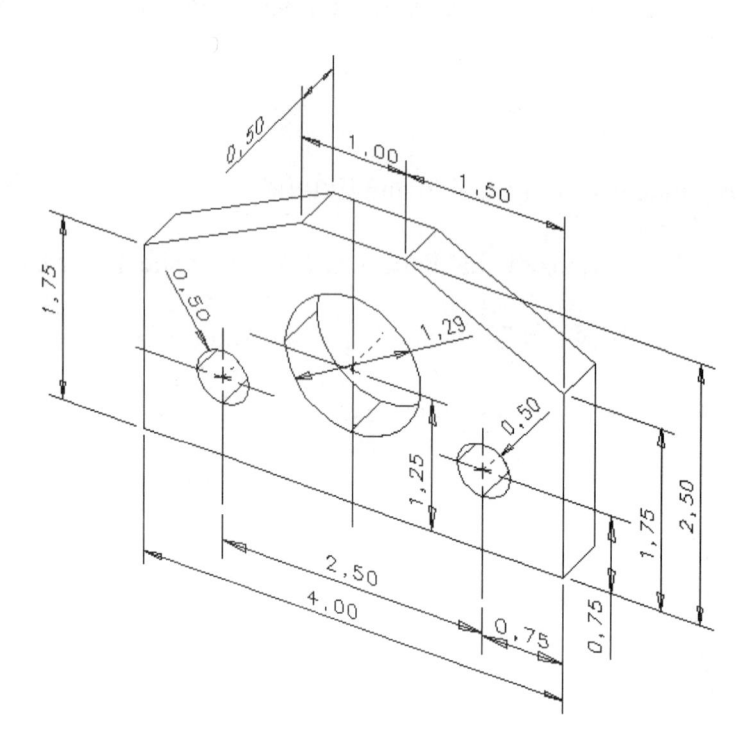

2.

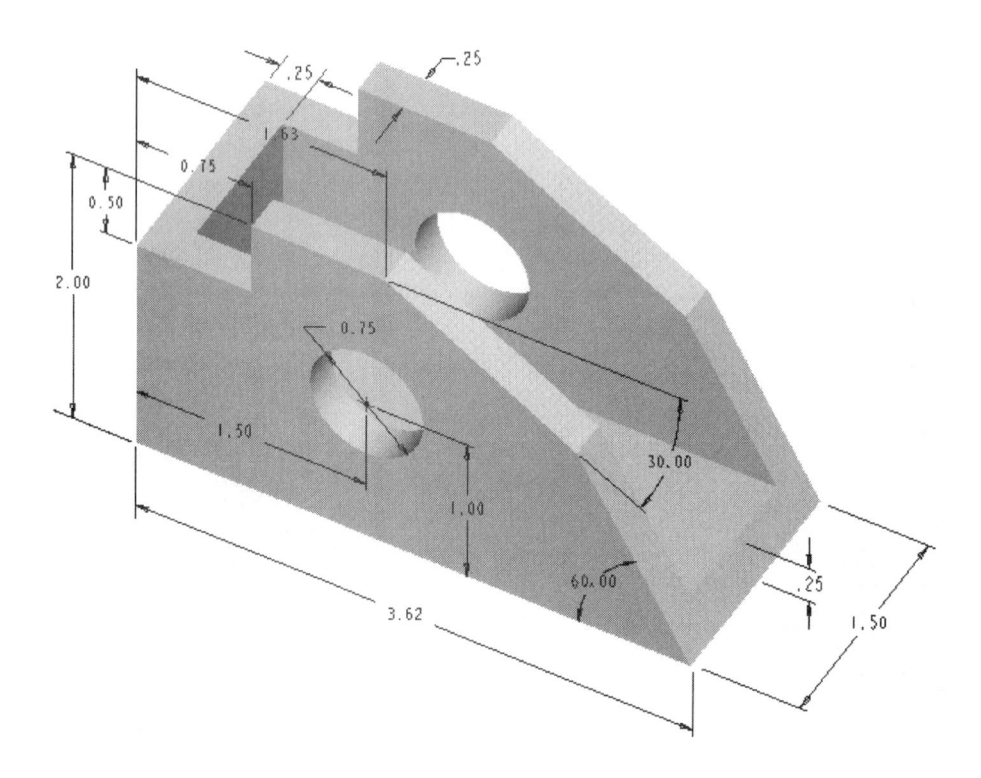

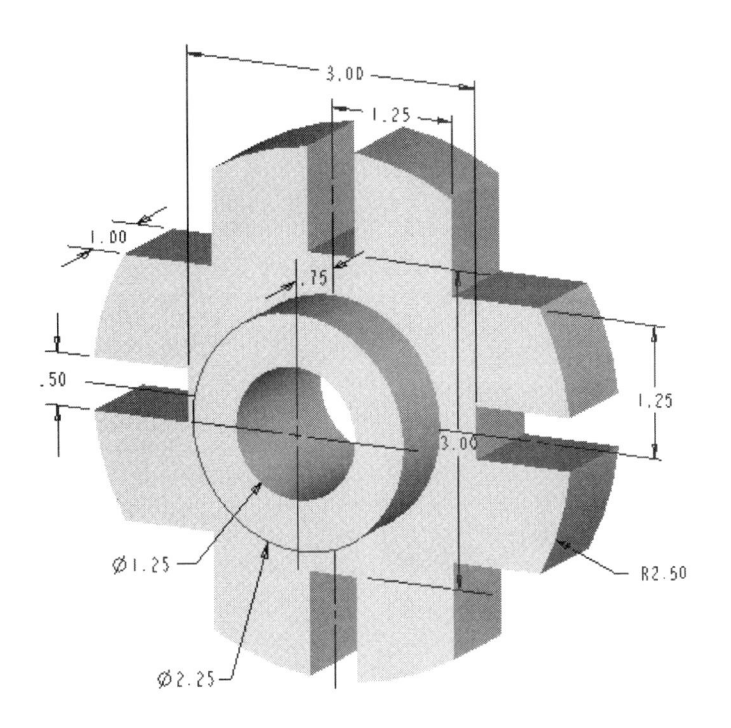

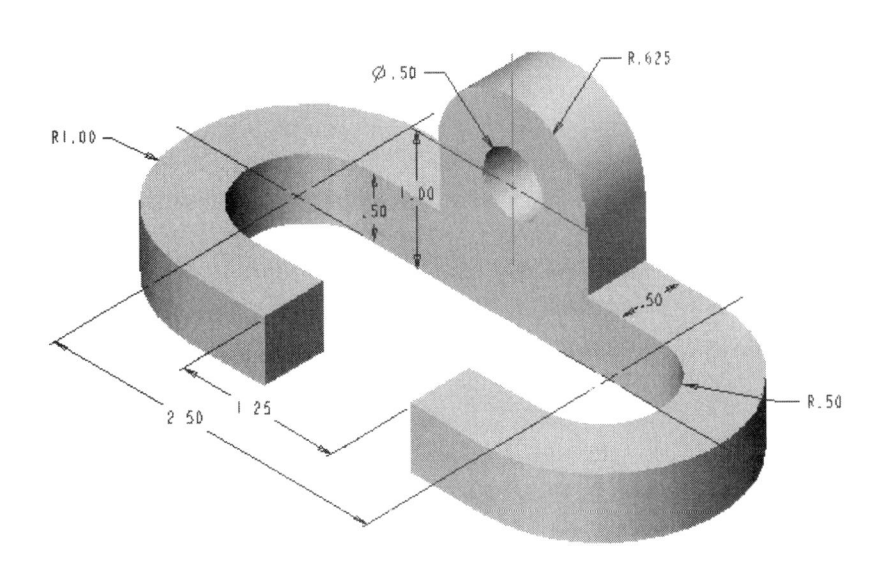

3.

4.

NOTES:

Chapter 5
3D Surface Modeling

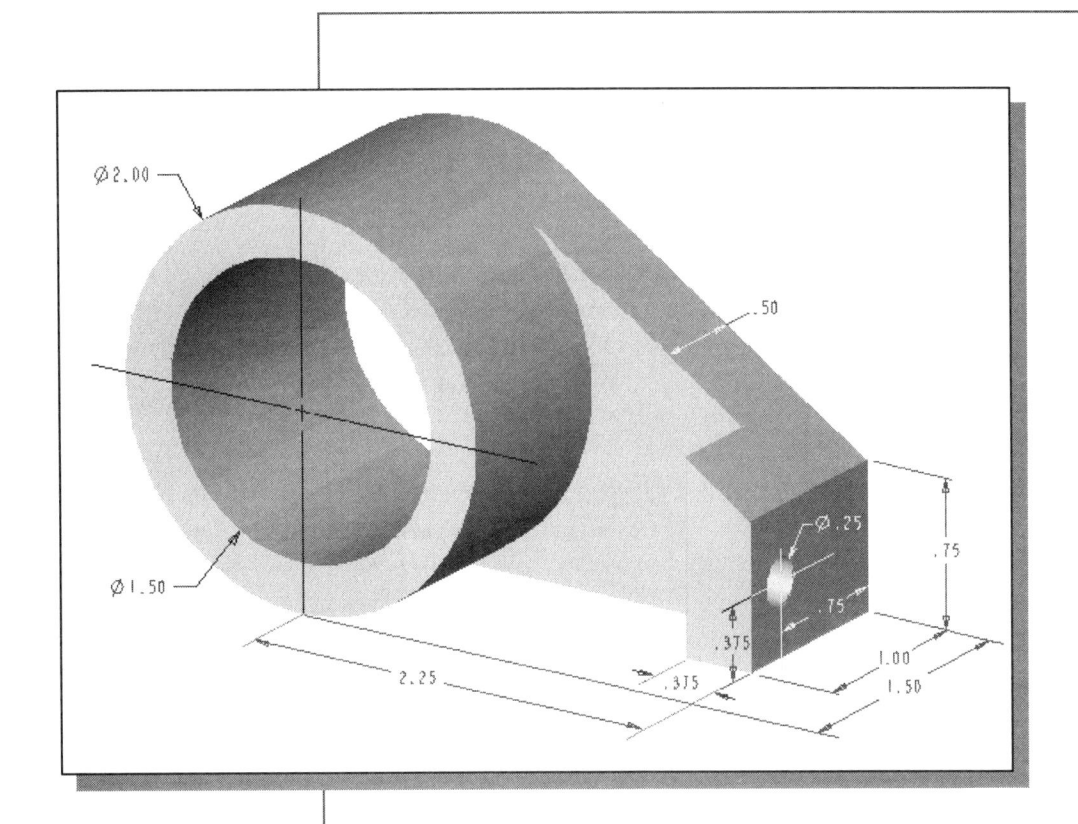

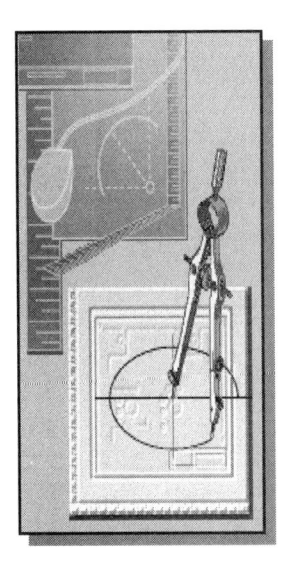

Learning Objectives

- Create 3D Surface Models
- Understand and Apply the Different Surface Modeling Techniques
- Understand the Use of the 2D SOLID Command
- Using the Predefined Surface Models
- Understand the Use of 3D FACE Command
- Use the Hidden Edge Option

Introduction

As illustrated in the previous chapters, there are no surfaces in a wireframe model; it consists only of points, lines, and curves that describe the edges of the object. Surface modeling was developed to provide the surface information that is missing in wireframe modeling. Essentially, defining the skin of a design creates a surface model. Although it is possible to create a surface model without using a wireframe model, in most cases it is much easier to create a surface model on top of a wireframe model. In surface modeling, a wireframe model can be used to provide information about the edges and corners so that the desired faces can be easily positioned and placed.

Surface modeling is more sophisticated than wireframe modeling in that surface modelers define not only the edges of 3D objects, but also the surfaces. Surface modeling provides hiding, shading, and rendering capabilities that are not available in wireframe modeling. Surface models do not provide the physical properties that solid models provide, such as mass, weight, center of gravity, and so on.

The AutoCAD surface modeler defines faceted surfaces using a filled polygon. The created faces of surface models are only planar, which means the surface models can only have approximate curved surfaces. It is important to note that the AutoCAD surface modeler does not create true curved surfaces. To differentiate these two types of surfaces, faceted surfaces are called **meshes** in AutoCAD. Because of the use of faceted approximation on true curved surfaces, the computer requirements of most faceted surface modelers are typically much less than that of solid modelers. Faceted surface modeling usually provides reasonably good representations of 3D designs with fast rendering and shading capabilities. Faceted surface models are also useful for creating geometry with unusual surface patterns, such as a 3D topographical model of mountainous terrain.

AutoCAD® 2006 provides three basic methods for creating surfaces – the **2D Solid**, **3D Face** and **Region** commands. The three commands were developed parallel to the historical development of the different types of computer modelers.

- **2D Solid**: The first generation surface command available in AutoCAD. Used mostly to fill an area in the sketch plane of the current UCS. This type of surface is not a true 3D surface.

- **3D Face**: Creates a true 3D planar surface (allowing X, Y and Z coordinates) of three-sided or four-sided shape. This is the type of surface developed primarily for creating faceted surface models.

- **Region**: Creates a 2D surface of arbitrary shape from existing 2D entities. This command creates the most flexible and the most complicated type of surface available in AutoCAD. This command was developed to allow manipulation of 2D surfaces using one of the solid modeling construction techniques, namely, the Constructive Solid Geometry method.

Although all three commands can be used to create planar surfaces, the resulting surfaces are not equal. In fact, the three commands are developed for specific tasks in mind. The **2D Solid** command is mostly used in 2D drawings to create 2D filled area and the **Region** command is designed so that general 2D shapes can be easily transformed into solid models. The **3D Face** command is the only one that is designed specifically for surface modeling and therefore it is the most suitable for such tasks. The use of the **2D Solid** and **Region** commands in 3D surface modeling can be somewhat awkward and at times very difficult. Note that the use of the **Region** command will be focused on in the solid modeling chapters of this text.

As one can imagine, sketching each surface manually can be very time consuming and tedious. AutoCAD also provides additional tools for surface modeling, such as **Predefined surfaces**, **Tabulated surfaces**, **Ruled surfaces** and **Revolved surfaces**. These tools are basically automated procedures, which can be used to define and create multiple copies of planar surfaces in specific directions. The principles and concepts used by these tools are also used in creating solid models, which are covered in chapter six through chapter eight of this text. You are encouraged to re-examine these commands after you have finished the solid modeling chapters.

In this chapter, the general procedures to create surface models are illustrated. The use of the **2D Solid** and **3D Face** commands are illustrated and differences discussed. We will also demonstrate the use of the more advanced surface modeling tools. Two wireframe models, which were created in the previous chapters, will be converted into surface models.

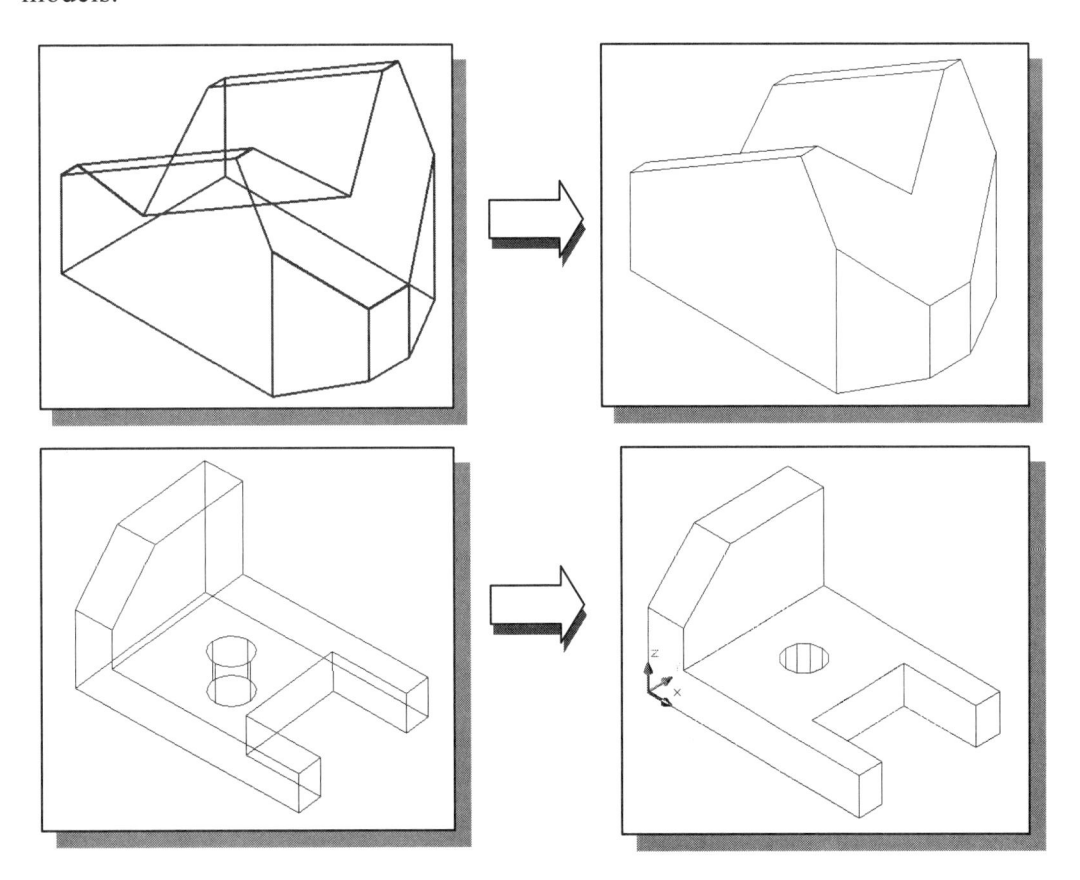

Starting Up AutoCAD® 2006

1. Select the **AutoCAD 2006** option on the *Program* menu or select the **AutoCAD 2006** icon on the *Desktop*.

2. In the *AutoCAD Today* startup dialog box, select the **Open a Drawing** icon with a single click of the left-mouse-button.

3. Click on the ***V-block.dwg*** filename to open the *V-block* wireframe model that was created in the previous chapter. (Use the Browse option to locate the file if it is not displayed.)

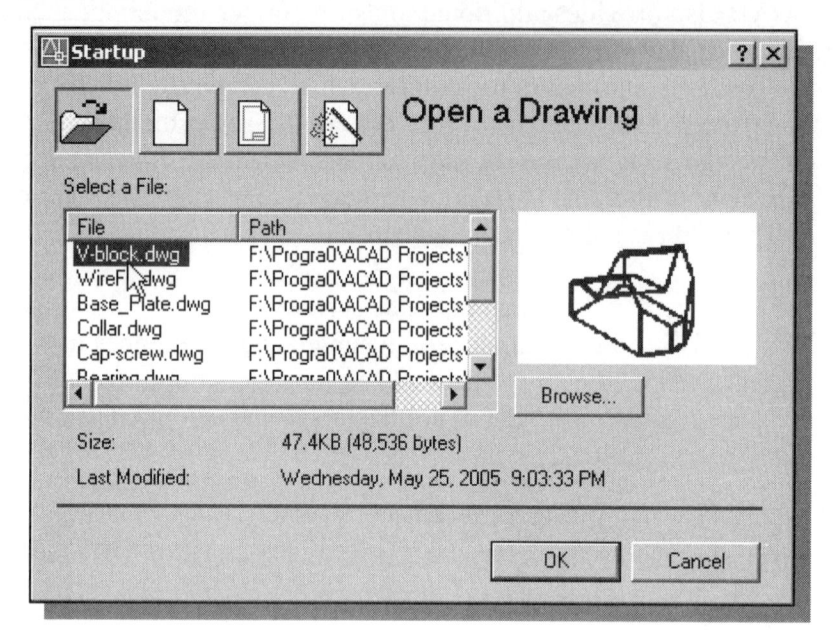

- The *V-block* wireframe model is retrieved and displayed in the graphics window.

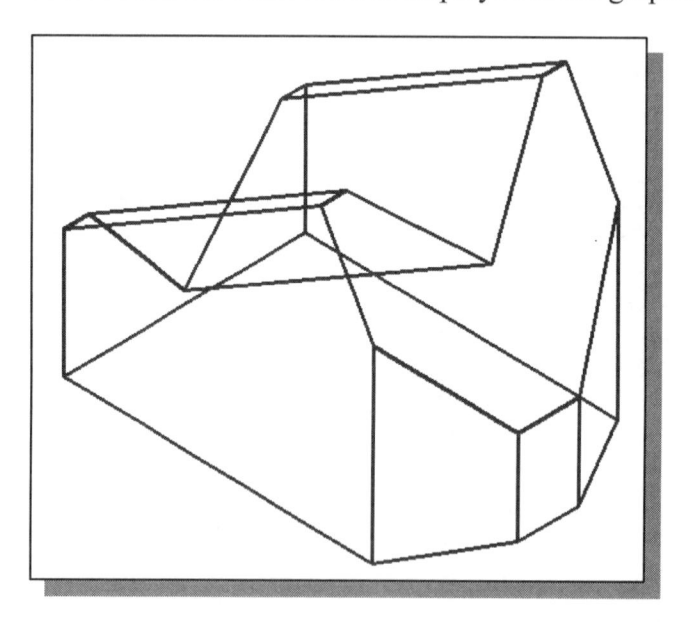

Using the UCS and Surfaces Toolbars

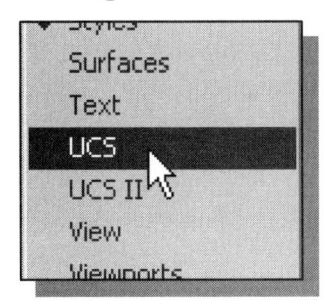

1. Move the cursor to the *Standard* toolbar area and **right-click** on any icon of the *Standard* toolbar to display a list of toolbar menu groups.

2. Select **UCS**, with the left-mouse-button, to display the *UCS* toolbar on the screen.

➢ The options available in the *UCS* toolbar allow us to quickly orient and align the UCS.

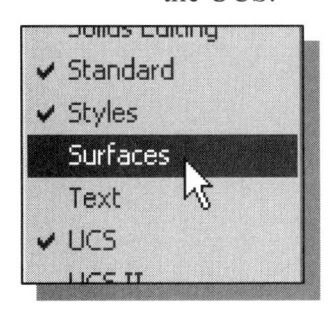

3. Move the cursor to the *Standard* toolbar area and **right-click** on any icon of the *Standard* toolbar to display a list of toolbar menu groups.

4. Select **Surfaces**, with the left-mouse-button, to display the *Surfaces* toolbar on the screen.

• Four groups of options are available in the *Surfaces* toolbar, which allow us to access the surface modeling commands quickly. On your own, move the cursor on top of the different icons and read the brief description of the individual commands in the help-line area.

Creating a Surface Using the 2D Solid Command

➢ The first generation surface available in AutoCAD is used to fill an area in the sketch plane of the current UCS. It is therefore necessary to properly orient the UCS prior to using the 2D Solid command.

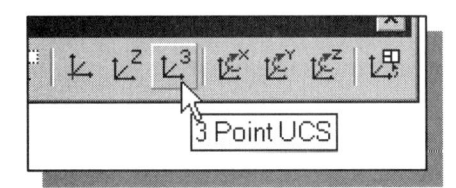

1. Select the **3 Point UCS** icon in the *UCS* toolbar.

• The **3 Point UCS** option allows us to specify the new X-axis and Y-axis directions to align the UCS.

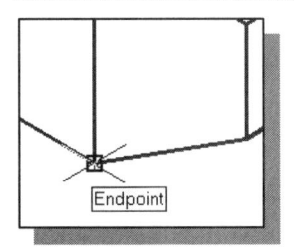

2. In the command prompt area, the message "*Specify new origin point<0,0,0>:*" is displayed. Pick the **lower right corner** of the front face of the wireframe model as shown.

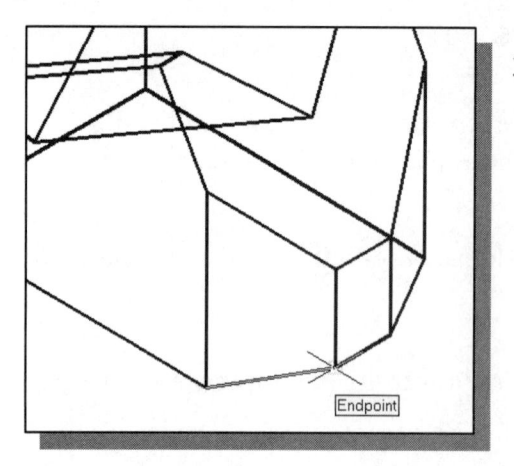

3. In the command prompt area, the message *"Specify point on positive portion of X-axis:"* is displayed. Pick the adjacent corner toward the right side of the model as shown.

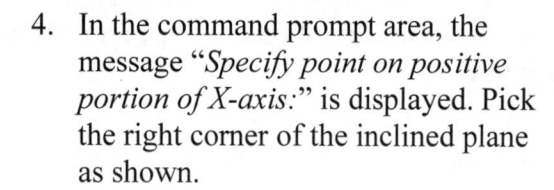

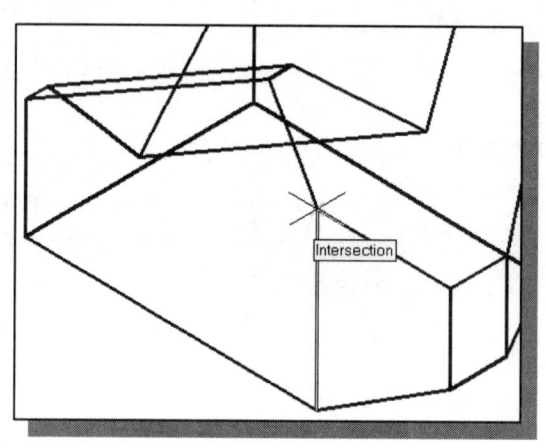

4. In the command prompt area, the message *"Specify point on positive portion of X-axis:"* is displayed. Pick the right corner of the inclined plane as shown.

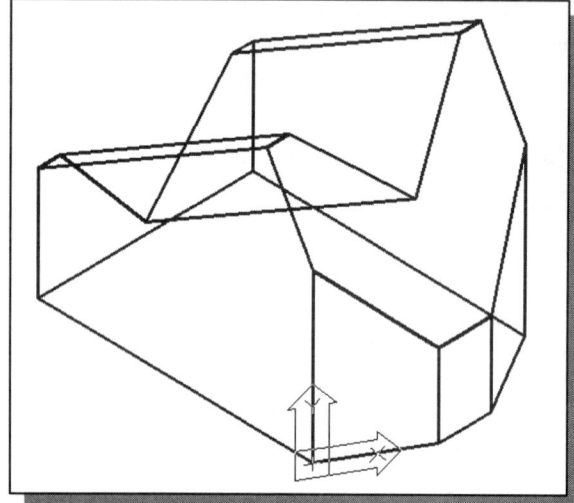

- The new UCS is aligned to the vertical inclined plane as shown.

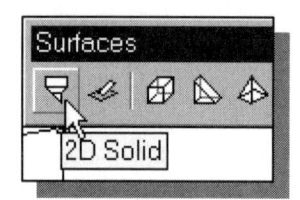

5. Select **2D Solid** in the *Surfaces* toolbar.

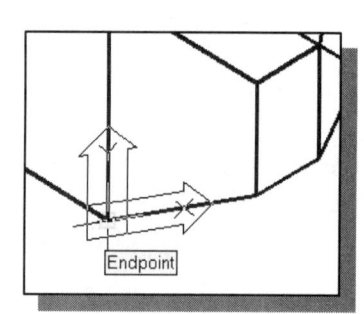

6. Place the first corner point of the 2D solid at the origin of the new UCS.

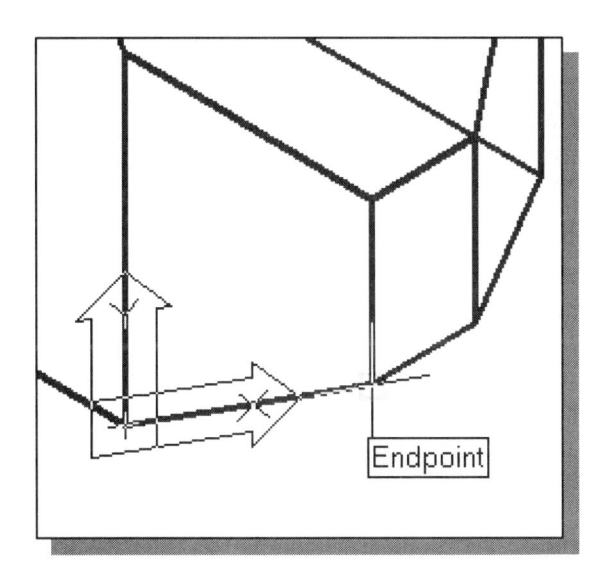

7. In the command prompt area, the message "*Specify second point:*" is displayed. Pick the bottom right corner of the inclined plane as shown.

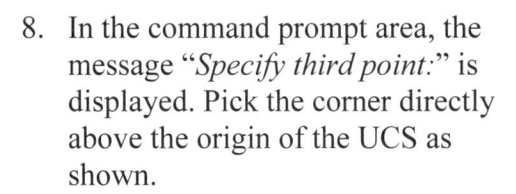

8. In the command prompt area, the message "*Specify third point:*" is displayed. Pick the corner directly above the origin of the UCS as shown.

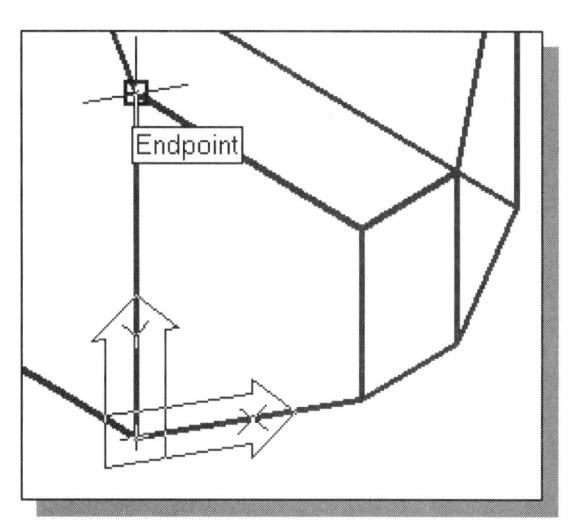

- The **2D Solid** command requires the third point to be specified diagonally opposite to the second point. This seemly strange way of specifying the third corner was established when the **2D Solid** command was first introduced back in the mid-1980s. Note that the **3D Face** command, the second-generation surface command in AutoCAD, does not follow this convention.

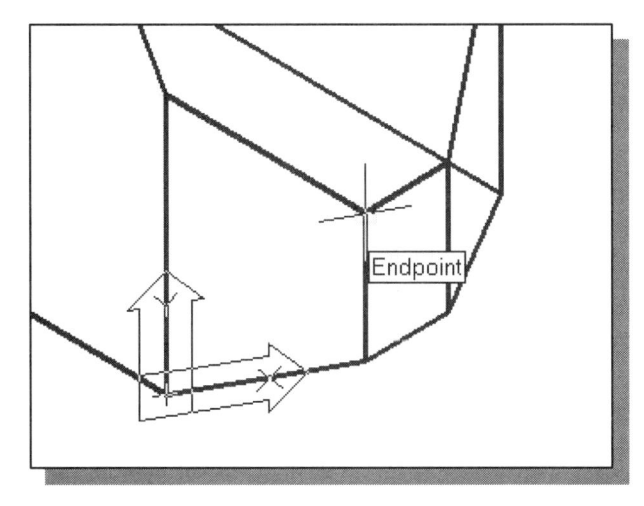

9. In the command prompt area, the message "*Specify fourth point or [Exit]:*" is displayed. Pick the corner directly above the second point we selected as shown in the figure.

10. Inside the graphics window, **right-mouse-click once** to end the 2D Solid command.

- The **2D Solid** command allows the creation of three-sided or four-sided filled polygons, which can be used to represent faces of surface models. Note that in the above steps, we could accept the three-sided polygon after defining the third corner.

Using the Shade Toolbar

1. Move the cursor to the *Standard* toolbar area and **right-click** on any icon of the *Standard* toolbar to display a list of toolbar menu groups.

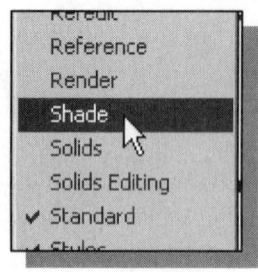

2. Select **Shade**, with the left-mouse-button, to display the *Shade* toolbar on the screen.

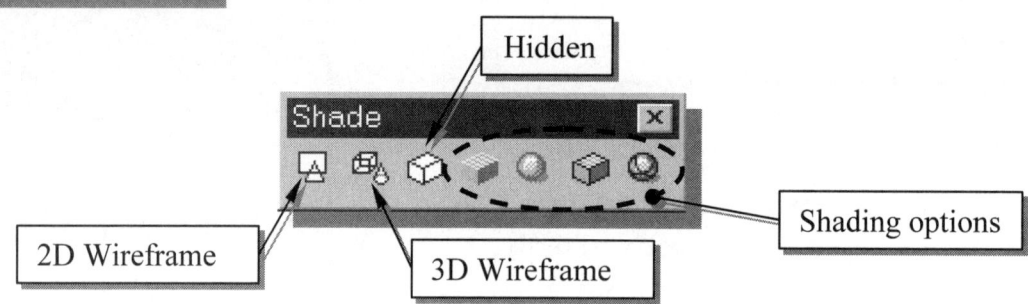

- **2D Wireframe**: Displays the objects using lines and curves to represent the boundaries of objects created. Linetypes and lineweights are visible with this option. Note that this is the default AutoCAD display mode.

- **3D Wireframe**: Displays the objects using lines and curves to represent the boundaries of objects created. Displays a shaded 3D user coordinate system (UCS) icon. Note that linetypes and lineweights are not visible with this option.

- **Hidden**: Displays the objects using the 3D wireframe representation with lines that are located behind surfaces and solids removed.

- **Flat Shaded**: Creates a shaded image of polygon faces and solids. The shaded objects appear flatter and less smooth than Gouraud Shaded objects.

- **Gouraud Shaded**: Creates a shaded image of polygon faces and solids using the Gouraud method. This mode generates an image that gives the objects a smooth and realistic appearance.

- **Flat Shaded, Edges On**: Combines the Flat Shaded and 3D Wireframe options. The objects are flat shaded with the wireframe edges showing through.

- **Gouraud Shaded, Edges On**: Combines the Gouraud Shaded and 3D Wireframe options. The objects are Gouraud shaded with the wireframe edges showing through.

3D Surface Modeling   5-9

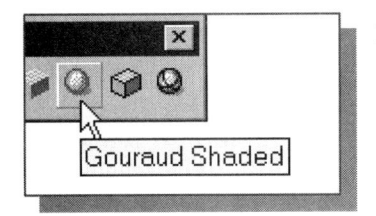

3. In the *Shade* toolbar, click on the **Gouraud Shaded** icon to display the shaded image of the model. There exists only one surface in our model. The surface was created with the **2D Solid** command.

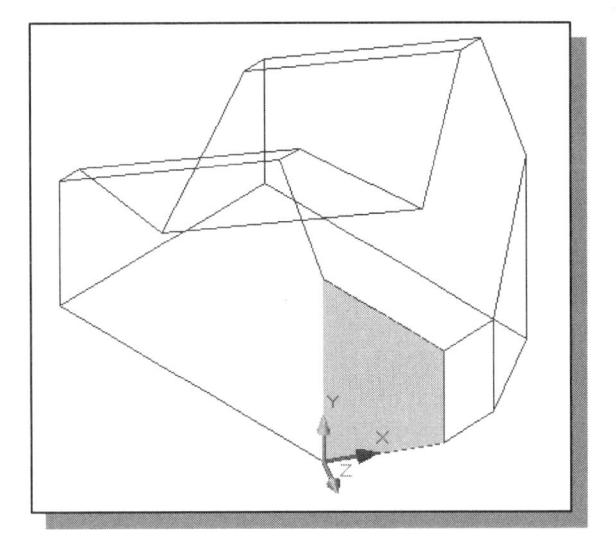

4. Select **3D Orbit** in the *View* pull-down menu.
   [View] → [3D Orbit]

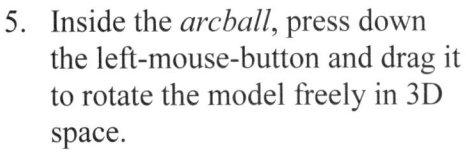

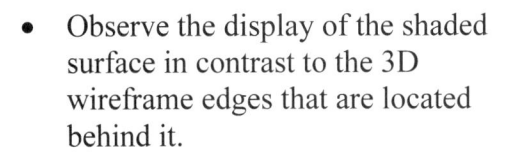

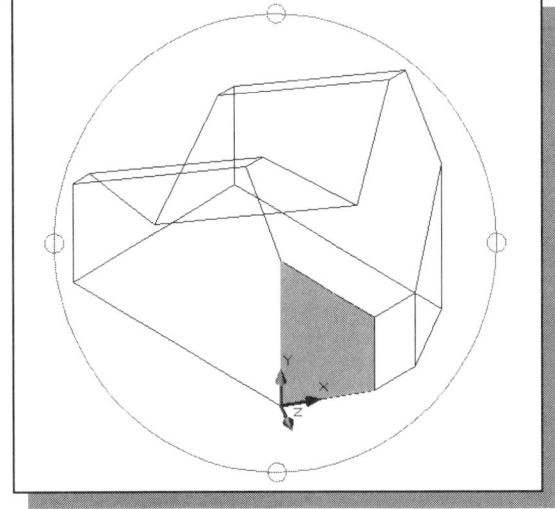

5. Inside the *arcball*, press down the left-mouse-button and drag it to rotate the model freely in 3D space.

- Observe the display of the shaded surface in contrast to the 3D wireframe edges that are located behind it.

6. On your own, reset the display to the **SE Isometric View** before continuing to the next section.

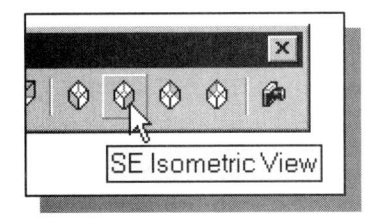

7. In the *UCS* toolbar, select the **World UCS**. This option resets the UCS to align to the world coordinate system.

Creating a Surface Using the 3D Face Command

- The second generation of surface command made available in AutoCAD was the **3D Face** command. The 3D Face command can be used to create true 3D planar surfaces by allowing the X, Y and Z coordinates of the corners to be selected independently of the current UCS. The created polygon can be a three-sided or four-sided shape. This command is the primary construction tool for surface modeling in AutoCAD.

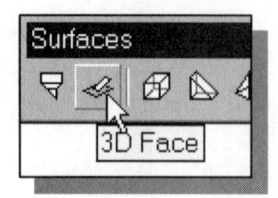

1. Select the **3D Face** in the *Surface* toolbar.

2. In the command prompt area, the message "*_3dface Specify first point or [invisible]:*" is displayed. Pick the lower right corner of the vertical inclined face of the model as shown.

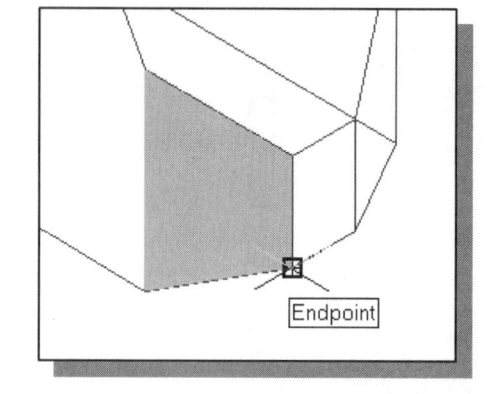

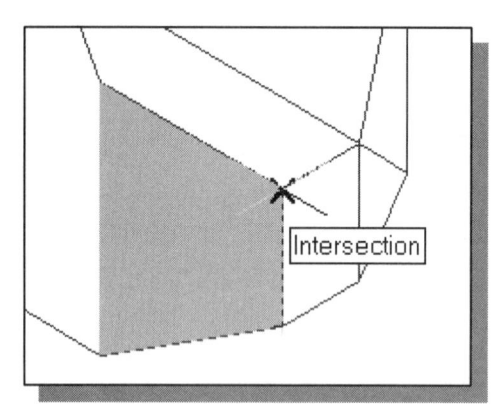

3. In the command prompt area, the message "*Specify second point or [invisible]:*" is displayed. Pick the adjacent corner above the previous selected corner of the vertical inclined face as shown.

4. In the command prompt area, the message "*Specify third point or [invisible]:*" is displayed. Pick the adjacent corner of the right vertical face of the model as shown.

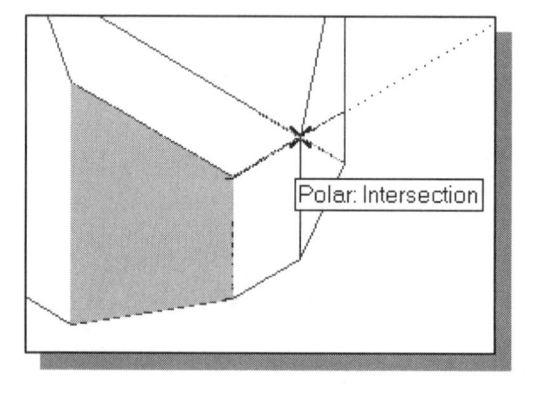

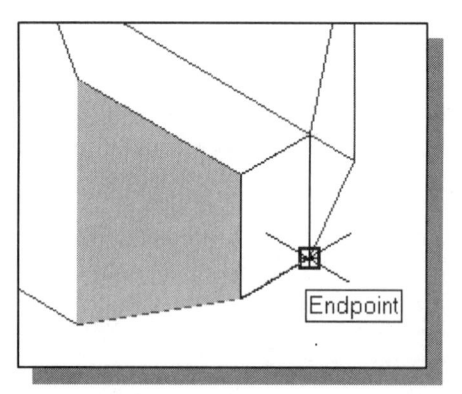

5. In the command prompt area, the message "*Specify fourth point or [invisible] <Create three-sided face>:*" is displayed. Pick the corner below the last selected corner as shown.

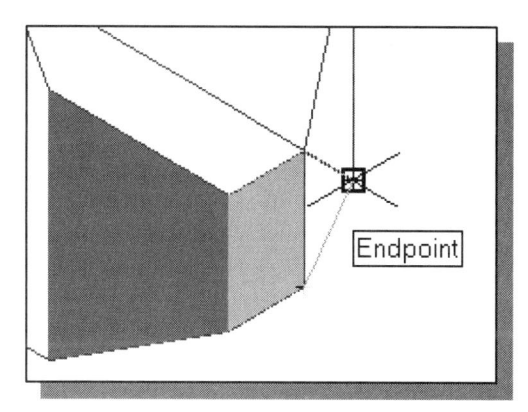

3D Surface Modeling   5-11

6. In the command prompt area, the message "*Specify third point or [invisible] <exit>:*" is displayed. Pick the back corner of the model as shown.

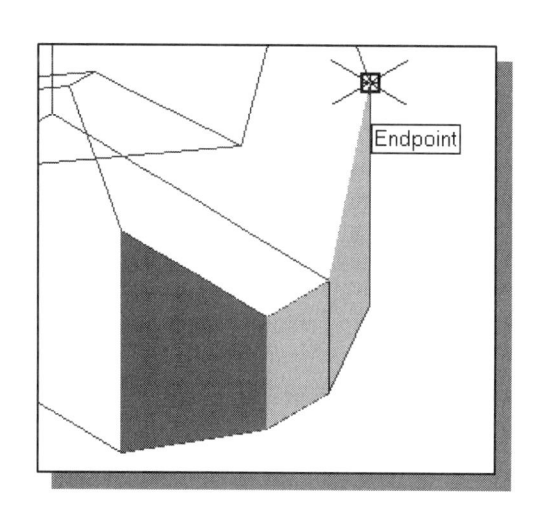

7. In the command prompt area, the message "*Specify fourth point or [invisible] <Create three-sided face>:*" is displayed. Pick the back corner of the model as shown.

- Note that this surface is created independent of the UCS and two corners of the previous face were used to position this face.

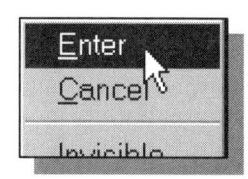

8. Inside the graphics window, right-mouse-click to activate the option menu and select **Enter** with the left-mouse-button to end the 3D Face command.

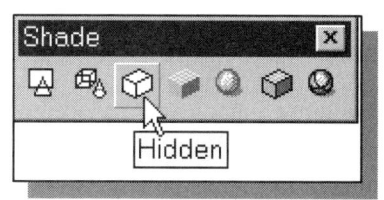

9. In the *Shade* toolbar, click on the **Hidden** icon to display the model with hidden lines removed.

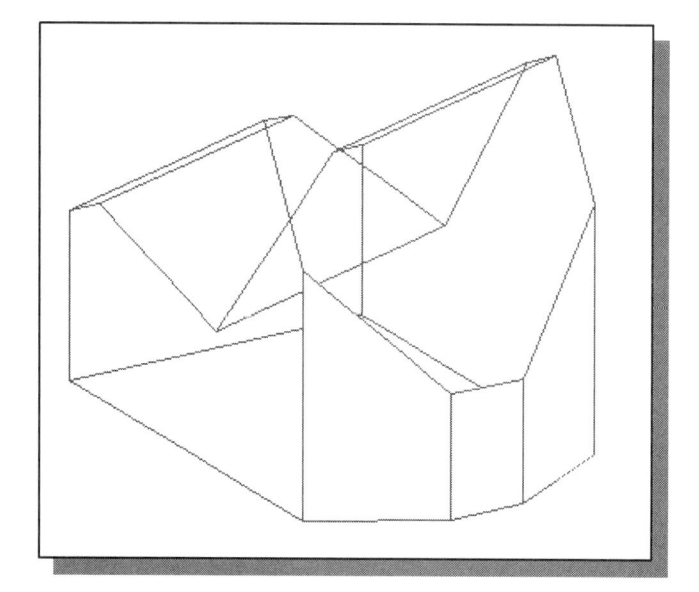

10. On your own, use the **3D Orbit** icon to rotate the model and examine the constructed surface model.

- Reset the display to **SE Isometric View** before proceeding to the next section.

Creating a Surface of Irregular Shape

- The **3D Face** command allows us to create three-sided or four-sided polygons. For surfaces of irregular shape, the Invisible Edge option is available in conjunction with the 3D Face command. Note that the Invisible Edge option cannot be applied to polygons created by the 2D Solid command.

1. Select the **2D Wireframe** in the *Shade* toolbar.

   - The 2D Wireframe command resets the display to the default AutoCAD display mode

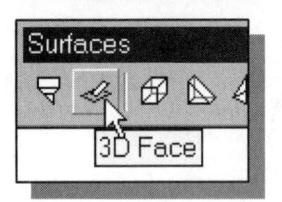

2. Select the **3D Face** in the *Surface* toolbar.

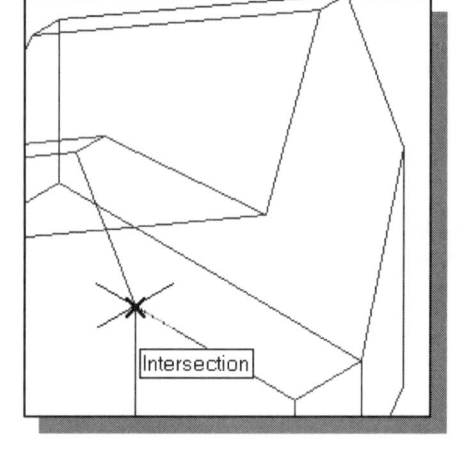

3. In the command prompt area, the message "*_3dface Specify first point or [invisible]:*" is displayed. Pick the top right corner of the model as shown.

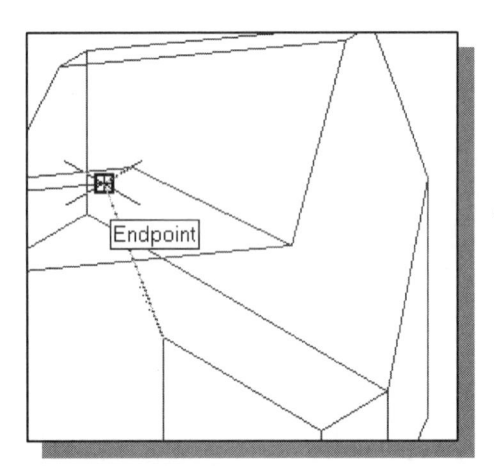

4. In the command prompt area, the message "*Specify second point or [invisible]:*" is displayed. Pick the top front corner of the model as shown.

5. In the command prompt area, the message "*Specify third point or [invisible] <exit>:*" is displayed. Pick the top corner of the model adjacent to the previously selected corner as shown.

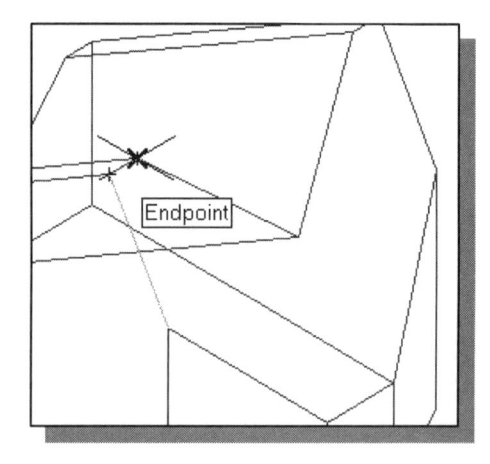

3D Surface Modeling   5-13

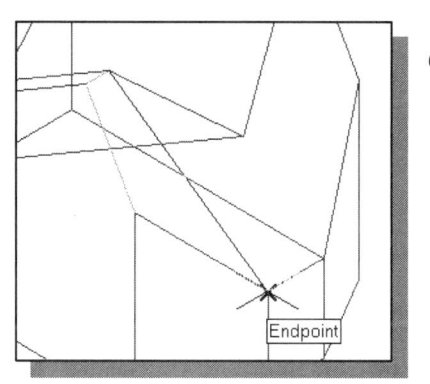

6. In the command prompt area, the message "*Specify fourth point or [invisible] <Create three-sided face>:*" is displayed. Pick the corner below the last selected corner as shown.

7. On your own, repeat the zigzagging pattern to define polygons until all corners of the inclined surface have been selected and additional polygons are created as shown in the figure. Note that the last polygon we created is a three-sided polygon.

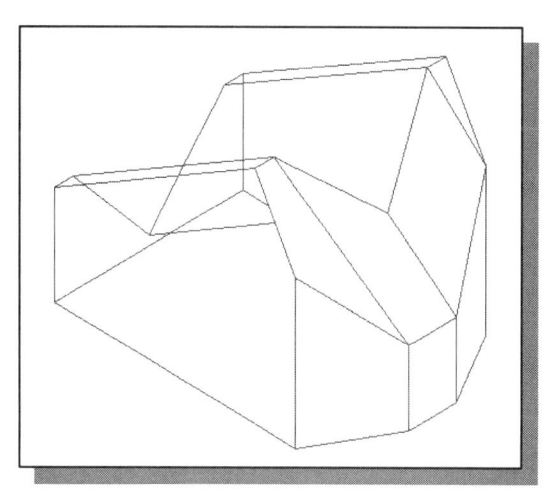

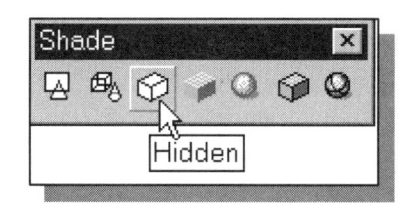

8. In the *Shade* toolbar, click on the **Hidden** icon to display the model with hidden lines removed. Note that the edges of the polygons are displayed as shown in the above figure.

9. In the *Shade* toolbar, click on the **Gouraud Shaded** icon to display the shaded image of the model. Note that the edges of the polygons are not visible when a shaded command is performed.

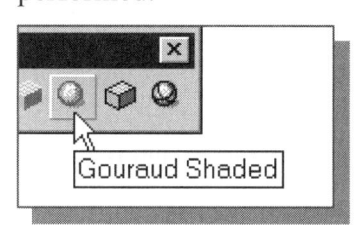

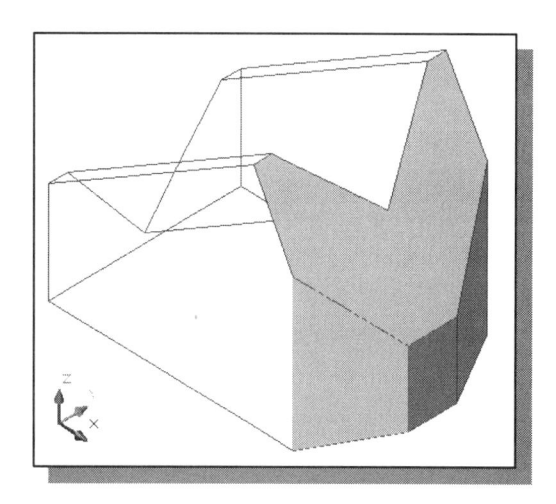

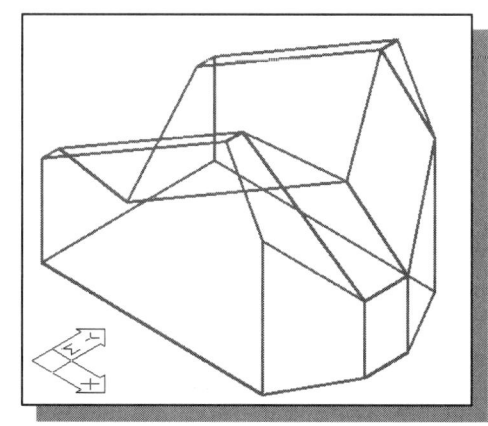

10. Select **2D Wireframe** in the *Shade* toolbar to reset the display to the default AutoCAD display mode.

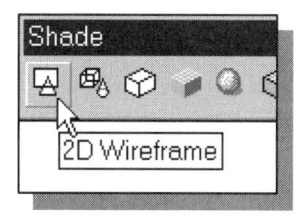

Using the Invisible Edge Option

- The **Invisible Edge** option is used to turn off the display of selected edges and therefore allow the adjacent polygons, created by the **3D Face** command, to appear as being joined together.

1. Select **Edge**, with the left-mouse-button, in the *Surfaces* toolbar.

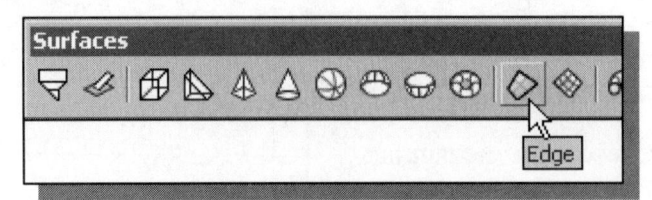

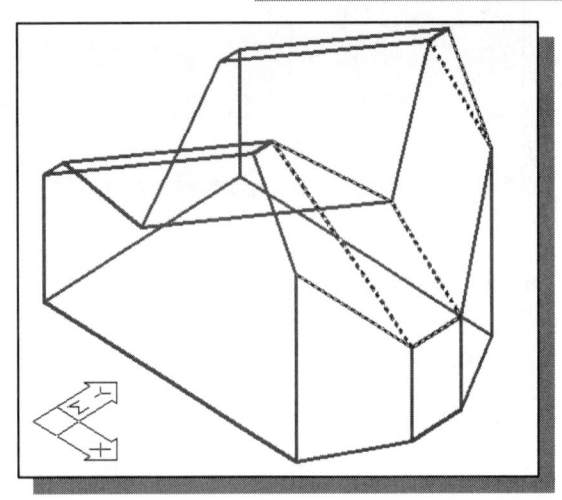

2. In the command prompt area, the message "*Specify edge of 3dface to toggle visibility or [Display]:*" is displayed. Pick the three edges inside the inclined surface as shown.

3. Inside the graphics window, right-mouse-click to activate the option menu and select **Enter** to end the **Edge** command.

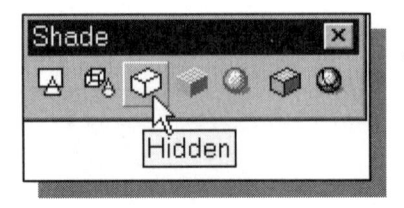

4. In the *Shade* toolbar, click on the **Hidden** icon to display the model with hidden lines removed.

- The selected edges are removed from the display so that the face of the inclined surface of the model appears to be more realistic.

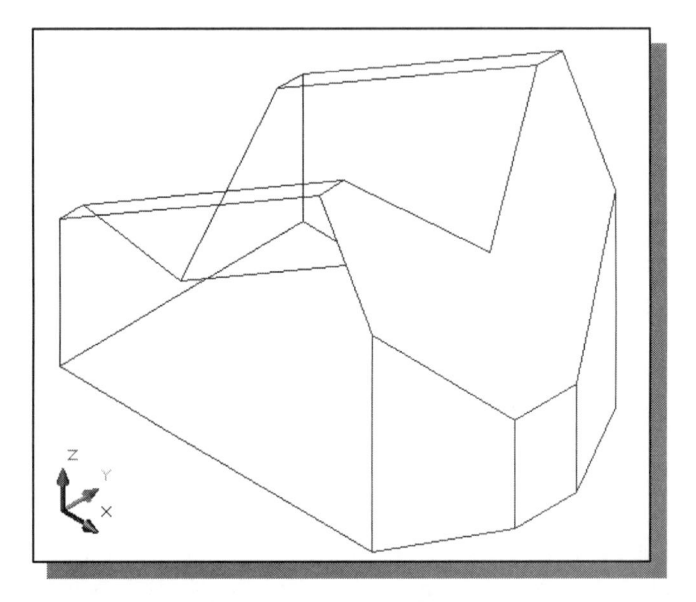

> On your own, repeat the above steps and complete the surface model of the design.

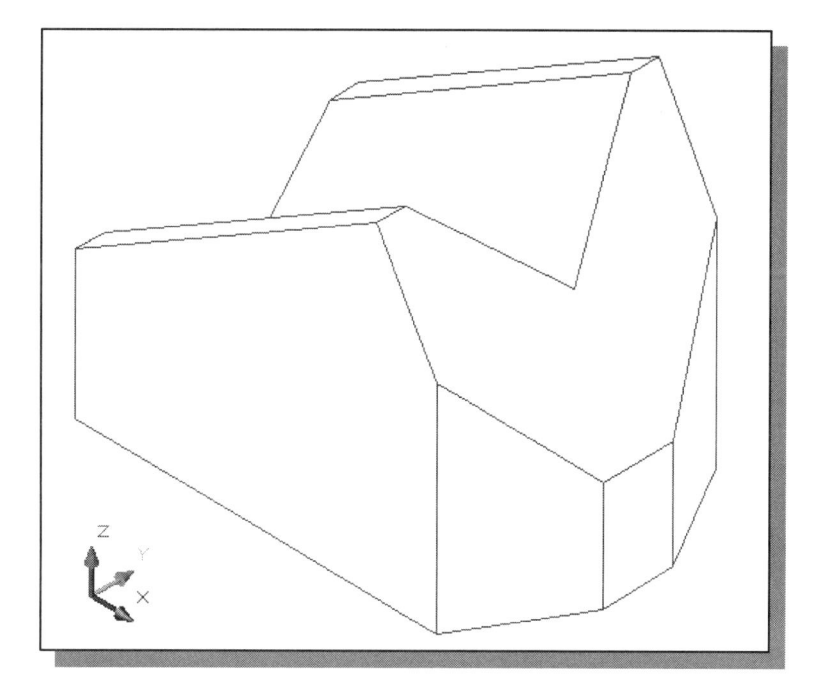

- On your own, examine the completed surface model by using the different shading commands.

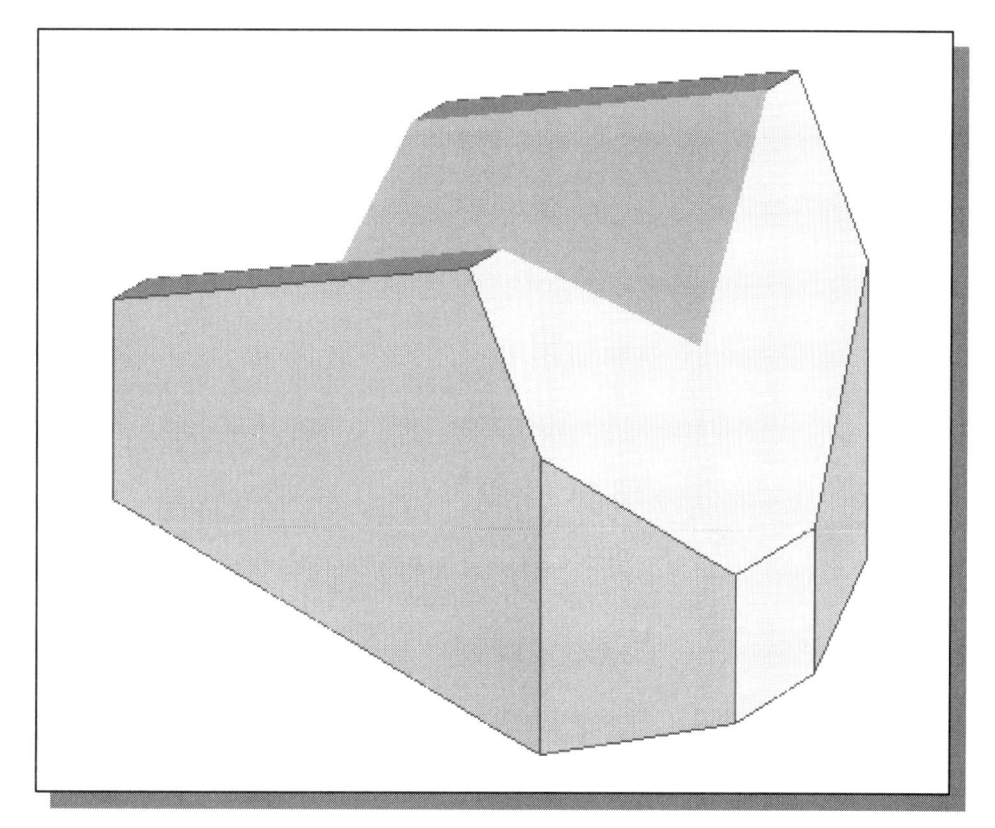

The *Locator* Wireframe Model

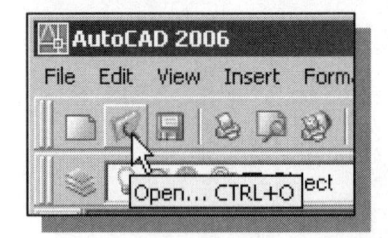

1. Click on the **Open** icon in the *Standard* toolbar.

2. In the *Select File* window, pick the **Locator** file that was created in Chapter 3.

3. Click **Open** to open the model file.

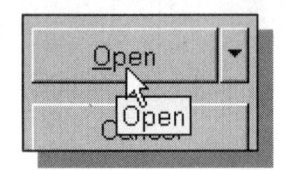

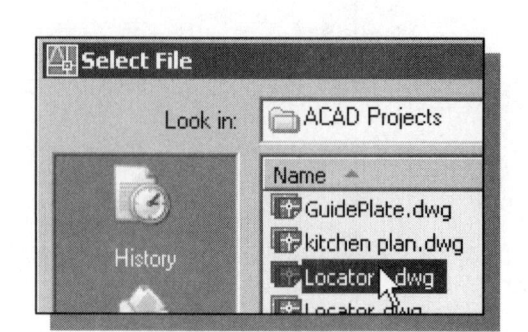

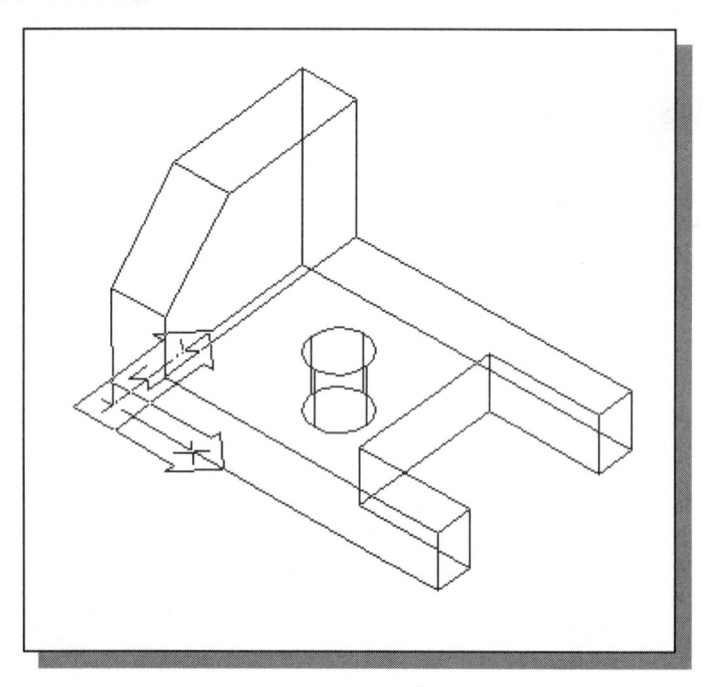

Moving Objects to a Different Layer

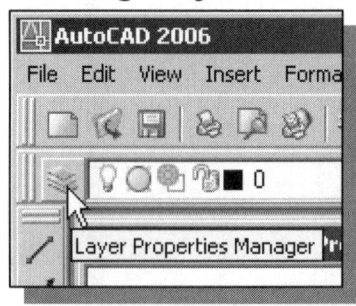

1. Pick **Layer Properties Manager** in the *Object Properties* toolbar.

2. Click on the **New** button to create new layers.

3. Create **two new layers** with the following settings:

Layer	*Color*	*LineType*	*LineWeight*
Construction	White	Continuous	Default
Wireframe	Blue	Continuous	0.30mm
Surface	Cyan	Continuous	Default

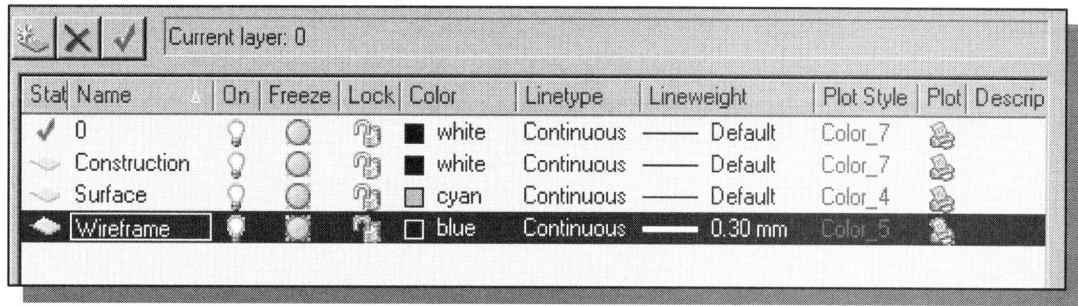

4. Click on the **OK** button to accept the settings and exit the *Layer Properties Manager* dialog box.

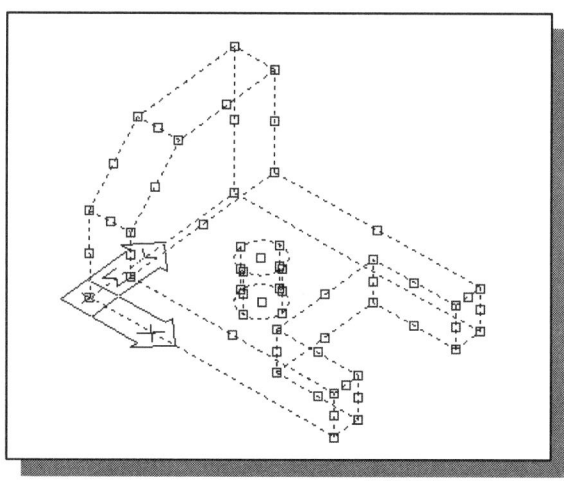

5. Inside the graphics window, pre-select all entities by using the left-mouse-button to create a *selection window* enclosing all entities.

- Note that all selected entities are highlighted.

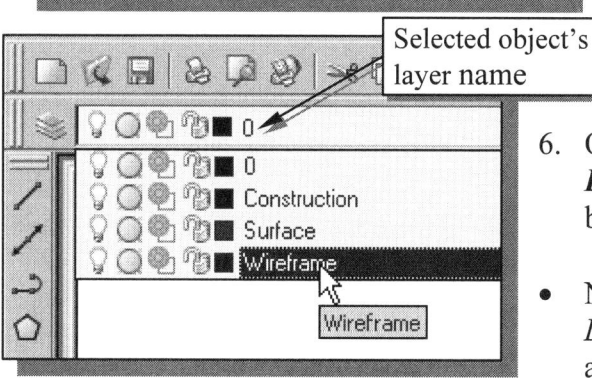

Selected object's layer name

6. On the *Object Properties* toolbar, choose the *Layer Control* box with the left-mouse-button.

- Notice the layer name displayed in the *Layer Control* box is the selected object's assigned layer and layer properties.

7. In the *Layer Control* box, click on the **Wireframe** layer name as shown.

❖ All entities of the *Locator* model are moved to the *Wireframe* layer.

8. Press the **[ESC]** key twice to deselect all highlighted entities.

Predefined Surface Models

- **Predefined surface models**: AutoCAD provides the following predefined 3D surface models: boxes, cones, dishes, domes, meshes, pyramids, spheres, tori (donuts), and wedges. AutoCAD creates these models by prompting the user to input specific dimensions for the desired models. For example, the length, width and height dimensions are required for the *Box* surface model.

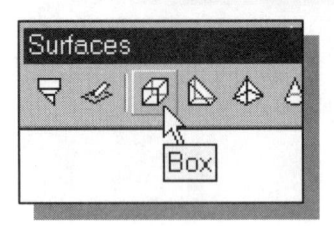

1. Select **Box** in the *Surfaces* toolbar.

2. In the command prompt area, the message "*Specify corner point of Box:*" is displayed. Pick a point that is toward the right of the *Locator* model.

3. In the command prompt area, the message "*Specify length of Box:*" is displayed.
   Specify length of Box: **4.5 [ENTER]**

4. In the command prompt area, the message "*Specify width of Box:*" is displayed.
   Specify width of Box: **3.0 [ENTER]**

5. In the command prompt area, the message "*Specify height of Box:*" is displayed.
   Specify height of Box: **2.5 [ENTER]**

6. In the command prompt area, the message "*Specify rotation angle about the Z axis or [Reference]:*" is displayed.
   Specify rotation angle about the Z axis or [Reference]: **0 [ENTER]**

➤ On your own, use the **Hidden**, **Shaded** and **Orbit** commands to examine the constructed surface model.

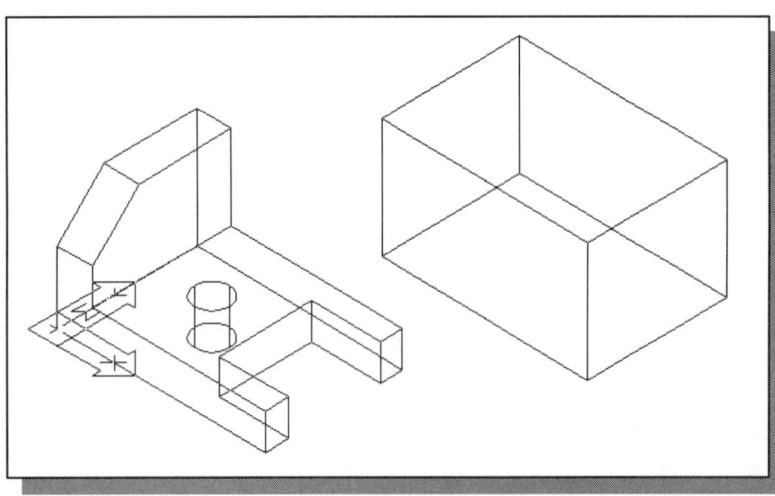

- The created *Box* is of the same size of the *Locator* model. Note that the surfaces of the box are created automatically and all faces of the model are grouped together as a single object.

> The predefined surface models can be used as a base for more complex designs. The main difficulty, and disadvantage, of surface modeling is that the constructed surfaces cannot be easily modified. In AutoCAD, we can use the grip-editing option to resize the constructed faces.

7. Select **Explode** in the *Modify* toolbar.

8. In the command prompt area, the message "*Select objects:*" is displayed. Select any edge of the surface model and note that the entire model is selected.

9. Inside the graphics window, right-mouse-click once to accept the selection and the faces are separated into individual objects.

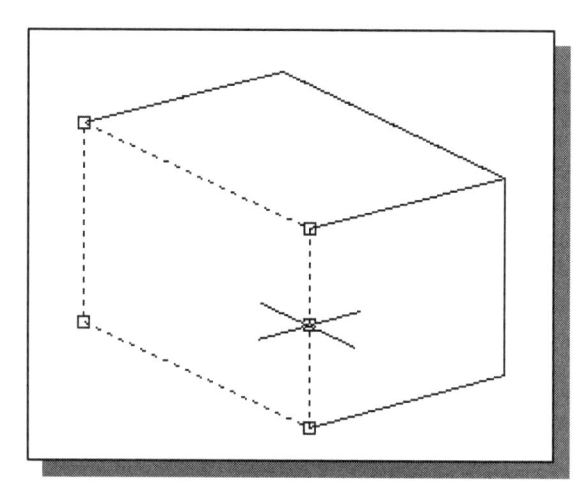

10. Select the front vertical face of the model by clicking on the edges of the front face as shown.

- Four grip points, the small rectangles displayed at the corners of the highlighted polygon, can be used to adjust the size and shape of the highlighted polygon.

11. Select the top right grip point with the left-mouse-button.

12. Move the cursor to the left, along the top edge of the model, of its current location as shown. Left-mouse-click once to reposition the grip point.

13. Press the [Esc] key to deselect the highlighted entities.

> On your own, experiment with creating other predefined surface models that are available in the *Surfaces* toolbar. Delete all surface models before proceeding to the next section.

Advanced Surface Modeling Commands

❖ Several of the advanced surface-modeling commands are displayed in the *Surfaces* toolbar. These commands allow us to quickly create and duplicate surfaces in specific manners.

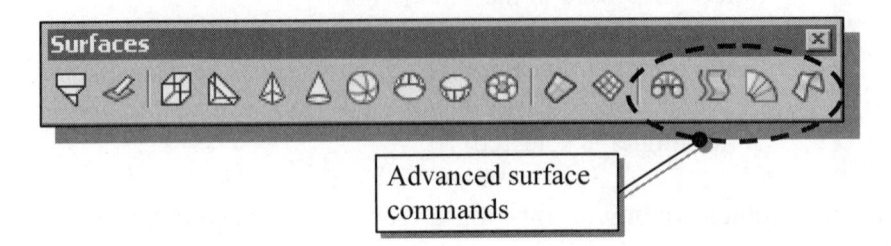

Advanced surface commands

- **Revolved Surface**: Creates a surface mesh by rotating a group of objects about an axis.

- **Tabulated Surface**: Creates a surface mesh representing a general tabulated surface defined by a path curve and a direction vector. The resulting mesh is a series of parallel polygons running along a specified path.

- **Ruled Surface**: Creates a surface mesh between two objects.

- **Edge Surface**: Creates a surface patch mesh from four edges.

Note that there are two AutoCAD system variables used to control the results of the Ruled Surface and Tabulated Surface commands: **SurfTab1** and **SurfTab2**.

SurfTab1 & SurfTab2 system variables: These two variables are used to set the number of increments used by the Ruled Surface and Tabulated Surface commands. The default values are set to six, which means any curve will be approximated with six straight lines.

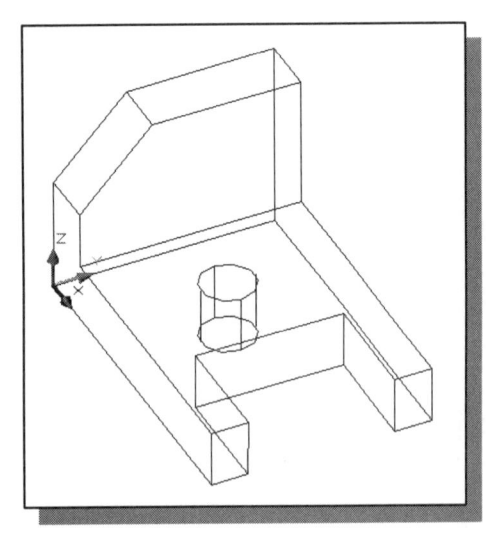

1. On your own, use the **3D Orbit** command and adjust the display of the wireframe model so that the four vertical lines connecting the two circles are visible as shown.

- To illustrate the use of the Ruled Surface and Tabulated Surface commands, we will first split the top circle into two arcs. Note that these commands allow us to construct a polygon mesh for different situations and regions. The split of the circle is necessary in creating a mapped surface on the top plane of the model.

3D Surface Modeling   5-21

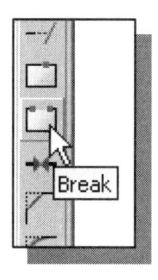

2. Select **Break** in the *Modify* toolbar.

- The Break command can be used to erase parts of objects or split an object in two. Note that we can only erase parts of a circle. We will therefore erase a portion of the circle and then split the circle into two arcs.

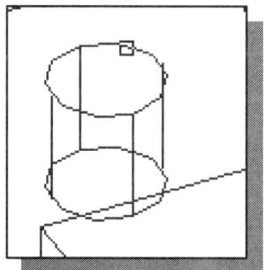

3. Select the top circle as shown. Note that the selected portion will be erased.

- By default, AutoCAD treats the selection point as the first break point. We can override the first point by choosing **First point** in the option menu.

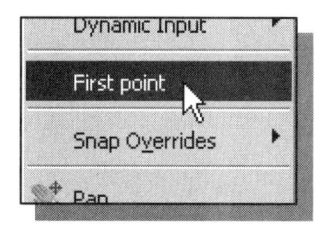

4. In the command prompt area, the message "*Specific second break point or [First point]:*" is displayed. Right-mouse-click once and select **First point** in the option menu.

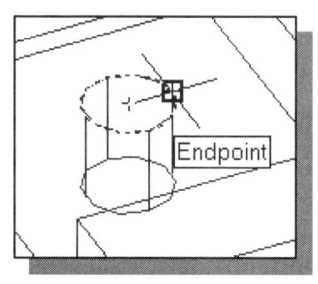

5. In the command prompt area, the message "*Specific first break point:*" is displayed. Choose the top endpoint of the vertical line as shown.

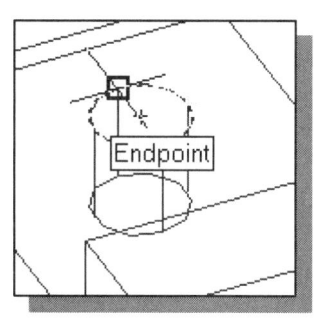

6. In the command prompt area, the message "*Specific second break point:*" is displayed. Choose the top endpoint of the vertical line as shown.

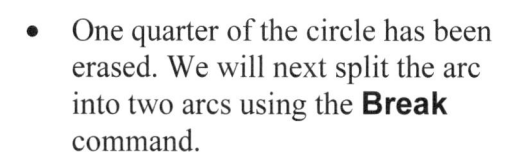

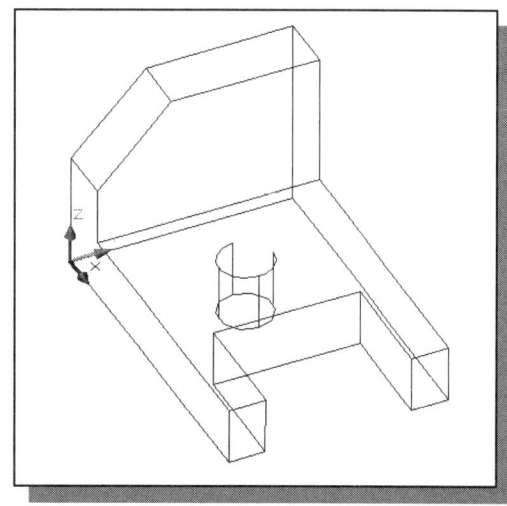

- One quarter of the circle has been erased. We will next split the arc into two arcs using the **Break** command.

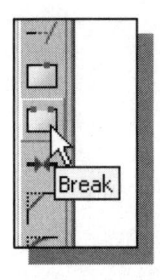

7. Select **Break** in the *Modify* toolbar.

8. Select the top circle as shown. Note that the selected portion will be erased.

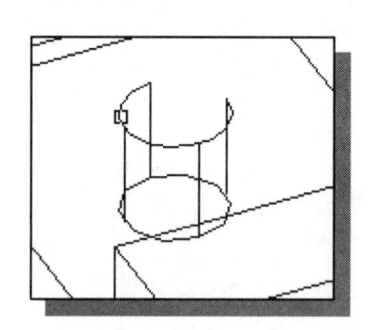

- By default, AutoCAD treats the selection point as the first break point. We can override the first point by choosing **First point** in the option menu.

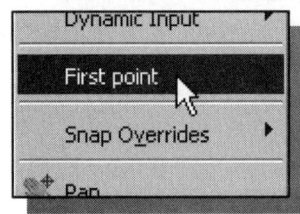

9. In the command prompt area, the message "*Specific second break point or [First point]:*" is displayed. Right-mouse-click once and select **First point** in the option menu.

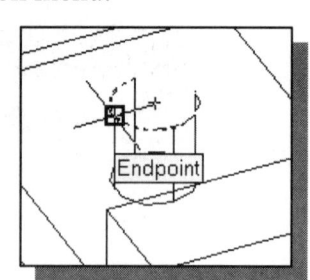

10. In the command prompt area, the message "*Specific first break point:*" is displayed. Choose the top endpoint of the left vertical line as shown.

11. In the command prompt area, the message "*Specific second break point:*" is displayed. To split an object, choose the same endpoint that was chosen as the first endpoint.

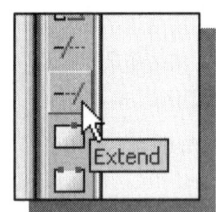

12. Select **Extend** in the *Modify* toolbar. Note the *Projection* type is set to **View**, as shown in the prompt window.

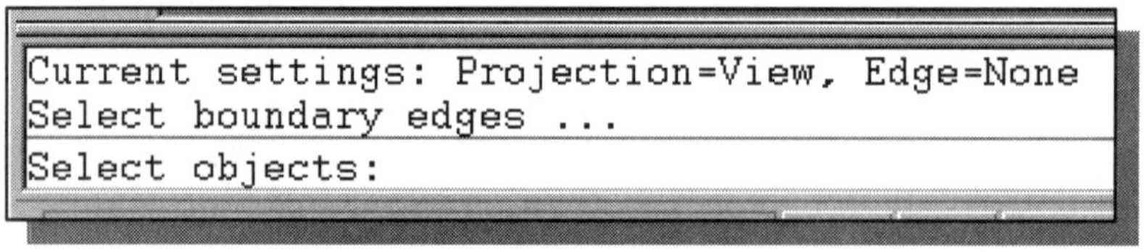

```
Current settings: Projection=View, Edge=None
Select boundary edges ...
Select objects:
```

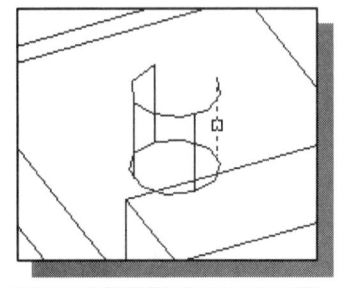

13. Select the right vertical edge to be the extending boundary as shown.

14. Inside the graphics window, right-mouse-click once to accept the selection.

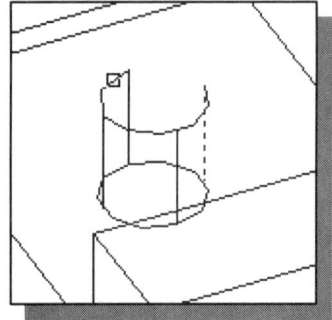

15. Pick the shorter arc near the right endpoint to extend the arc in that direction.

16. Inside the graphics window, right-mouse-click once to display the option menu and select **Enter** to end the **Extend** command.

Using the Tabulated Surface Option

- In AutoCAD, the **TABSURF** command allows us to construct a polygon mesh representing a general tabulated surface defined by a path curve and a direction vector.

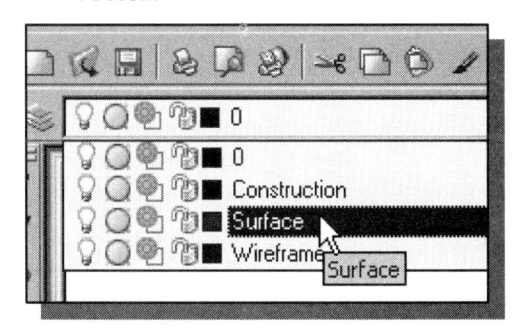

1. Set the *Surface* layer as the *Current Layer* by choosing the layer name in the *Layer Control* box as shown.

- We will place the surface model on a different layer than the wireframe model.

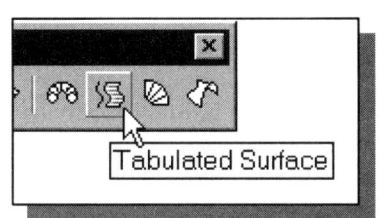

2. Pick **Tabulated Surface** in the *Surfaces* toolbar as shown.

3. In the command prompt area, the message "*Select object for path curve:*" is displayed. Choose the upper arc as shown.

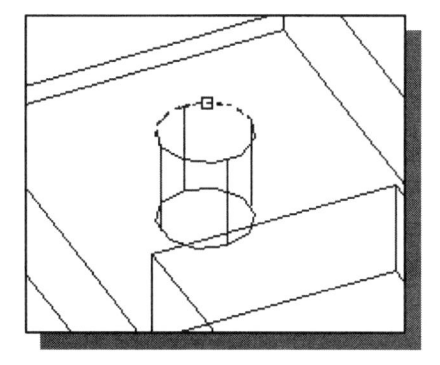

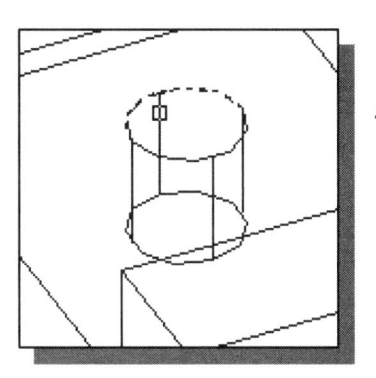

4. In the command prompt area, the message "*Select object for direction vector:*" is displayed. Choose the vertical line near the top endpoint as shown. Note that the endpoint of the line is used as a reference point to determine the direction of the polygon mesh.

5. On your own, use the **Orbit** and **Shaded** commands to examine the constructed polygons.

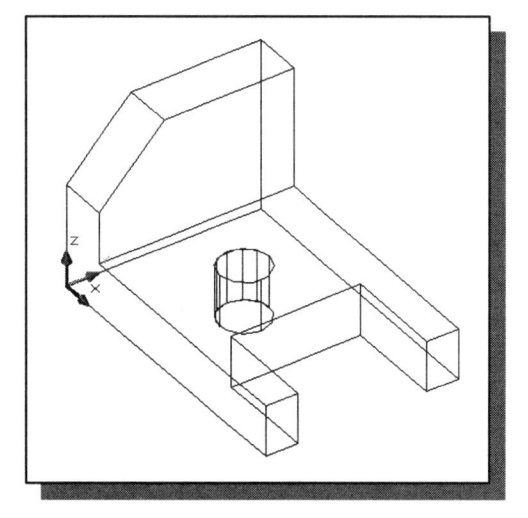

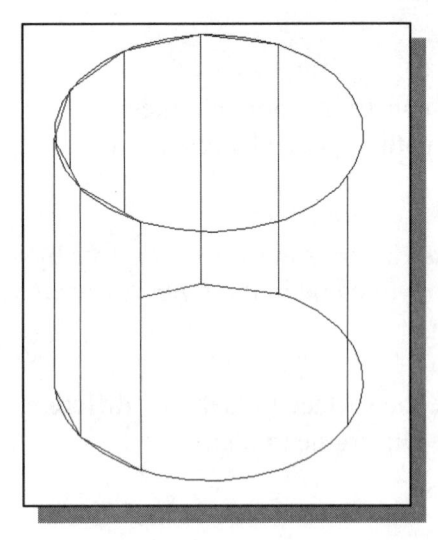

- Exactly six polygons are created and positioned along the selected *path curve*, the upper arc, with the **Tabulated Surface** command. This is set by the **SurfTab1** variable. We can adjust number of segments to use by typing the word, **surftab1**, at the command prompt.

❖ Note that a surface modeler using polygons to approximate true curves is called a *faceted surface modeler*.

Using the Ruled Surface Option

- In AutoCAD, the **RULESURF** command allows us to construct a polygon mesh between two objects. We can use two different objects to define the edges of the ruled surface: lines, points, arcs, circles, ellipses, elliptical arcs, polylines, or splines. The two objects to be used as the *rails* of a ruled surface mesh must both be either open or closed. For open curves, AutoCAD starts construction of the ruled surface based on the locations of the specified points on the curves.

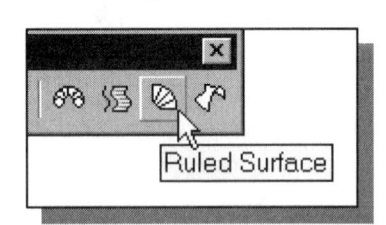

1. Pick **Ruled Surface** in the *Surfaces* toolbar as shown.

2. In the command prompt area, the message "*Select first defining curve:*" is displayed. Choose the lower arc by clicking on the right side as shown.

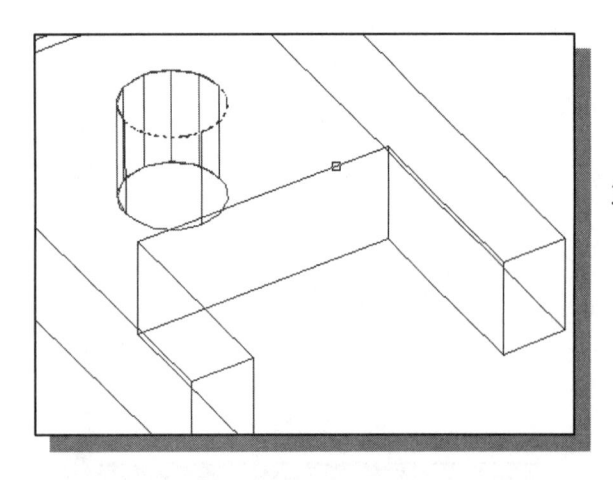

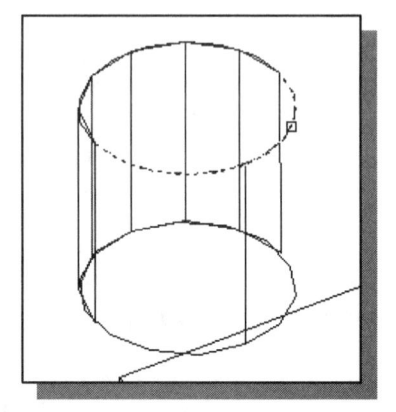

3. In the command prompt area, the message "*Select second defining curve:*" is displayed. Choose the inside straight edge on the right side as shown.

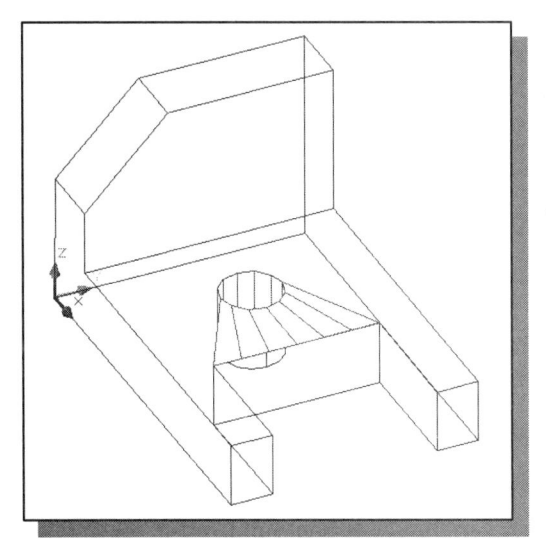

❖ Exactly six polygons are created and positioned along the two *defining curves* with the **Ruled Surface** command.

• Note that the created polygons are grouped together and all polygons are being treated as one object. To hide the edges of the polygons, it is necessary to separate the polygon into individual entities.

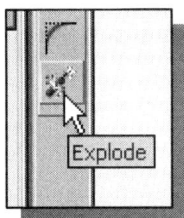

4. Select **Explode** in the *Modify* toolbar.

5. In the command prompt area, the message "*Select objects:*" is displayed. Select any edge of the ruled surface and notice that all six polygons are selected.

6. Inside the graphics window, right-mouse-click once to accept the selection and the polygons are separated into individual objects.

7. Select **2D Wireframe** in the *Shade* toolbar to display the entities in wireframe mode.

8. Select **Edge**, with the left-mouse-button, in the *Surfaces* toolbar.

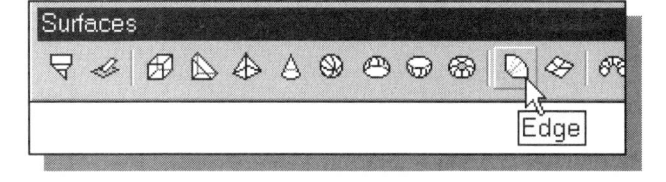

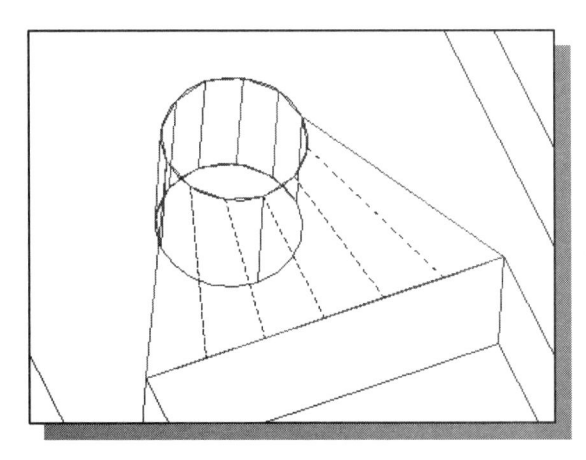

9. In the command prompt area, the message "*Specify edge of 3dface to toggle visibility or [Display]:*" is displayed. Pick the five edges inside the ruled surface as shown.

10. Inside the graphics window, right-mouse-click to activate the option menu and select **Enter** to end the **Edge** command.

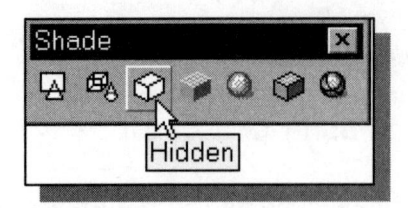

11. In the *Shade* toolbar, click on the **Hidden** icon to display the model with hidden lines removed.

- The selected edges are removed from the display so that the face of the ruled surface appears to be more realistic.

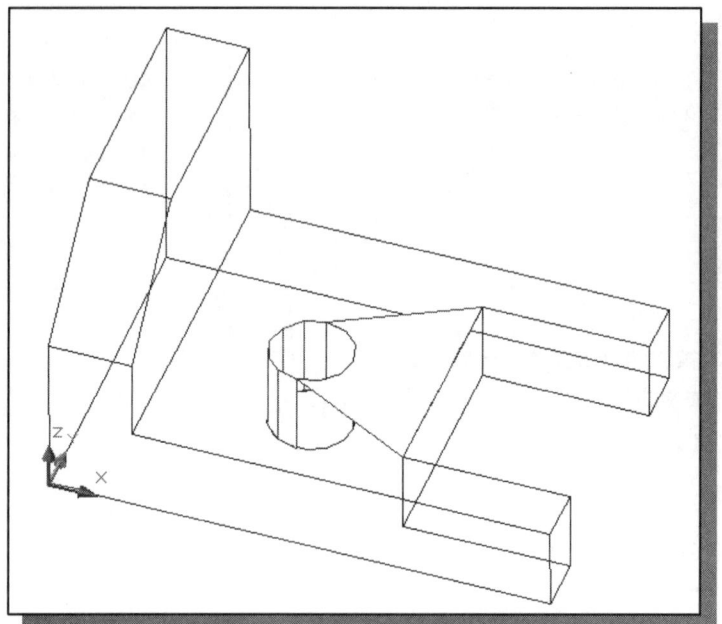

12. On your own, repeat the above steps and create another ruled surface as shown. (Hint: Use the **Realtime Zoom** function to assist the selection of the arc.)

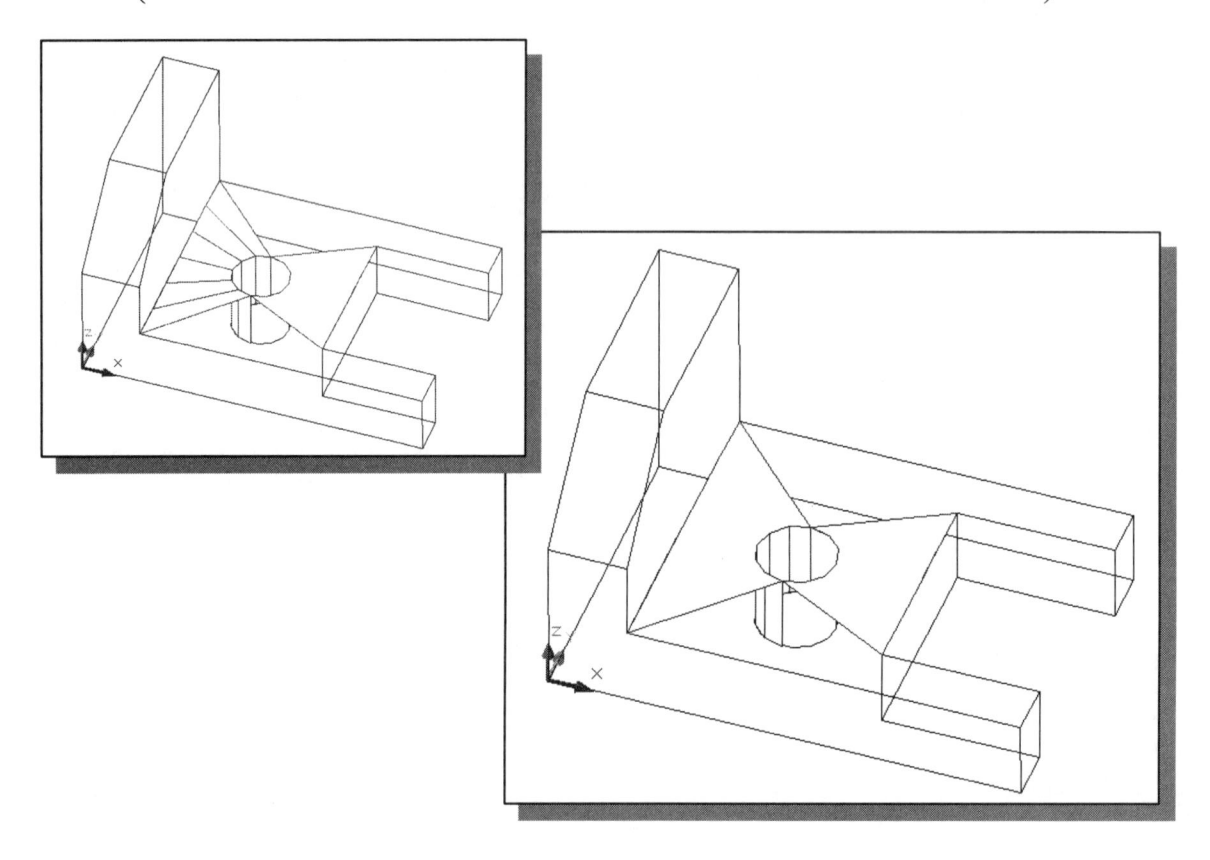

13. On your own, use the **Ruled Surface** option to create a polygon mesh for the other arc on the top surface. What is required to perform such an operation?

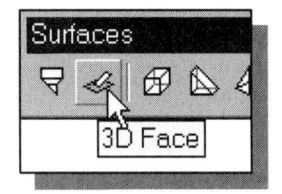

14. Select **3D Face** in the *Surface* toolbar.

- On your own, complete the surface model as shown.

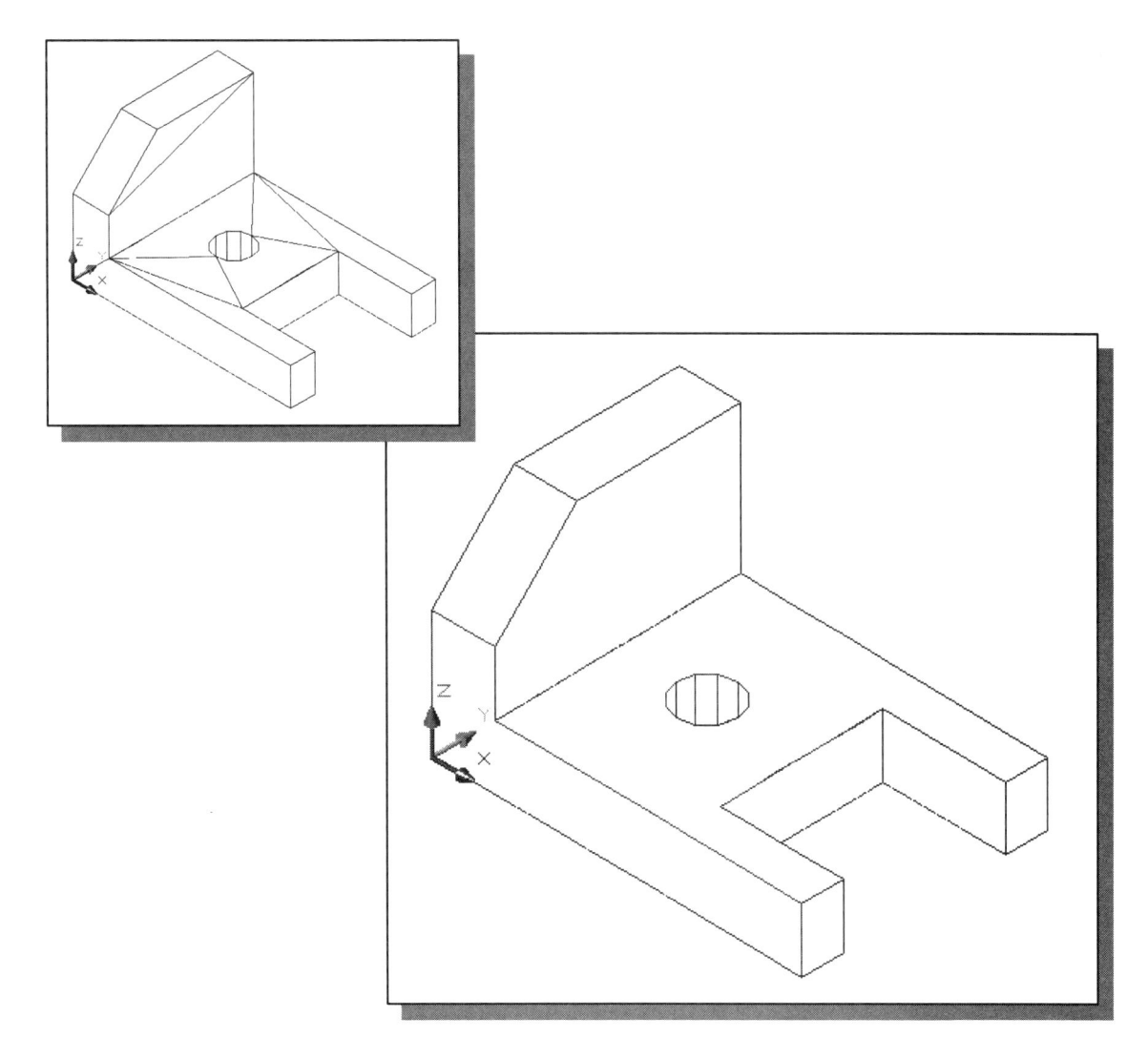

Questions:

1. List and describe three differences between *Wireframe models* and *Surface models*?

2. List and describe the three types of surfaces available in AutoCAD.

3. List and describe two different shading options in AutoCAD.

4. What is the difference between the **2D Solid** and **3D Face** commands in AutoCAD?

5. List and describe two predefined surface models in AutoCAD.

6. What is the difference between **Tabulated Surface** and **Ruled Surface** in AutoCAD?

7. Identify and describe the following commands:

    a)

    b)

    c)

    d)

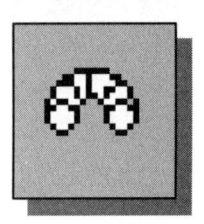

Exercises: All dimensions are in inches.

1.

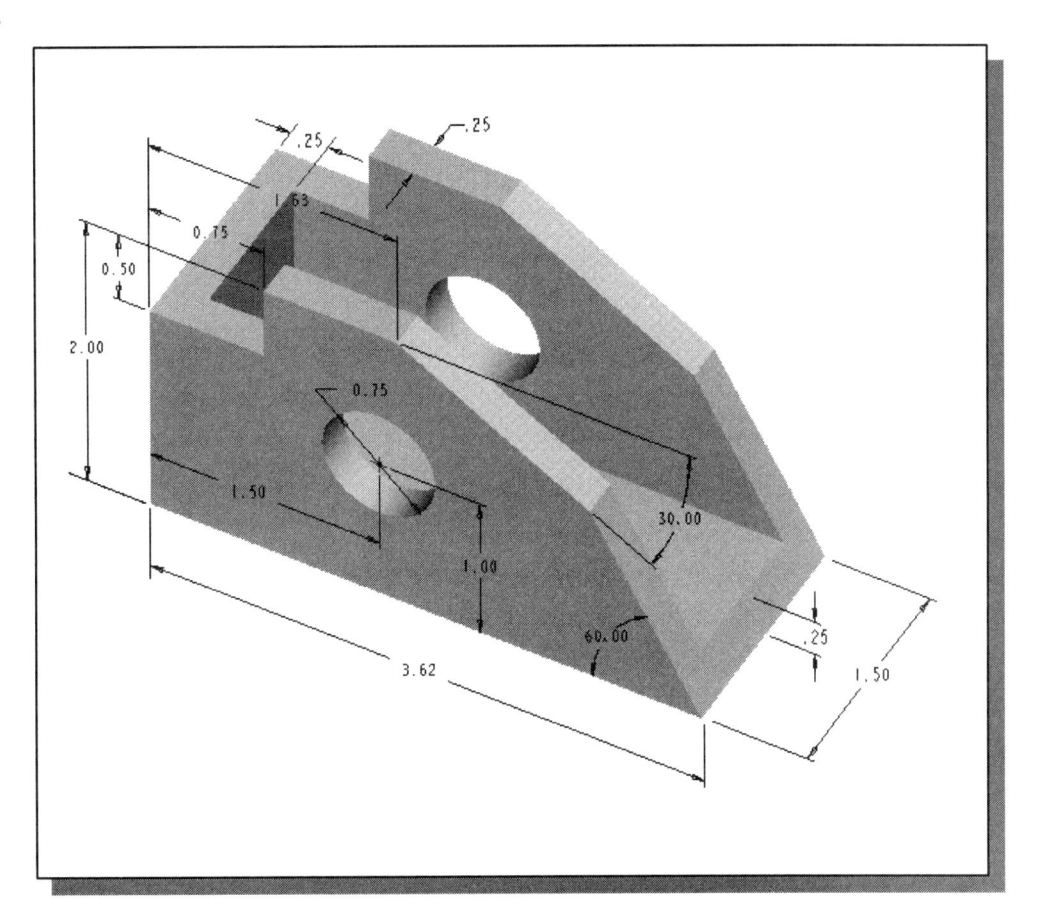

2.

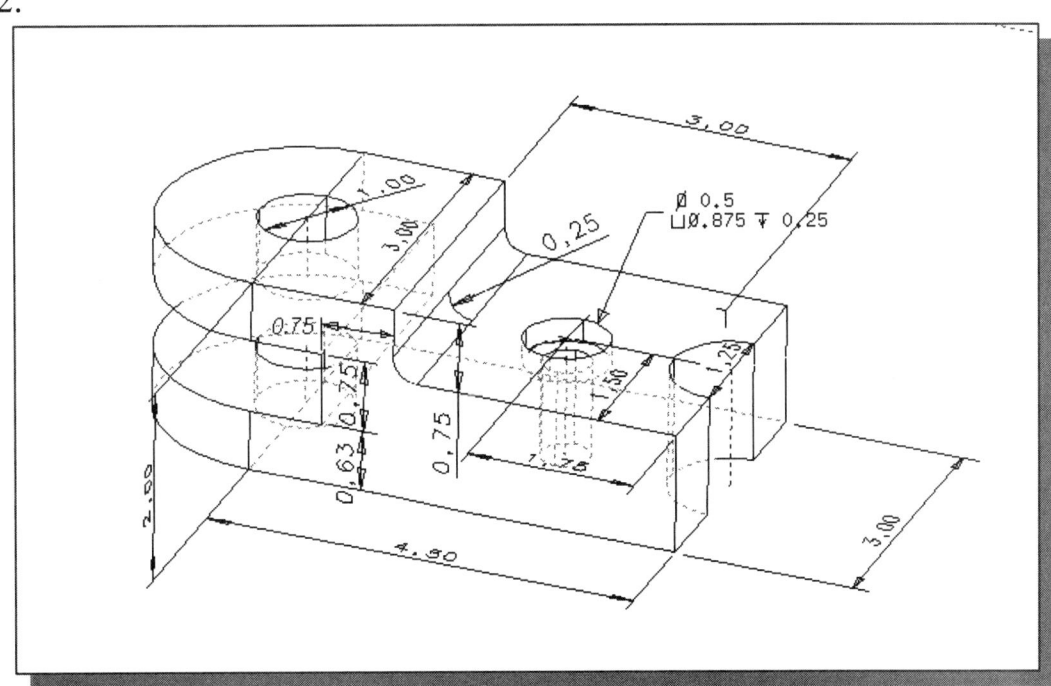

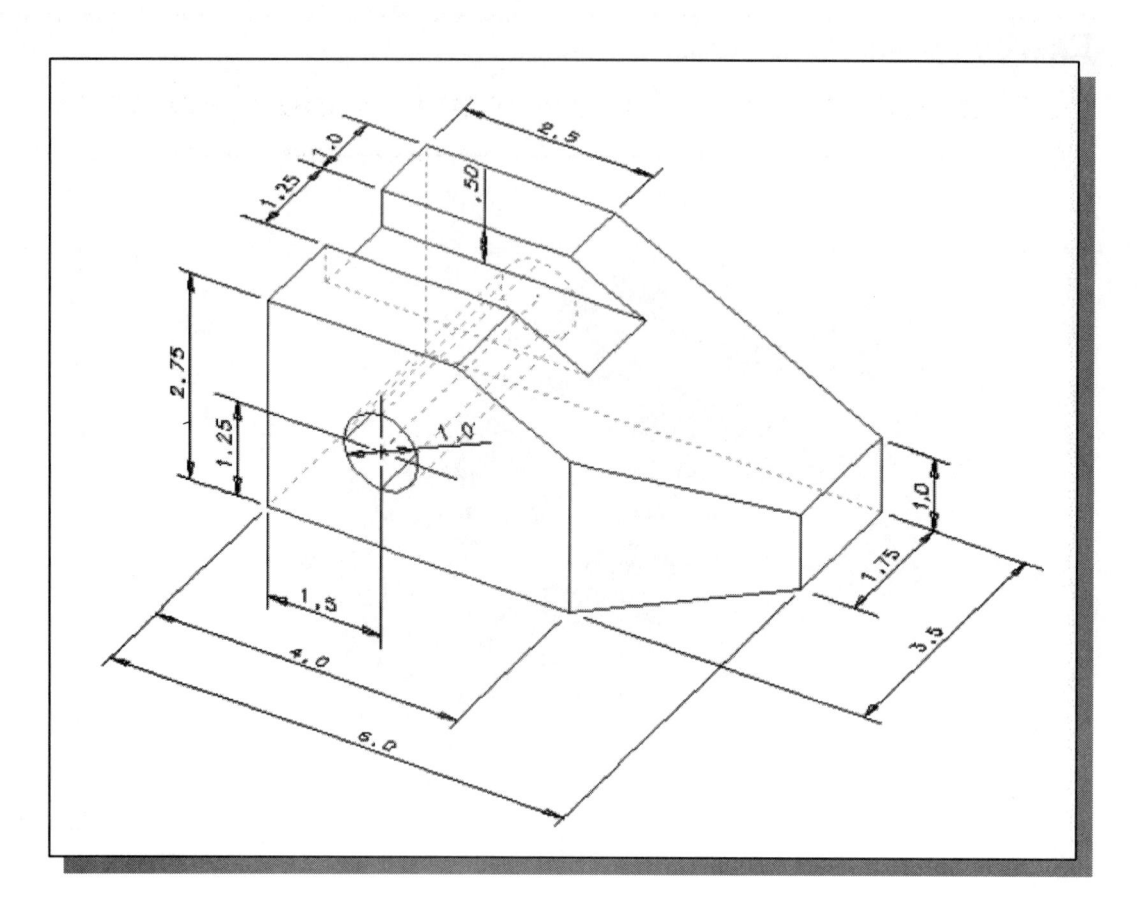

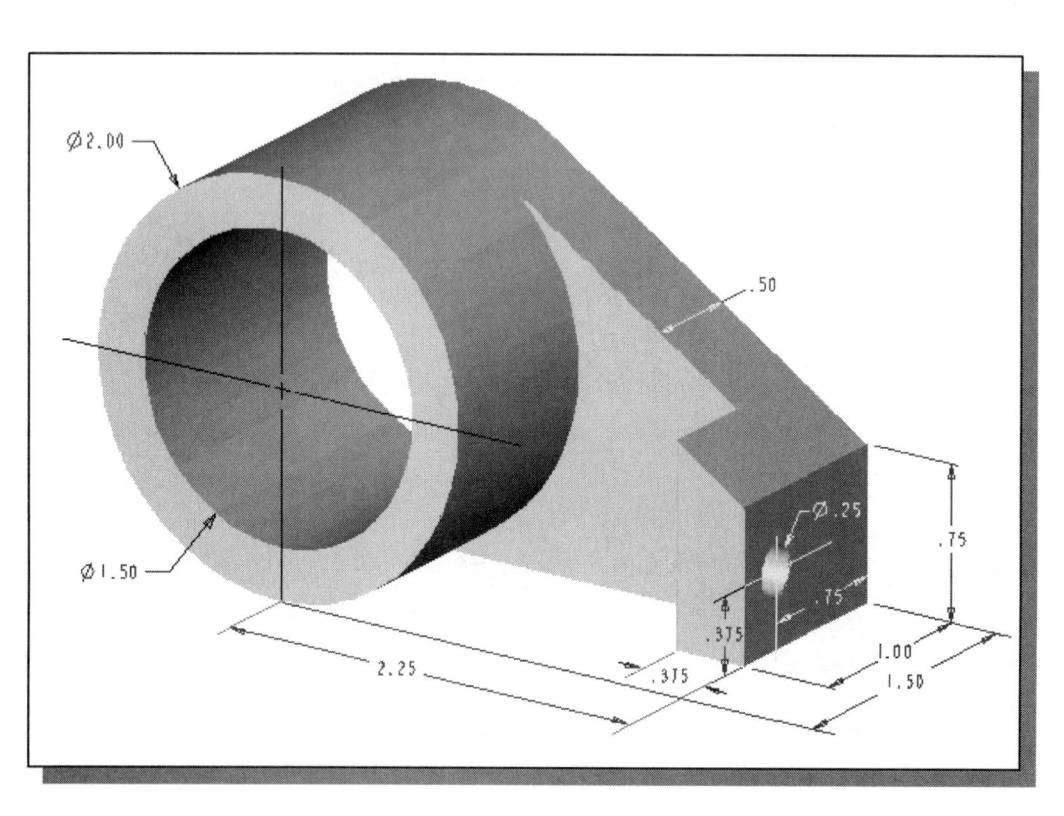

3.

4.

Chapter 6
Solid Modeling – Constructive Solid Geometry

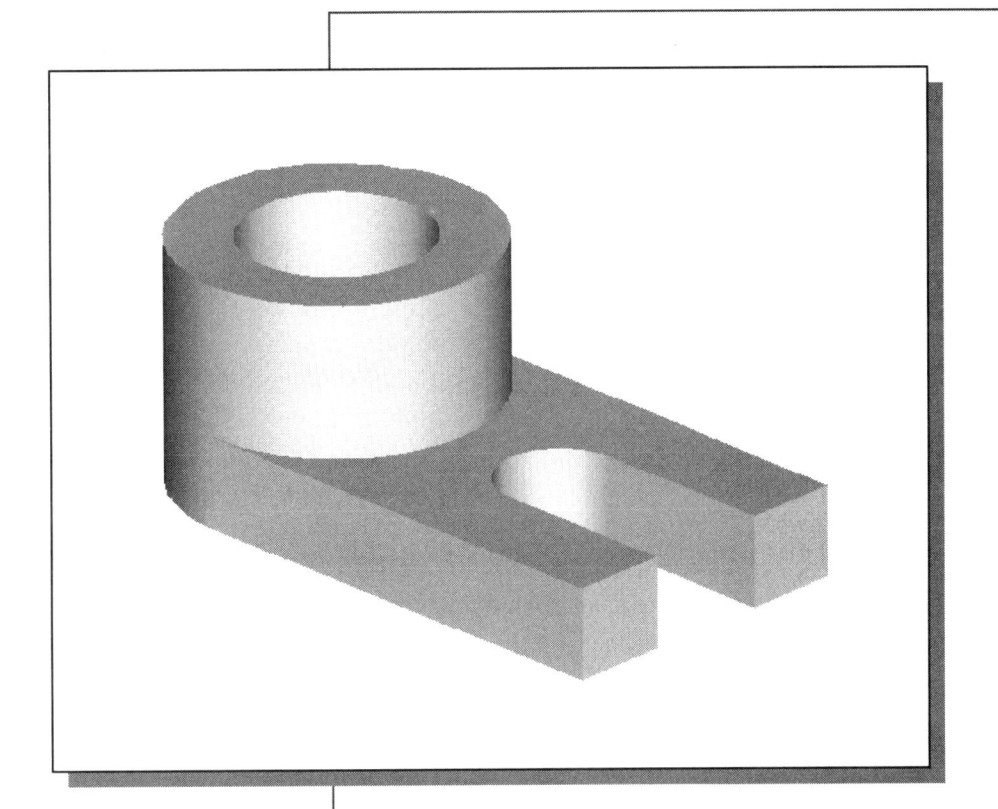

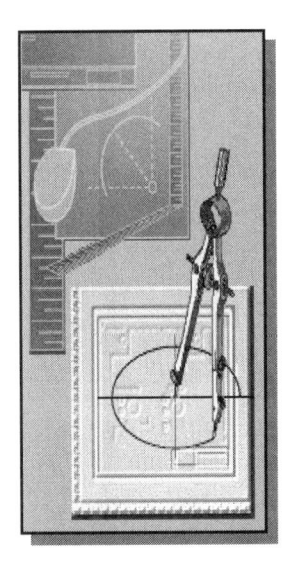

Learning Objectives

- **Understand the Constructive Solid Geometry Concept**
- **Create a Binary Tree**
- **Understand the Basic Boolean Operations**
- **Create Solids Using AutoCAD Primitive Solids**
- **Generate Shaded SOLID Images**

Introduction

In the previous chapters, we have looked at wireframe modeling and surface modeling. Wireframe modeling describes only the corners and edges of designs. Surface modeling describes part surfaces but not interiors. Although surface models are fairly good representations of the actual designs, designers are still required to interactively examine surface models to insure that the various surfaces on a model are contiguous throughout. Many of the concepts used in 3D wireframe and surface modelers are incorporated in the solid modeling scheme, but it is solid modeling that offers the most advantages as a design tool.

In the solid modeling presentation scheme, the solid definitions include nodes, edges, and surfaces, and it is a complete and unambiguous mathematical representation of a precisely enclosed and filled volume. Two predominant methods for representing solid models are **constructive solid geometry** (CSG) representation and **boundary representation** (B-rep). CSG defines a model in terms of combining basic solid shapes and B-rep defines a model in terms of its edges and surfaces. AutoCAD's solid modeler is a hybrid modeler; the user interface and construction techniques are CSG-based, whereas the B-rep capabilities are invoked automatically and are transparent to the user.

In this chapter, we will discuss the fundamental concepts of solid modeling. We will demonstrate the procedure required to construct a solid model using AutoCAD's CSG-based user interface.

The *Guide-Block* Design

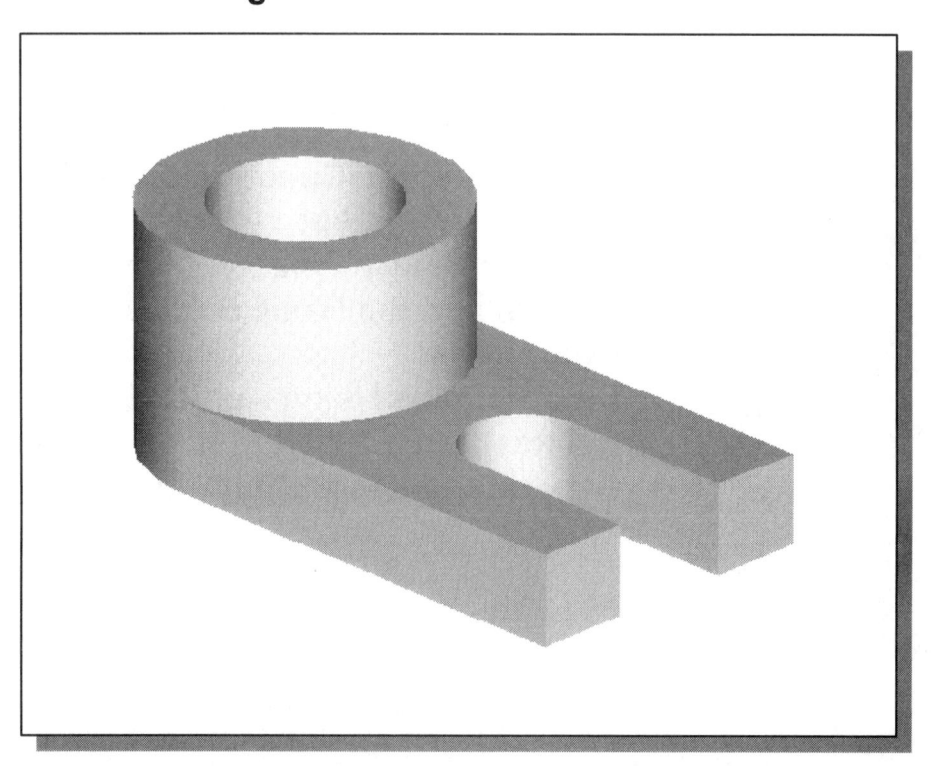

Constructive Solid Geometry Concept

In the 1980s, one of the main advancements in **Solid Modeling** was the development of the **Constructive Solid Geometry** (CSG) method. CSG describes the solid model as the combination of basic three-dimensional shapes (**primitive solids**). The basic primitive solid set typically includes: Rectangular-prism (Block), Cylinder, Cone, Sphere, and Torus (Donut). Two solid objects can be combined into one object in various ways, these operations are known as **Boolean operations**. There are three basic Boolean operations: **UNION (Join)**, **SUBTRACT (Cut)**, and **INTERSECT**. The UNION operation combines the two volumes included in the different solids into a single solid. The SUBTRACT operation subtracts the volume of one solid object from the other solid object. The INTERSECT operation keeps only the volume common to both solid objects. The CSG method is also known as the **Machinist's Approach**, as the method is parallel to machine shop practices.

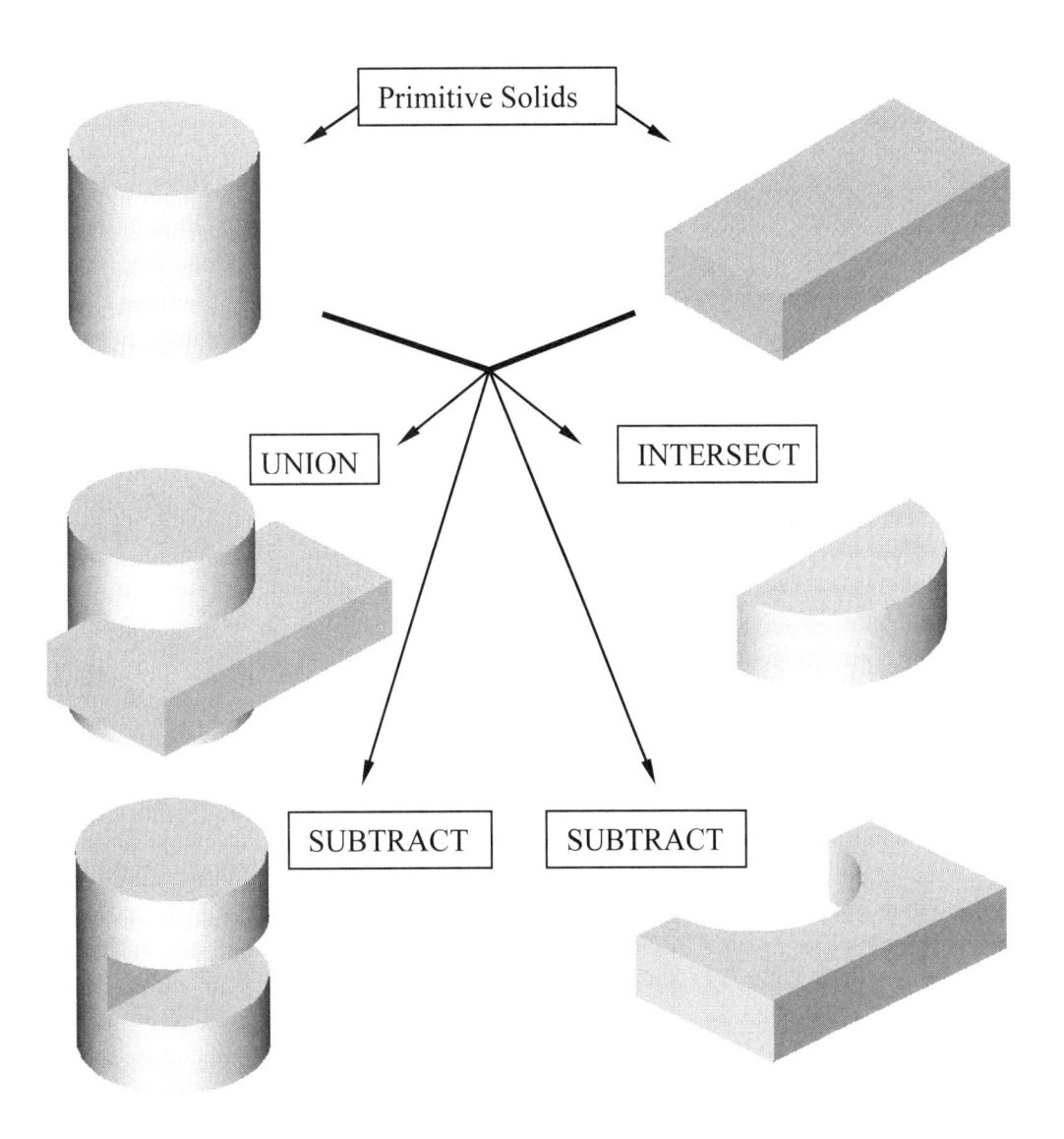

Binary Tree

CSG is also referred to as the method used to store a solid model in the database. The resulting solid can be easily represented by what is called a **binary tree**. In a binary tree, the terminal branches (leaves) are the various primitives that are linked together to make the final solid object (the root). The binary tree is an effective way to represent the steps required to construct the solid model. Complicated solid models can be modeled by considering the different combinations of Boolean operations required in the binary tree. This provides a convenient and intuitive way of modeling that imitates the manufacturing process. A binary tree is an effective way to plan your modeling strategy before you start creating anything.

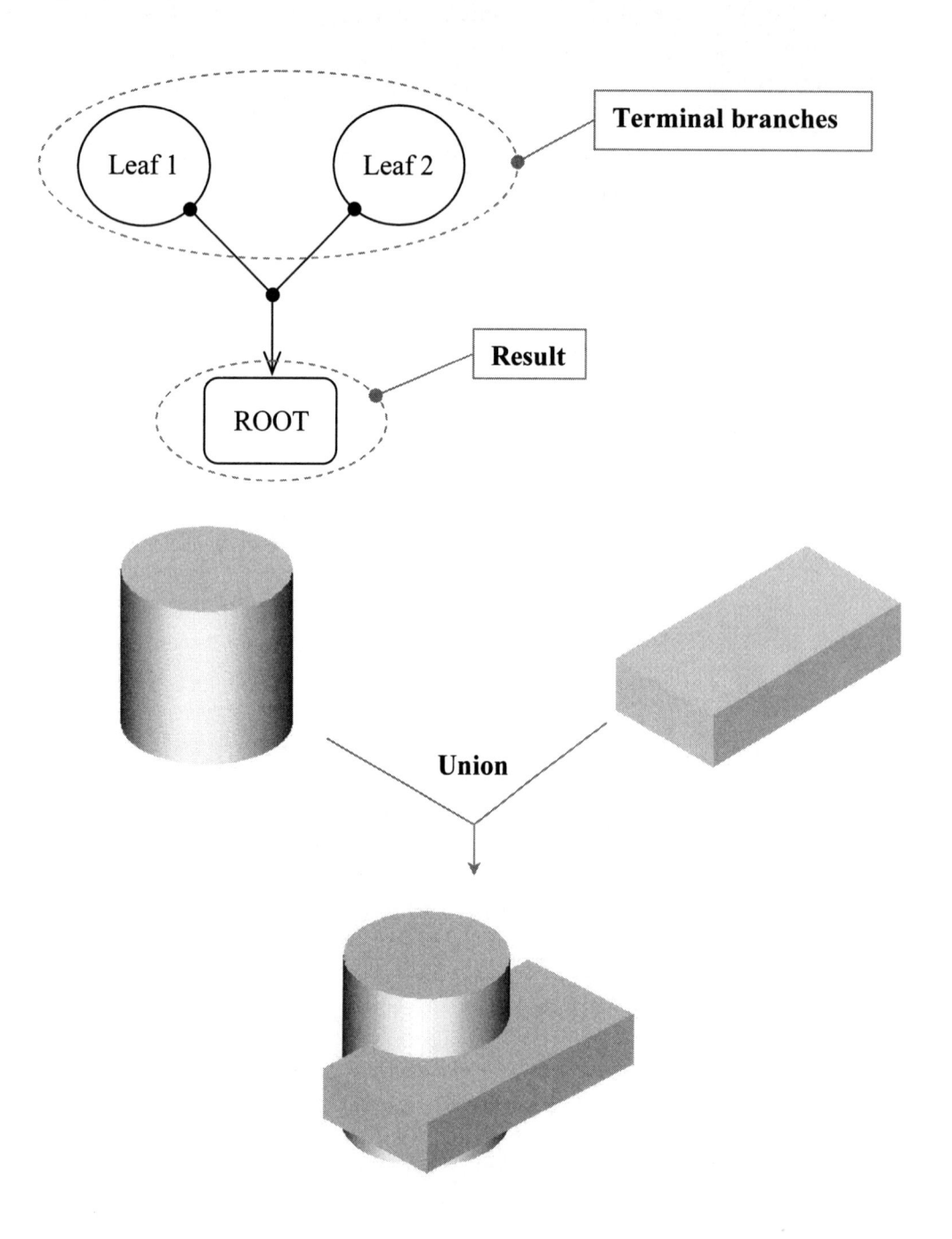

The *Guide-Block* CSG Binary Tree

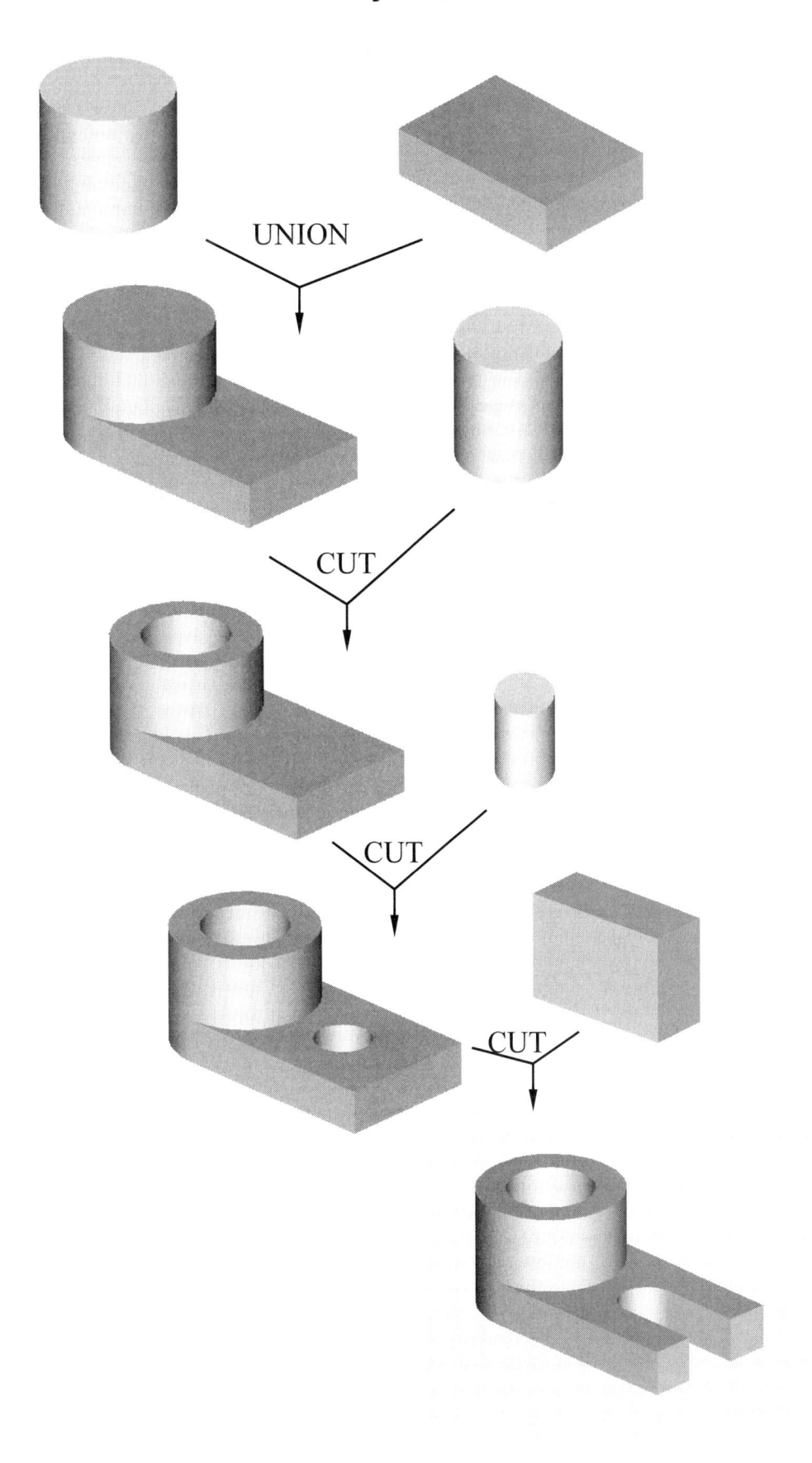

Starting Up AutoCAD 2006

1. Select the **AutoCAD 2006** option on the *Program* menu or select the **AutoCAD 2006** icon on the *Desktop*.

2. In the startup dialog box, select the **Start from Scratch** option with a single click of the left-mouse-button.

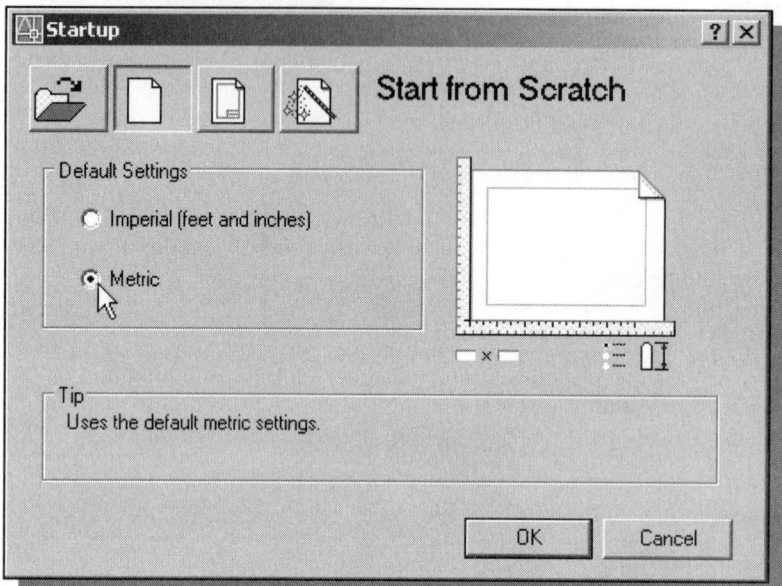

3. In the *Default Settings* section, pick **Metric** as the drawing units.

4. Click **OK** to accept the settings and exit the *Startup* window.

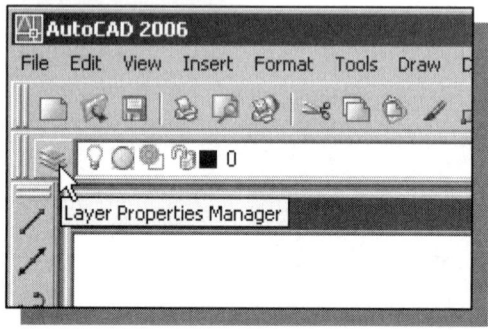

5. Pick **Layer Properties Manager** in the *Object Properties* toolbar.

6. Create a **new layer** with the following settings:

Layer	Color	LineType
Solid_Objects	**Cyan**	**Continuous**

7. Set layer *Solid_Objects* as the *Current Layer*.

8. Click on the **OK** button to accept the settings and exit the *Layer Properties Manager* dialog box.

Creating the First 3D Object

1. In the *Status Bar* area, reset the option buttons so that only *POLAR*, *OSNAP*, *OTRACK*, and *MODEL* are switched **ON**.

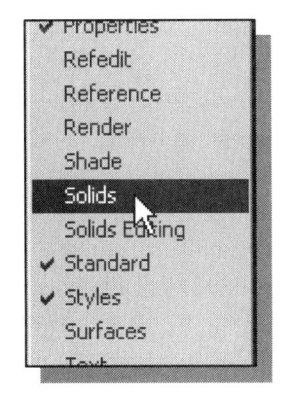

2. Move the cursor to the *Standard* toolbar area and **right-click** on any icon of the *Standard* toolbar to display a list of toolbar menu groups.

3. Select **Solids**, with the left-mouse-button, to display the *Solids* toolbar on the screen.

➢ The first section of the *Solids* toolbar contains the basic set of pre-defined solids available in **AutoCAD® 2006**.

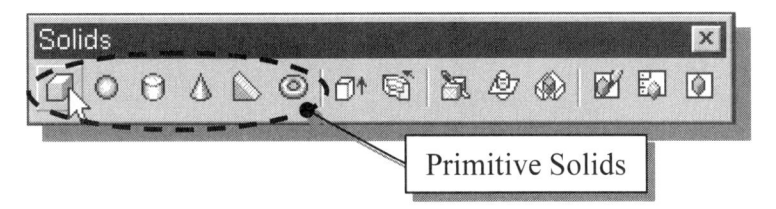

Primitive Solids

- We will first create a **75 × 50 × 15** rectangular block.

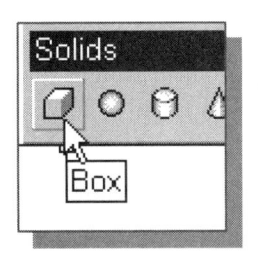

4. In the *Solids* toolbar, click on the **Box** icon. In the command prompt area, the message "*Specify corner of box or [Center] <0,0,0>:*" is displayed.

5. Pick a location that is near the lower left corner of the graphics window.

6. In the command prompt area, the message "*Specify corner or [Cube/Length]:*" is displayed. Enter: **@75,50 [ENTER]**.

7. In the command prompt area, the message "*Specify Height:*" is displayed. Enter: **15 [ENTER]**.

Viewing the 3D Block

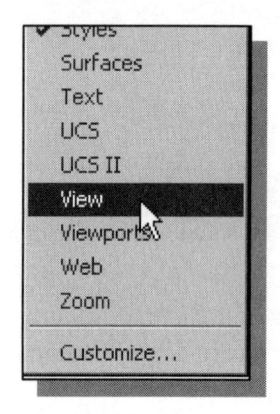

1. Move the cursor to the *Standard* toolbar area and **right-click** on any icon of the *Standard* toolbar to display a list of toolbar menu groups.

2. Select **View**, with the left-mouse-button, to display the *View* toolbar on the screen.

➢ The *View* toolbar contains two sections of icons that allow us to quickly switch to standard 2D and 3D views.

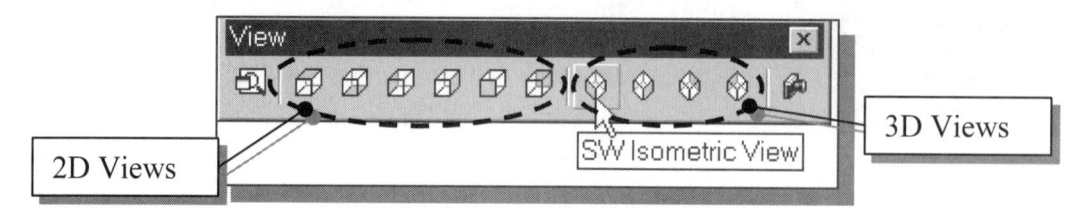

3. In the *View* toolbar, click on the **SE Isometric View** icon.

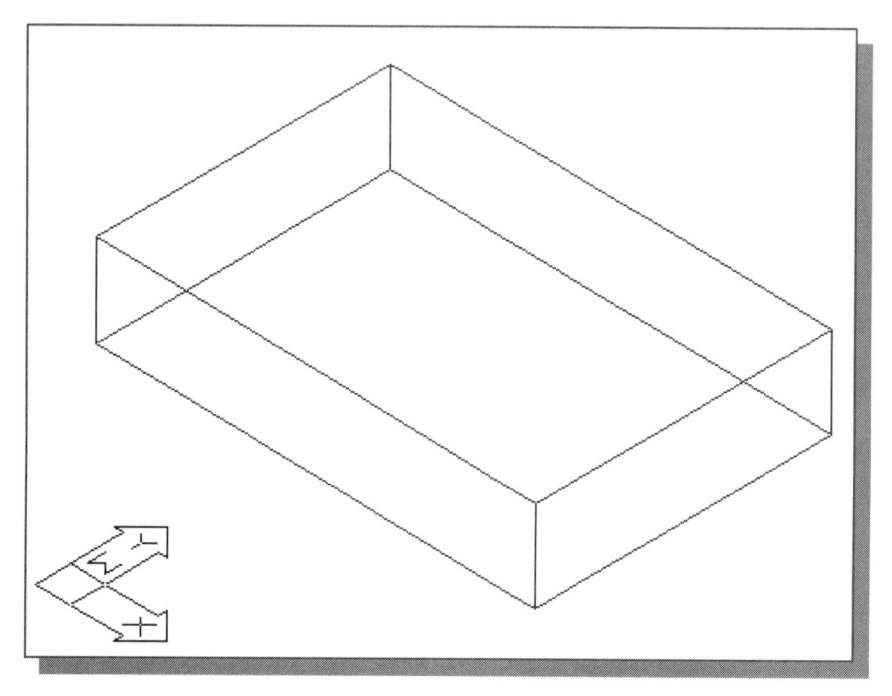

Creating the Second Solid Feature

- We will next create a cylinder block as the second solid feature.

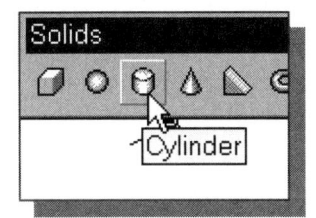

    1. In the *Solids* toolbar, click on the **Cylinder** icon. In the command prompt area, the message "*Specify center point for base cylinder or [Elliptical] <0,0,0>:*" is displayed.

    2. Inside the graphics window, press down the **[SHIFT]** key and **right-mouse-click** once to bring up the *Object Snap* shortcut menu.

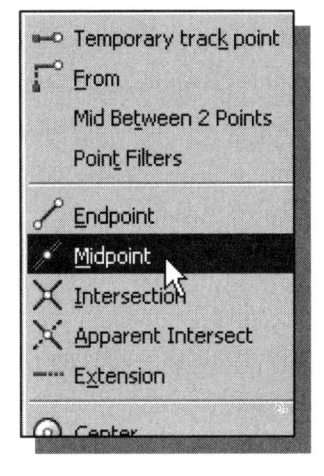

    3. Select the **Midpoint** option in the shortcut menu. Snap to and select the midpoint of the left-bottom edge of the rectangular block as shown.

    4. In the command prompt area, the message "*Specify radius for base of cylinder or [Diameter]:*" is displayed. Snap to and select one of the endpoints of the bottom edge of the rectangular block.

    5. In the command prompt area, the message "*Specify height of cylinder or [Center of other end]:*" is displayed. Enter: **40 [ENTER]**.

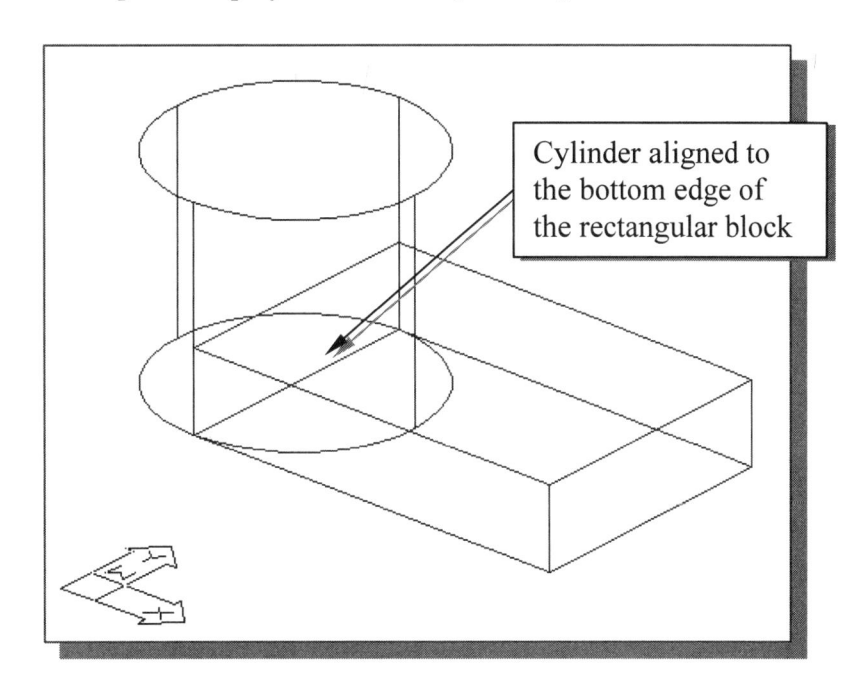

    Cylinder aligned to the bottom edge of the rectangular block

- Two solid objects, one rectangular box and one cylinder, are displayed on the screen.

Boolean Operation – *UNION*

1. Move the cursor to the *Standard* toolbar area and **right-click** on any icon of the *Standard* toolbar to display a list of toolbar menu groups.

2. Select **Solids Editing**, with the left-mouse-button, to display the *Solids Editing* toolbar on the screen.

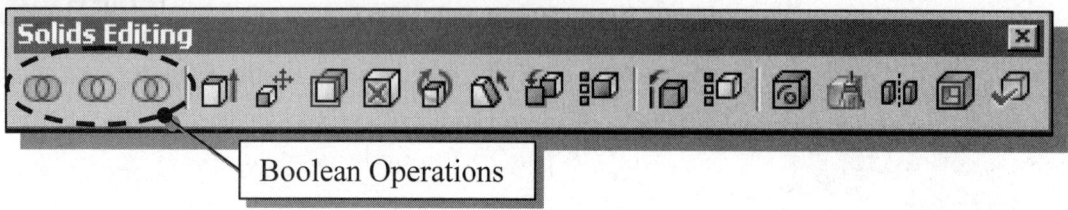

Boolean Operations

➢ The first section of the *Solids Editing* toolbar contains the basic set of Boolean operation icons.

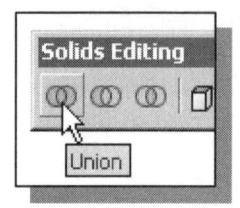

3. In the *Solids Editing* toolbar, click on the **Union** icon. In the command prompt area, the message "*Select Objects:*" is displayed.

4. Pick both solids by clicking on their edges.

5. Inside the graphics window, right-mouse-click to accept the selection and proceed with the Union command.

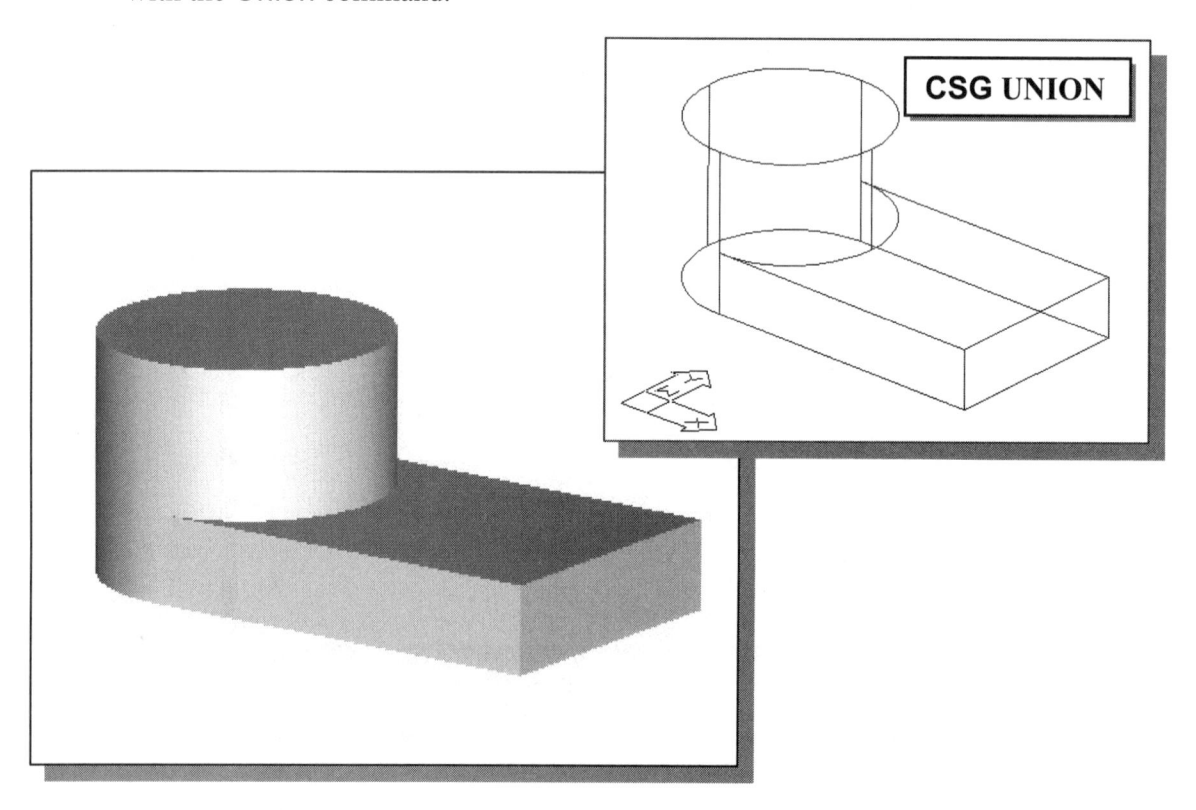

CSG UNION

Creating the Second Cylinder Feature

- We will create another cylinder as the next solid feature.

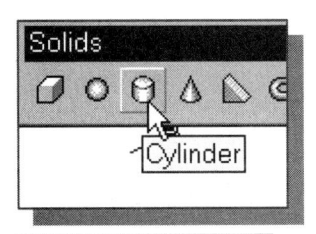

1. In the *Solids* toolbar, click on the **Cylinder** icon. In the command prompt area, the message *"Specify center point for base cylinder or [Elliptical] <0,0,0>:"* is displayed.

2. Inside the graphics window, press down the **[SHIFT]** key and **right-mouse-click** once to bring up the *Object Snap* shortcut menu.

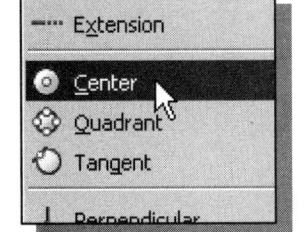

3. Select the **Center** option in the shortcut menu.

4. Snap to and select the **center of the circle** that is at the **top** of the solid block.

5. In the command prompt area, the message *"Specify radius for base of cylinder or [Diameter]:"* is displayed. Enter: **15 [ENTER]**.

6. In the command prompt area, the message *"Specify height of cylinder or [Center of other end]:"* is displayed. Enter: **-50 [ENTER]**.
   (The negative value will extrude the cylinder downward, in the negative Z direction.)

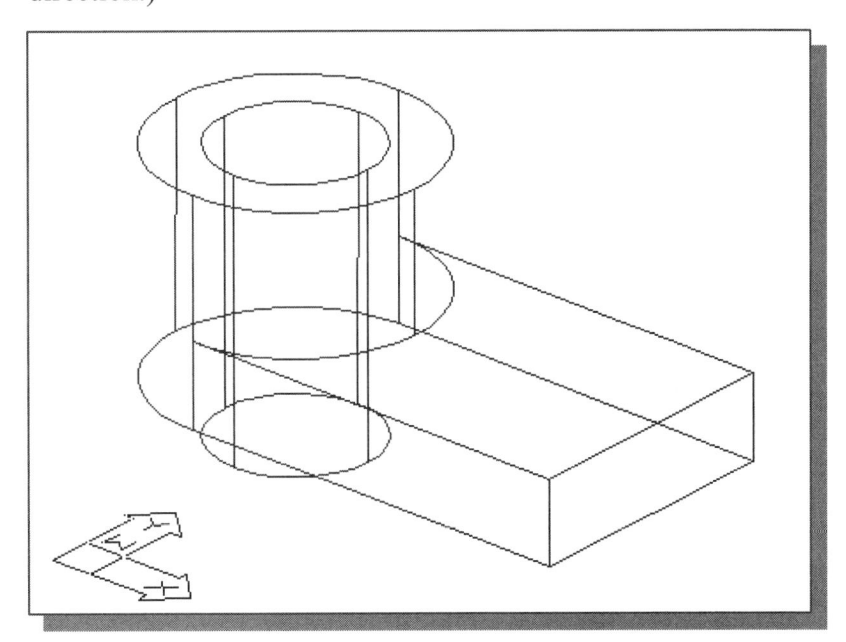

- Two solid objects, the base block and a cylinder, are displayed on the screen. On your own, use the Front View and Top View options to confirm the top of the two solids are aligned. (Before continuing, restore the view as shown; including the Coordinate icon.)

Boolean Operation – *SUBTRACT*

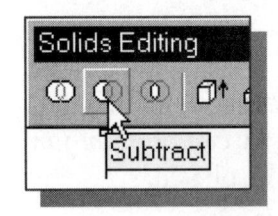

1. In the *Solids Editing* toolbar, click on the **Subtract** icon. In the command prompt area, the message "*_subtract Select solids and regions to subtract from .. Select Objects:*" is displayed.

2. Pick the base block solid by clicking on one of the straight edges.

3. Inside the graphics window, **right-mouse-click** to accept the selection and proceed with the Subtract command.

4. In the command prompt area, the message "*Select solids and regions to subtract .. Select Objects:*" is displayed.

5. Pick the cylinder block by clicking on one of the circular edges.

6. Inside the graphics window, **right-mouse-click** to accept the selection and proceed with the Subtract command.

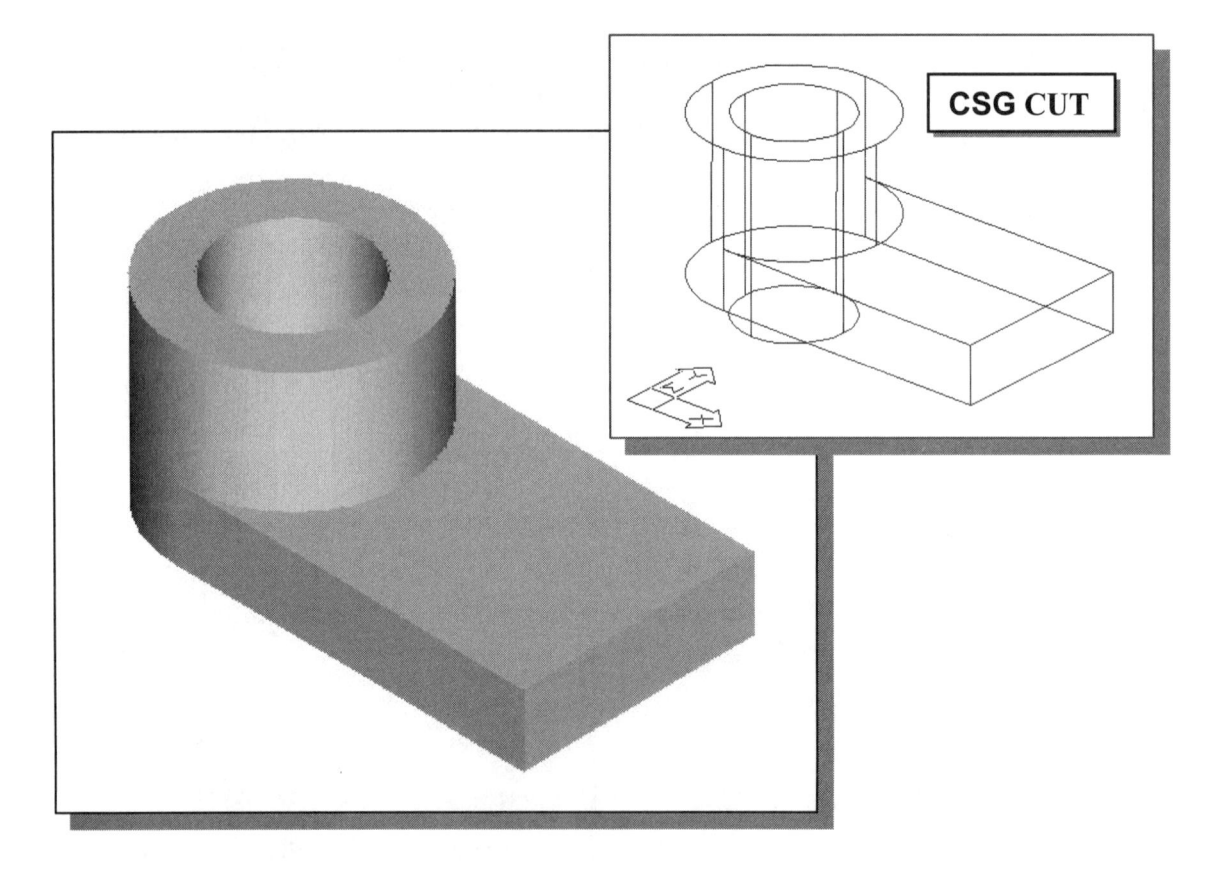

CSG CUT

Creating Another Solid Feature

- As can be seen, the CSG construction approach is quite straightforward. The main task is in positioning and aligning the solid blocks prior to applying a Boolean operation.

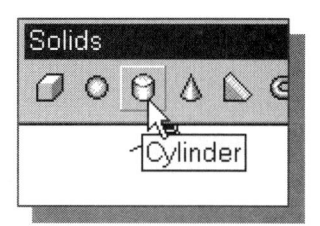

1. In the *Solids* toolbar, click on the **Cylinder** icon. In the command prompt area, the message "*Specify center point for base cylinder or [Elliptical] <0,0,0>:*" is displayed.

2. Inside the graphics window, press down the **[SHIFT]** key and **right-mouse-click** once to bring up the *Object Snap* shortcut menu.

3. Select the **From** option in the shortcut menu.

4. Snap to the **top front** corner of the base block and enter: **@-30,25 [ENTER]**.

5. In the command prompt area, the message "*Specify radius for base of cylinder or [Diameter]:*" is displayed. Enter: **10 [ENTER]**.

6. In the command prompt area, the message "*Specify height of cylinder or [Center of other end]:*" is displayed. Enter: **-30 [ENTER]**.

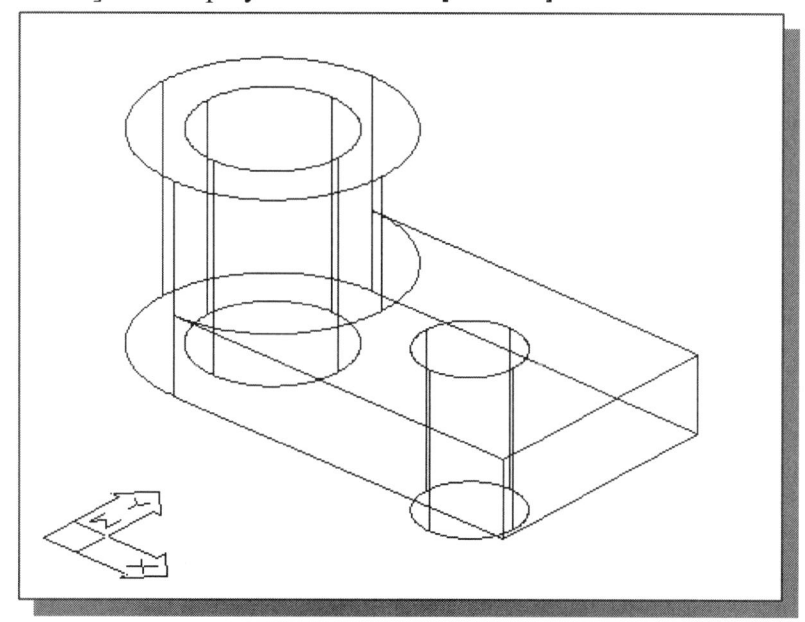

7. On your own, perform the Boolean **Subtract** operation.

Shaded Options

1. Move the cursor to the *Standard* toolbar area and right-click on any icon of the *Standard* toolbar to display a list of toolbar menu groups.

2. Select **Shade**, with the left-mouse-button, to display the *Shade* toolbar on the screen.

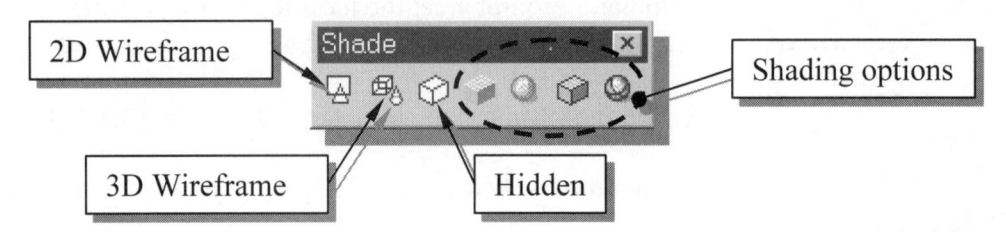

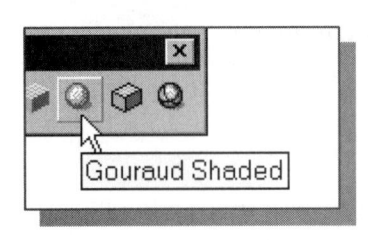

- 2D Wireframe
- 3D Wireframe
- Hidden
- Shading options

3. In the *Shade* toolbar, click on the **Gouraud Shaded** icon. This method produces a relatively smooth shaded solid.

➢ On your own, examine the effects of the other *Shading* options.

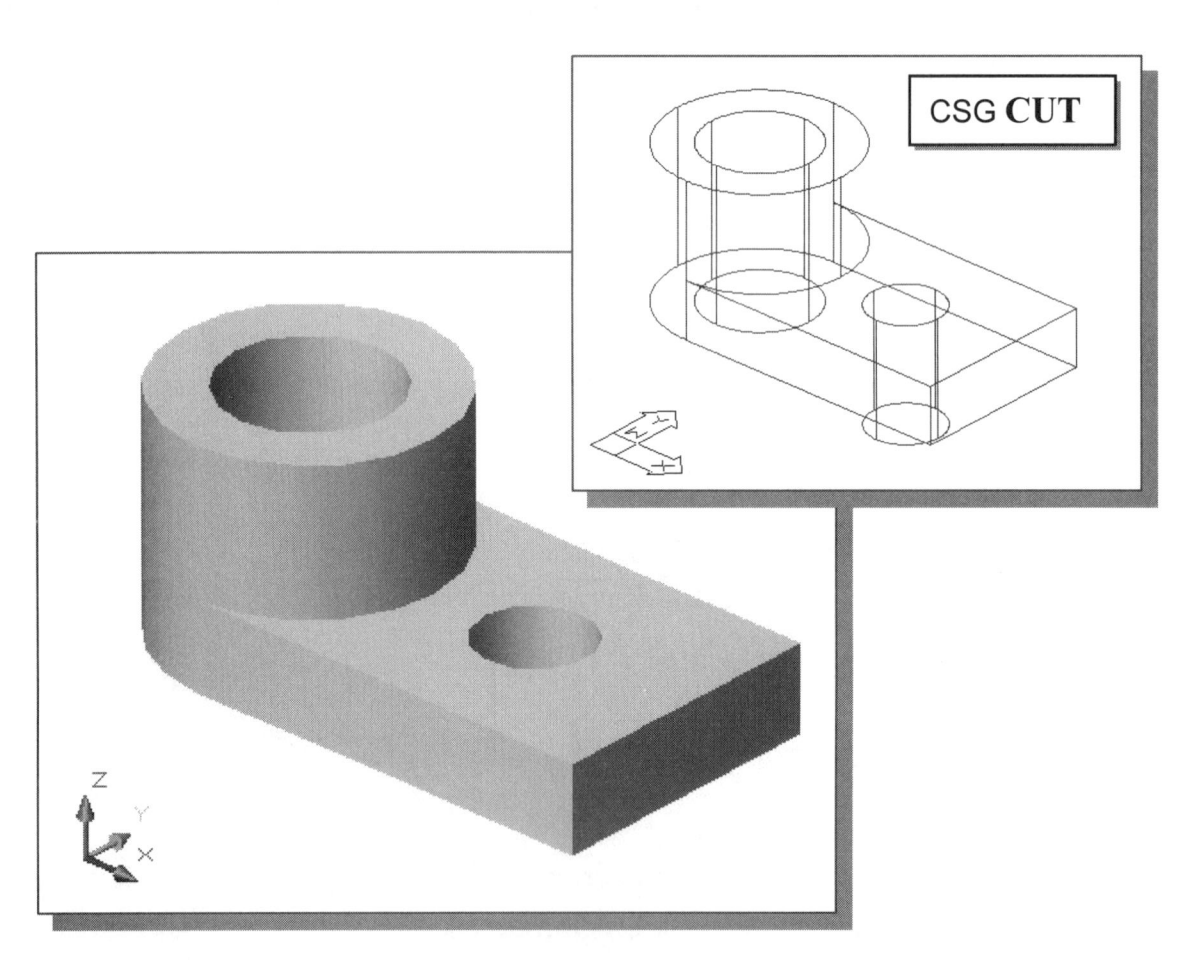

CSG CUT

Creating the Final Feature

- We will create a rectangular cut as the final solid feature of the part. In the previous sections, with some careful planning, we positioned the solid blocks at the correct space locations while we were creating the solids. It is also possible to take a more general approach to orient and position solid blocks in space coordinates.

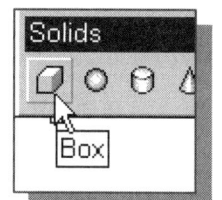

    1. In the *Solids* toolbar, click on the **Box** icon. In the command prompt area, the message "*Specify corner of box or [Center] <0,0,0>:*" is displayed.

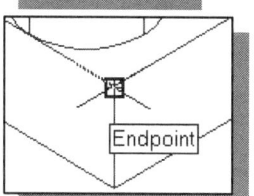

    2. Snap to and select the **top front corner** of the base block.

    3. In the command prompt area, the message "*Specify corner or [Cube/Length]:*" is displayed. Enter: **@-30,30** [**ENTER**].

    4. In the command prompt area, the message "*Specify Height:*" is displayed. Enter: **-20** [**ENTER**].

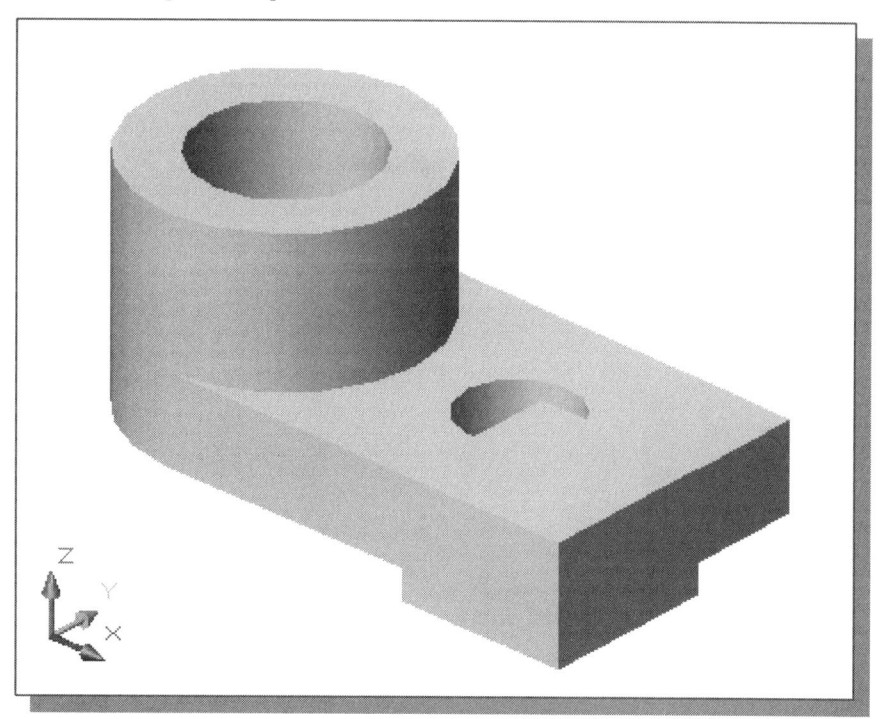

> Two overlapping solids appear on the screen. We will use the *editing tools* to orient the cutter, the rectangular block, to its correct location.

Rotating the Rectangular Block

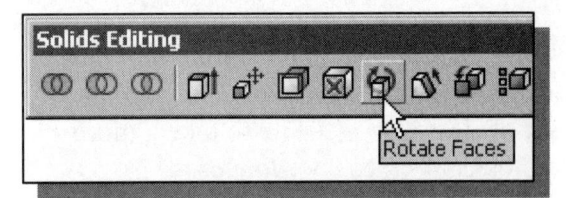

1. In the *Solids Editing* toolbar, click on the **Rotate Faces** icon. In the command prompt area, the message "*Select Faces or [Undo/Remove/All]:*" is displayed.

2. Select the edges of the rectangular block until **all edges are highlighted** as shown.

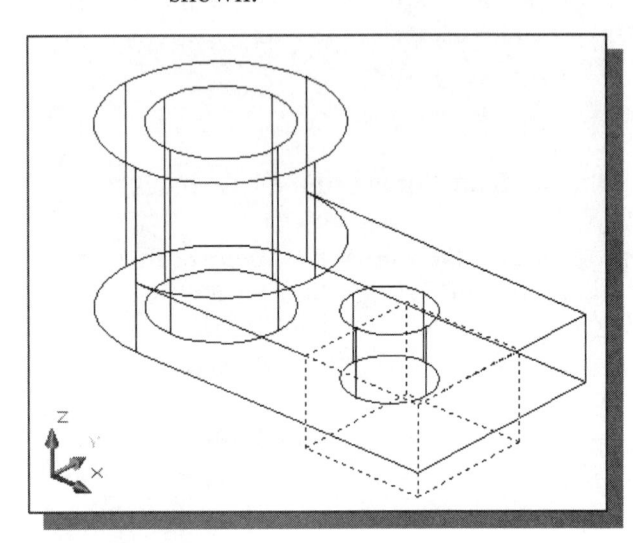

3. Inside the graphics window, **right-mouse-click** and select **Enter** in the popup option menu to proceed with the **Rotate** command.

4. In the command prompt area, the message "*Specify an axis point or [Axis by object /View /Xaxis /Yaxis /Zaxis]<2 points>:*" is displayed. Inside the graphics window, **right-mouse-click** and select **X axis** in the popup option menu to proceed with the **Rotate** command.

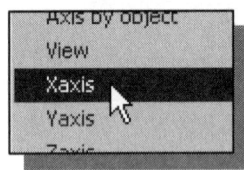

5. In the command prompt area, the message "*Specify the origin of the rotation <0,0,0>:*" is displayed. Pick the **top front** corner of the rectangular block as the rotation origin.

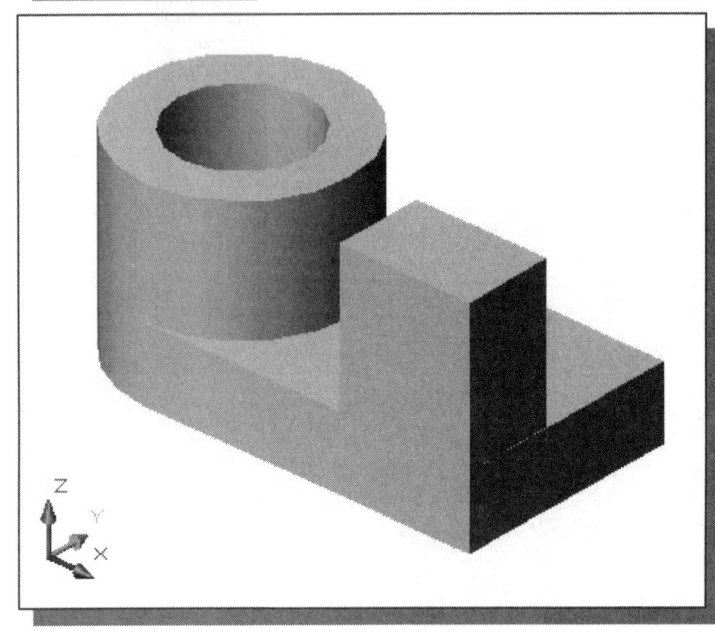

6. In the command prompt area, the message "*Specify a rotation angle or [Reference]:*" is displayed. Enter: **90 [ENTER]**.

7. Inside the graphics window, **right-mouse-click** and select **Exit** in the popup menu.

8. **Right-mouse-click** inside the graphics window and select **Exit** to exit the **Rotate** command.

Moving the Rectangular Block

1. **Pre-select** the rectangular block by left-clicking on any edge of the block.

2. Inside the graphics window, right-mouse-click and select **Move** in the popup menu. In the command prompt area, the message "*Specify base point or displacement:*" is displayed.

3. Snap to and select the **top front corner** of the rectangular block as the reference point.

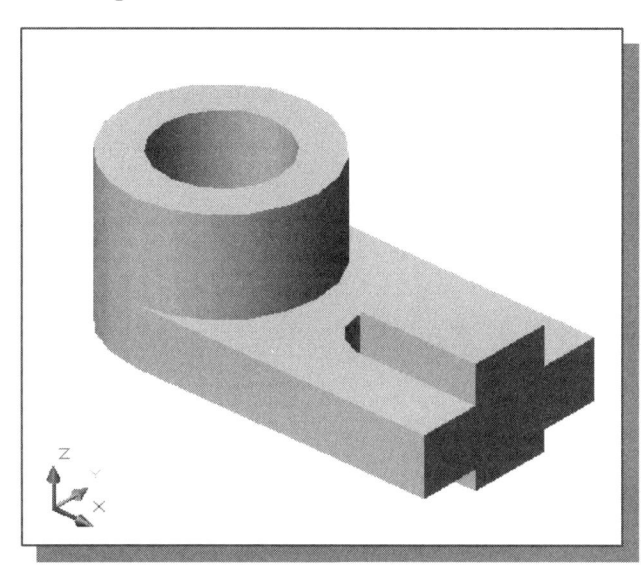

4. In the command prompt area, the message "*Specify second point of displacement or <use first point as displacement>:*" is displayed. Enter: **@0,15,-20** [ENTER].

5. On your own, use the **3D Orbit** command to examine the model.

6. On your own, perform the **Subtract** Boolean operation and complete the model.

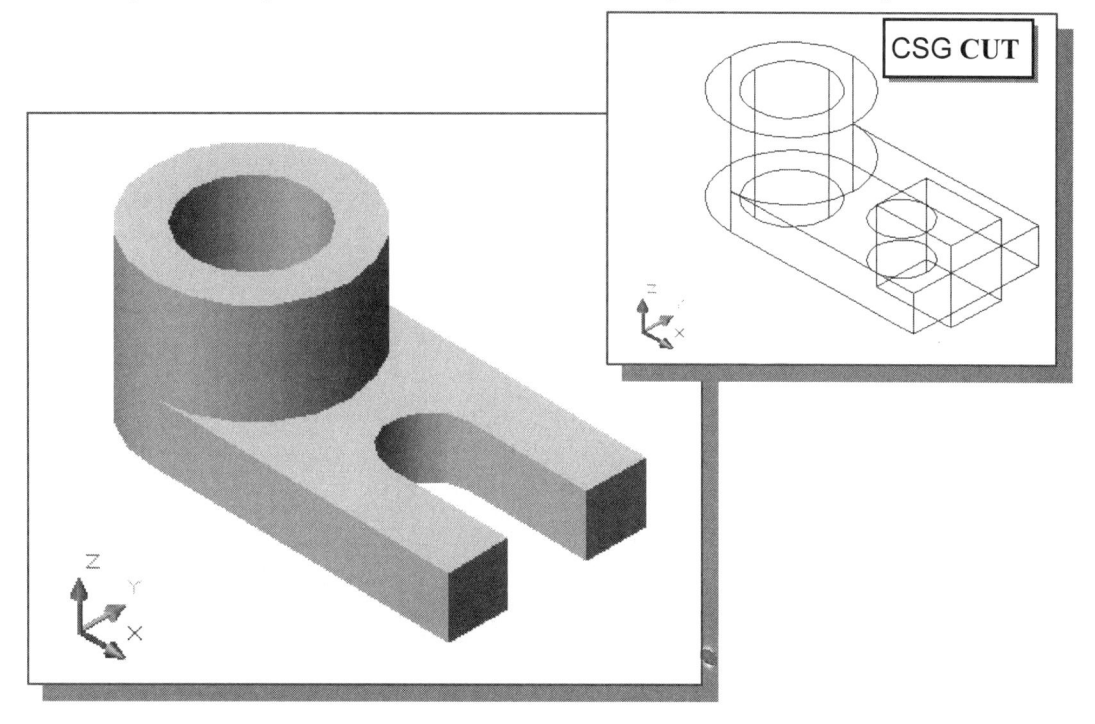

Questions:

1. What are the three basic Boolean operations commonly used in computer geometric modeling software?

2. What is a *primitive solid*?

3. What does *CSG* stand for?

4. Which Boolean operation keeps only the volume common to the two solid objects?

5. What is the difference between a *CUT* feature and a *UNION* feature in AutoCAD?

6. Using the CSG concept, create Binary Tree sketches showing the steps you plan to use to create the two models shown on the next page:

Ex.1)

Ex.2)

Exercises: All dimensions are in inches.

1.

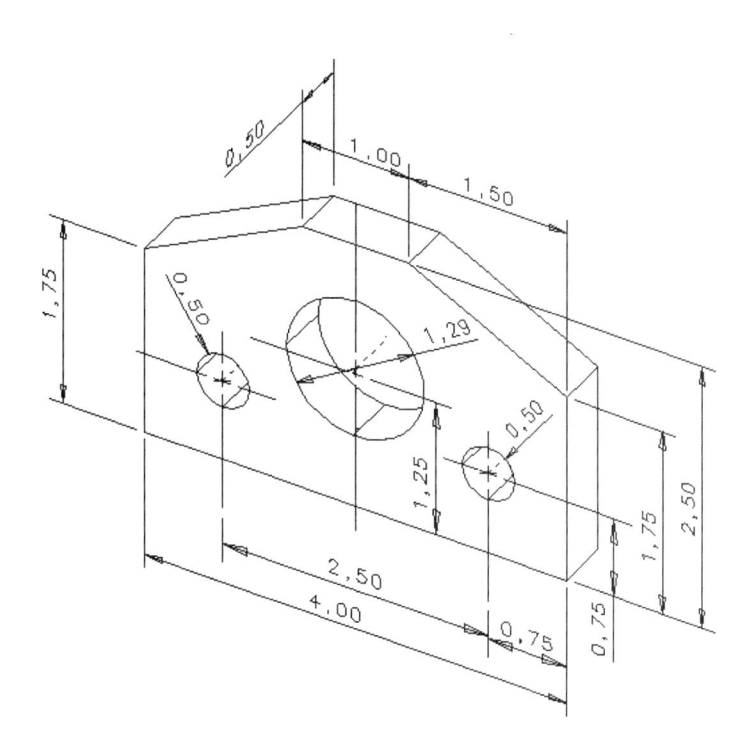

2.

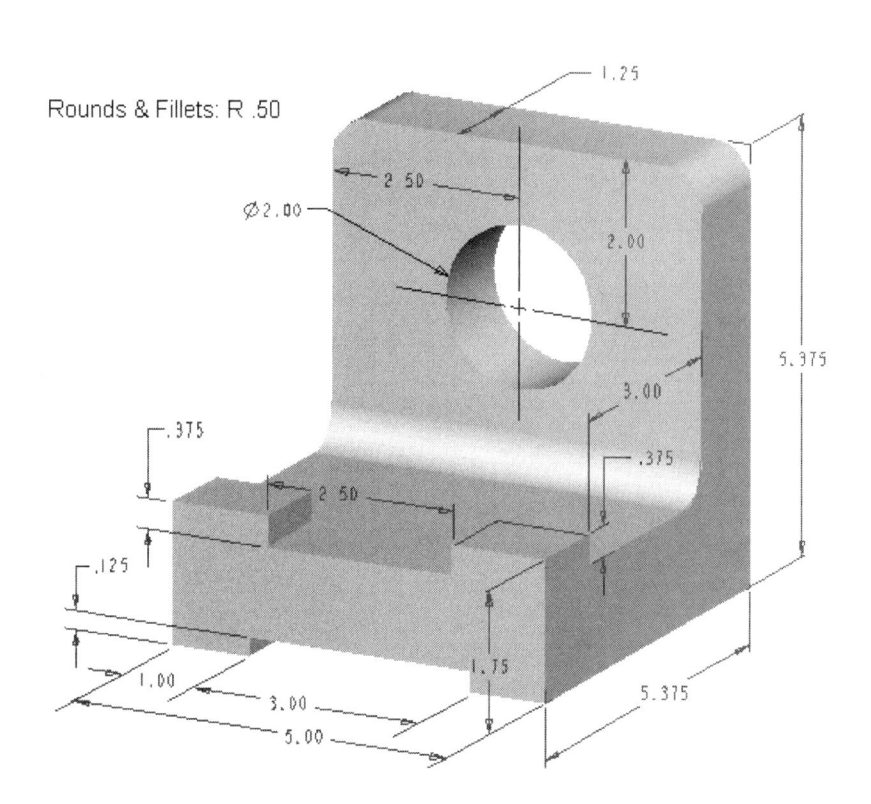

Rounds & Fillets: R .50

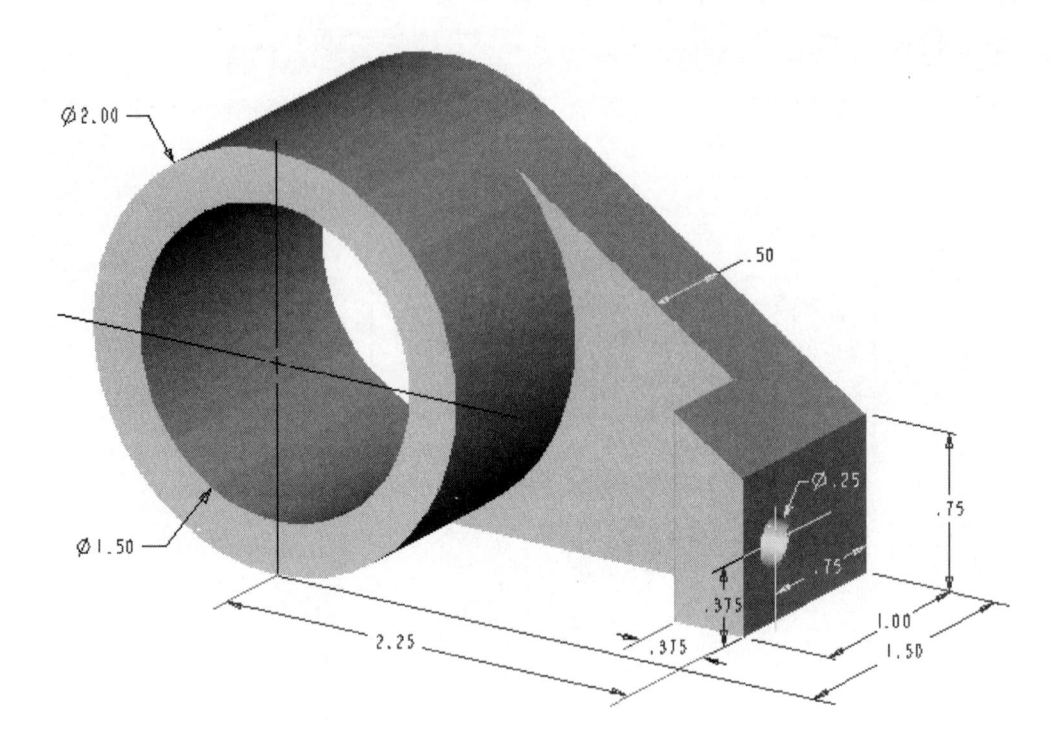

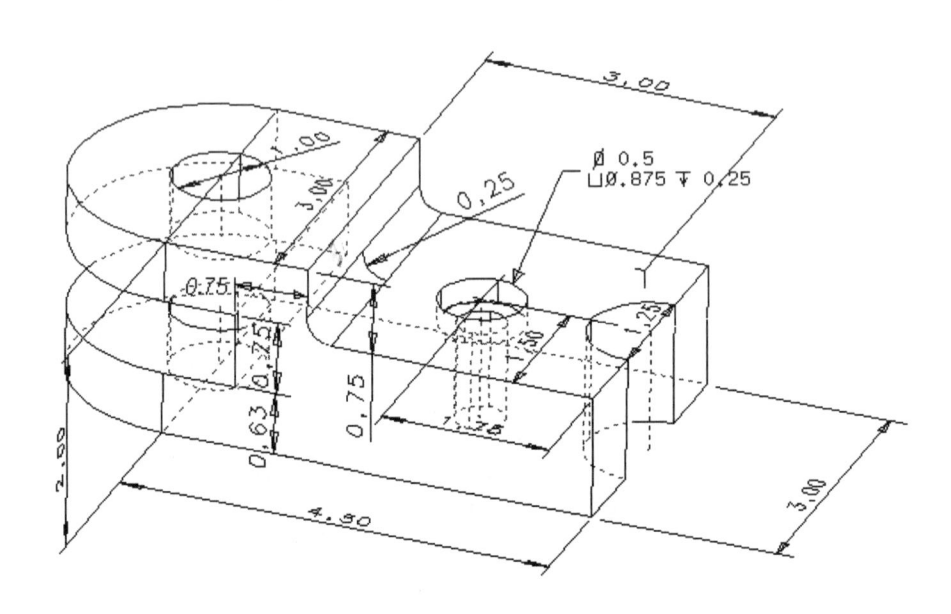

3.

4.

Chapter 7
Regions, Extrude and Solid Modeling

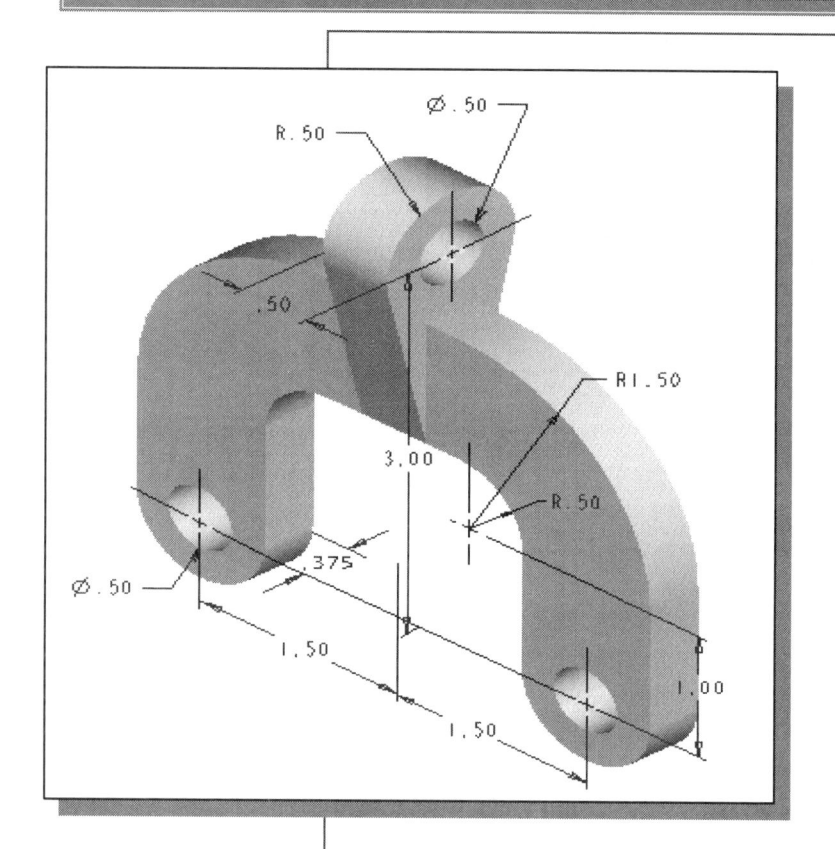

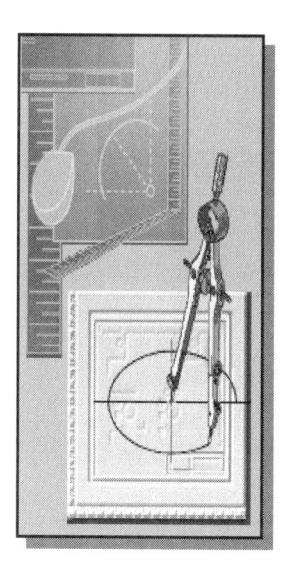

Learning Objectives

- **Use 2D Extrusion Method to Create Solid Models**
- **Understand and Create Regions**
- **Edit 3D Geometry with the Rotate and Move Commands**
- **Examine the Mass Properties of Solids**
- **Understand the Concept of Boundary-Representation Modeling Approach**

Introduction

AutoCAD provides many powerful modeling and design-tools, and there are many different approaches to accomplish solid modeling tasks. The CSG approach, as illustrated in the previous chapter, allows designers to create solid models by using a set of predefined shapes to simulate the manufacturing process. Using only a limited set of predefined solids becomes less efficient as more complex geometric definitions are involved. Another solid modeling method, parallel to the development of the CSG approach, is the **Boundary Representation (B-rep)** approach. CSG defines a model in terms of combining basic solid shapes and B-rep defines a model in terms of its edges and surfaces. Two of the most important solid construction operations based on the B-rep approach are **Extrude** and **Revolve**. The Extrude and Revolve options can be used to convert two-dimensional filled polygons into 3D solid models. This concept of using two-dimensional sketches of the three-dimensional features proves to be an effective way to construct solid models. Many designs are in fact the same shape in one direction. Computer input and output devices we use today are largely two-dimensional in nature, which makes this modeling technique quite practical. AutoCAD's solid modeler is a hybrid modeler; the user interface and construction techniques are CSG-based, whereas the B-rep capabilities are invoked automatically and are transparent to the user.

It is important to point out that solid modeling provides us with very powerful design tools. With the abundant tools available in solid modeling software, the modeling techniques are limited only to the imagination of the designer. Instead of the individual 2D views, one should concentrate on the 3D features of designs. In this chapter, the *V-Block* design from chapter four is revisited to introduce the solid modeling tools available in *AutoCAD*. The procedure to creating solid models using the **Extrude** command is demonstrated.

The *V-BlockS* Design

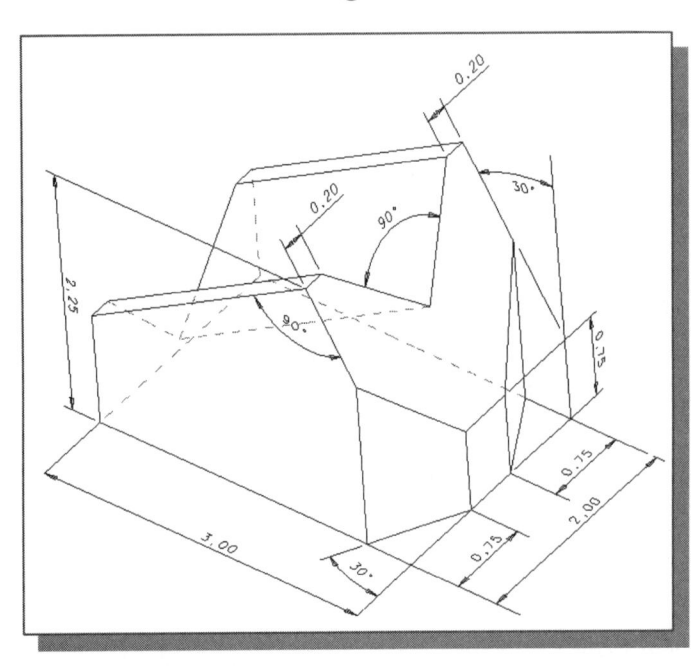

> Make a rough sketch of a CSG binary tree showing the steps that can be used to construct the design. What are the advantages and limitations of using the CSG approach in creating the design? What other tools could be helpful to the modeling task at hand? Take a few minutes to consider these questions and you are encouraged to construct the design on your own prior to going through the tutorial.

Starting Up AutoCAD® 2006

1. Select the **AutoCAD 2006** option on the *Program* menu or select the **AutoCAD 2006** icon on the *Desktop*.

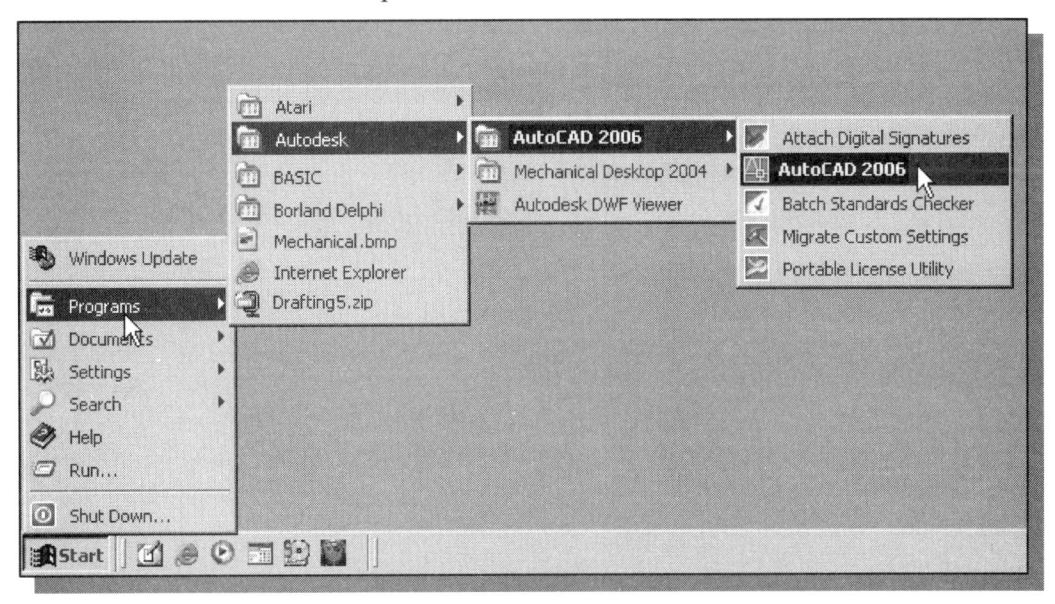

2. In the startup window, select **Start from Scratch**, as shown in the figure below.

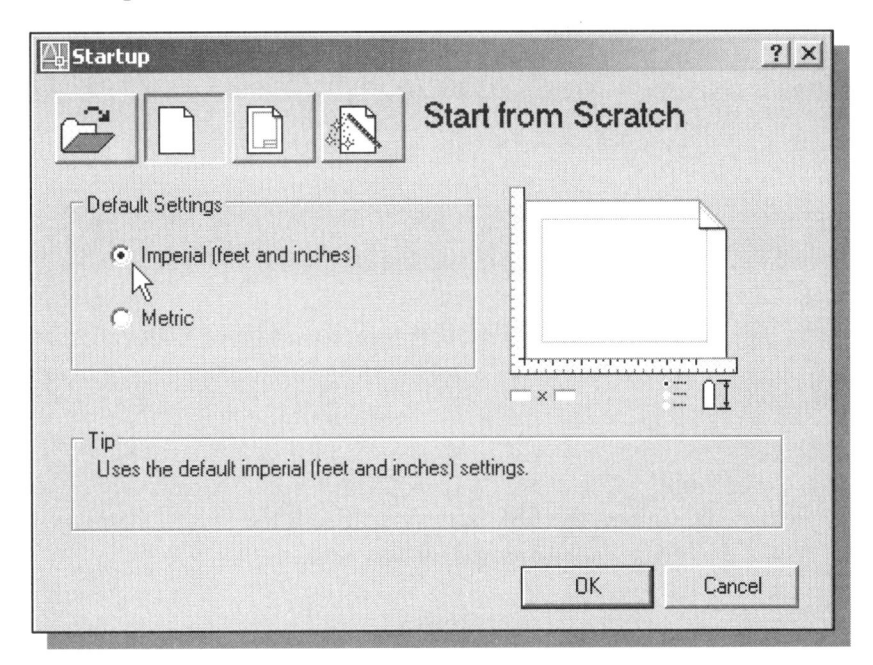

3. In the *Default Settings* section, pick **Imperial (feet and inches)** as the drawing units.

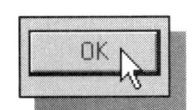

4. Pick **OK** in the startup dialog box to accept the selected settings.

Layers Setup

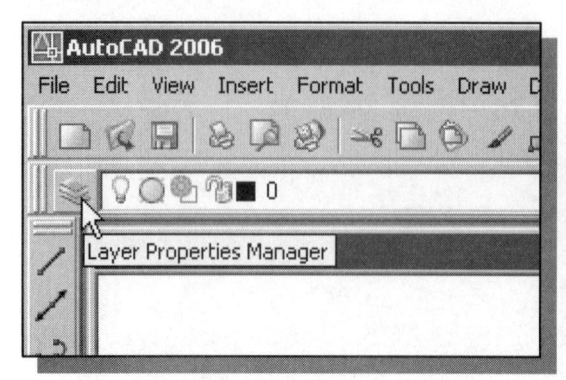

1. Pick **Layer Properties Manager** in the *Object Properties* toolbar.

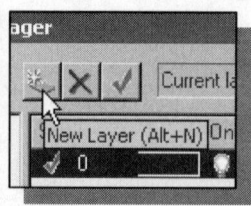

2. Click on the **New** button to create new layers.

3. Create **two new layers** with the following settings:

Layer	Color	LineType	LineWeight
Construction	White	Continuous	Default
Object	Cyan	Continuous	0.30mm

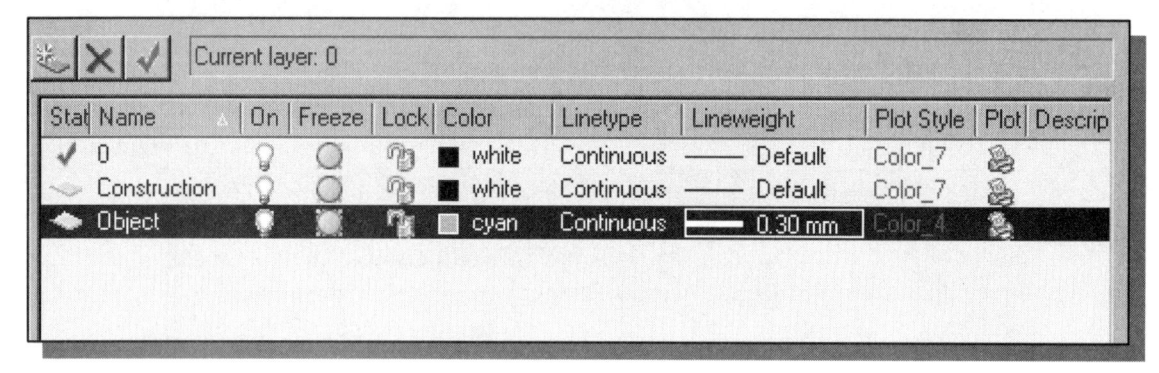

4. Highlight the layer *Construction* in the list of layers.

5. Click on the **Current** icon to set layer *Construction* as the *Current Layer*.

6. Click on the **OK** button to accept the settings and exit the *Layer Properties Manager* dialog box.

7. In the *Status Bar* area, reset the options and turn **ON** the *POLAR, OSNAP, OTRACK, DYN, LWT* and *MODEL* options.

Setting Up a 2D Sketch

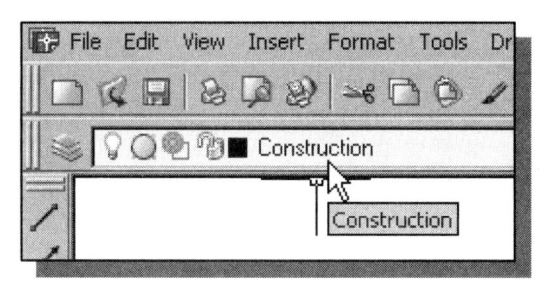

1. In the *Layer Control* box, confirm layer *Construction* is set as the *Current Layer*.

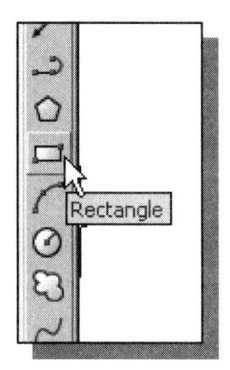

2. Select the **Rectangle** icon in the *Draw* toolbar. In the command prompt area, the message "*Specify first corner point or [Chamfer/Elevation/Fillet/Thickness/Width]:*" is displayed.

3. Place the first corner point of the rectangle near the lower left corner of the screen. Do not be overly concerned about the actual coordinates of the selected location; the CAD drawing space is as big as you can imagine.

4. We will create a 3" × 2.25" rectangle. Enter **@3,2.25 [ENTER]**.

❖ The Rectangle command creates rectangles as *polyline* features, which means all segments of a rectangle are created as a single object.

5. We will next make a copy of the rectangle. Pick **Copy Object** in the *Modify* toolbar.

6. Pick any edge of the rectangle we just created.

7. Inside the graphics window, **right-mouse-click** to accept the selection.

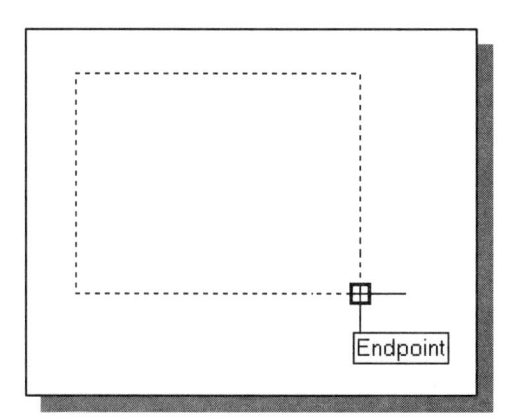

8. In the command prompt area, the message "*Specify base point or displacement, or [Multiple]:*" is displayed. Pick the lower right corner of the rectangle as the base point. A copy of the rectangle is attached to the cursor at the selected base point.

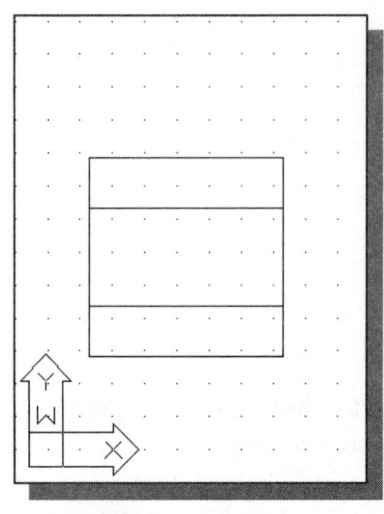

9. In the command prompt area, the message "*Specify second point of displacement, or <use first point as displacement>:*" is displayed. Enter: **@0,0.75 [ENTER]**.

❖ This will position the second rectangle at the location for the 30-degree angle that is needed in the design.

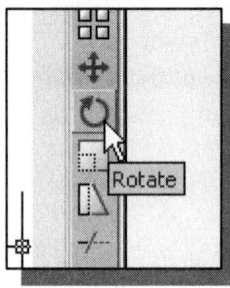

10. Pick **Rotate** in the *Modify* toolbar as shown.

11. Pick the top horizontal line on the screen. The second rectangle, the copy we just created, is selected.

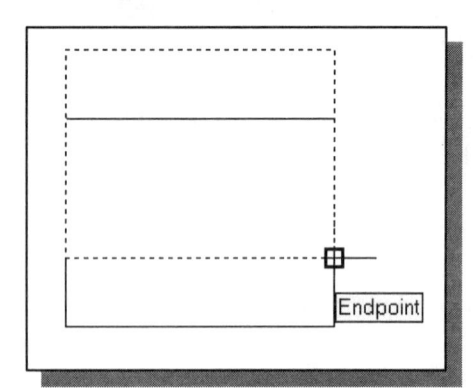

12. Inside the graphics window, right-mouse-click to accept the selection.

13. In the command prompt area, the message "*Specify base point:*" is displayed. Pick the lower right corner of the selected rectangle as the base point.

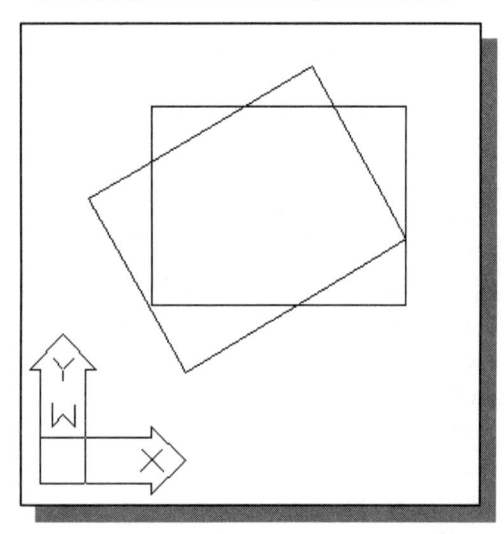

14. In the command prompt area, the message "*Specify the rotation angle or [Reference]:*" is displayed. Enter: **30 [ENTER]**.

Defining the Front Edges of the Design

1. In the *Status Bar* area, right-mouse-click on any of the *POLAR* options to bring up the option menu.

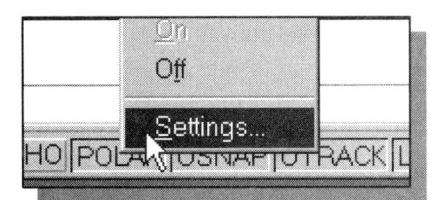

2. Select **Settings** in the option menu as shown.

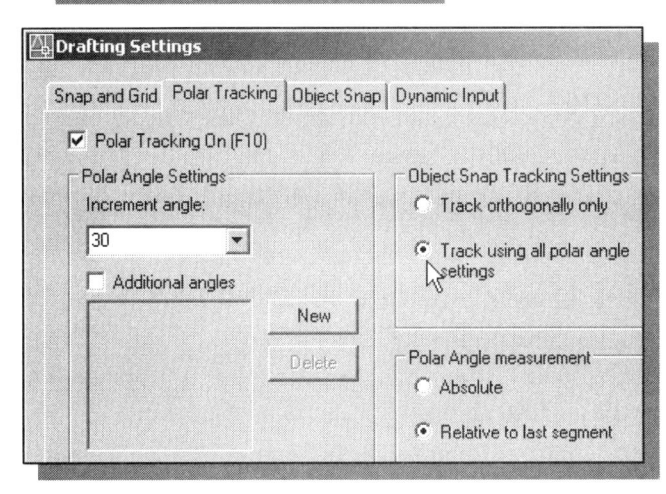

3. In the *Drafting Settings* window, confirm the *Increment angle* is set to **30** in the *Polar Angle Settings* area.

4. Switch on the ***Track using all polar angle settings*** and ***Relative to last segment*** options as shown.

5. In the *Drafting Settings* window, click on the **Object Snap** tab to display and set the *Object Snap* options.

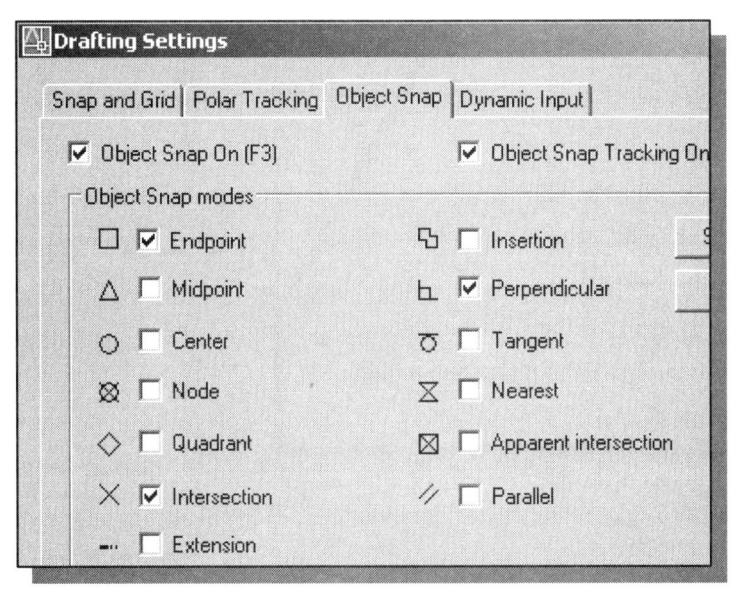

6. On your own, reset the *Object Snap modes* so that only the *Endpoint*, *Intersection* and *Perpendicular* are switched *ON* as shown in the figure.

7. In the *Drafting Settings* window, click **OK** to accept the settings.

- The AuotCAD *AutoTrack*^{TM} feature includes two tracking options: *POLAR* tracking and *OBJECT SNAP* tracking. When the *POLAR* option is turned on, alignment markers are displayed to help us create objects at precise positions and angles.

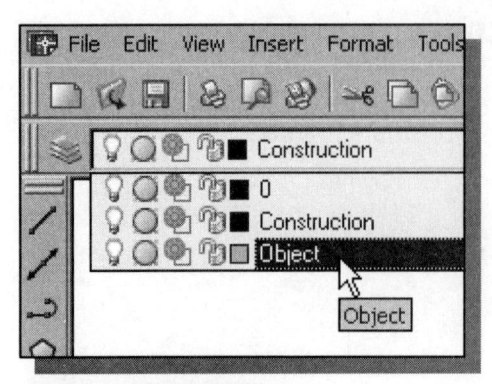

8. On the *Object Properties* toolbar, choose the *Layer Control* box with the left-mouse-button.

9. Move the cursor over the name of layer *Object*; the tool tip "*Object*" appears.

10. **Left-mouse-click once** and layer *Object* is set as the *Current Layer*.

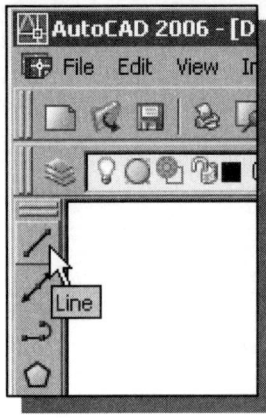

11. Select the **Line** command icon in the *Draw* toolbar.

12. In the command prompt area, the message "*_line Specify first point:*" is displayed. Pick the **lower left corner** of the bottom horizontal line in the front view as the starting point of the line segments.

13. Pick the **lower right corner** of the bottom horizontal line in the front view as the second point.

14. Select the **third** and **fourth** points as shown in the below figure.

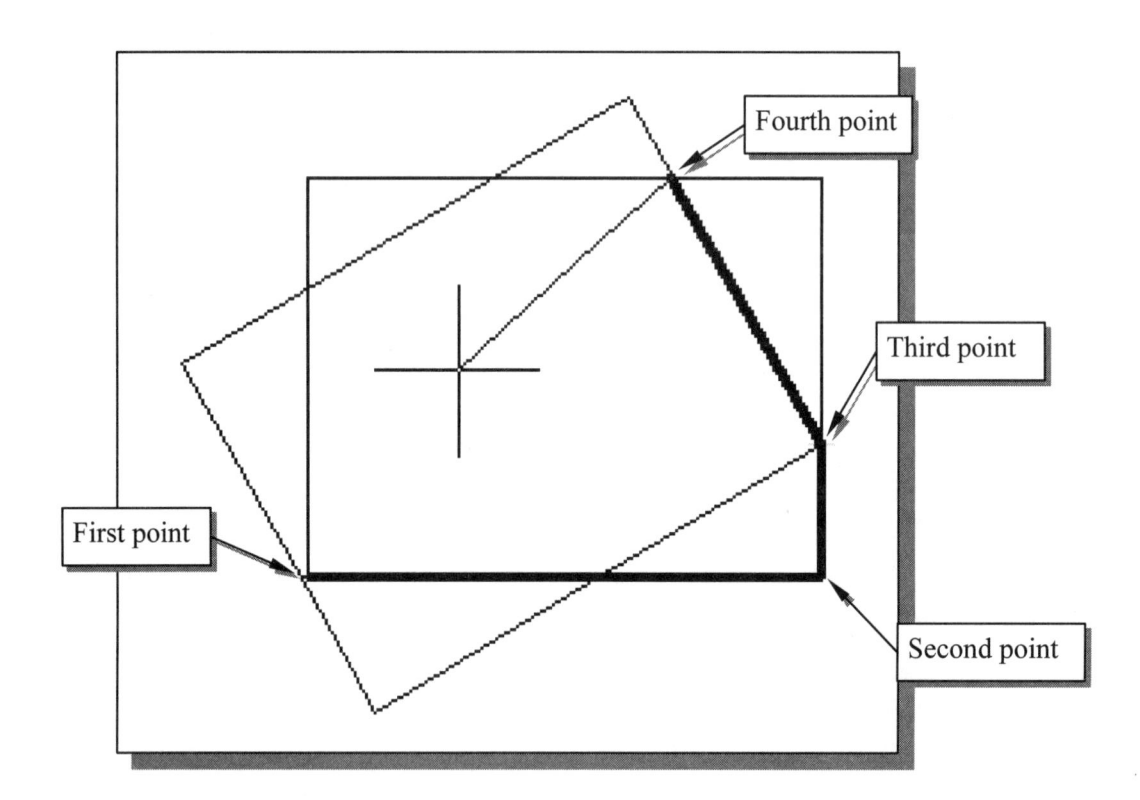

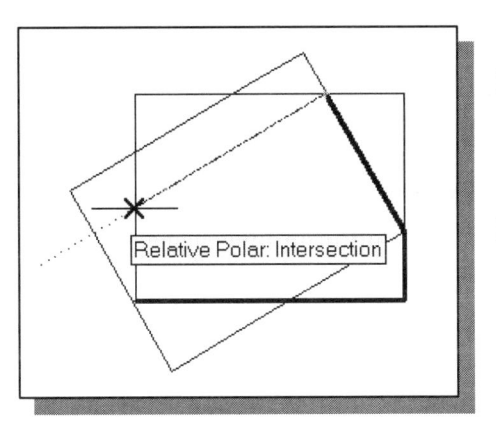

15. Move the cursor near the left vertical line and notice the *AutoTrack* feature automatically snaps the cursor to the intersection point.

16. Left-click at the intersection point as shown.

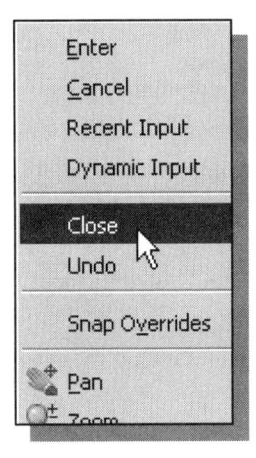

17. Inside the graphics window, right-mouse-click once to bring up the popup menu.

18. Pick **Close** in the popup menu. The Close option will create a line segment connecting the last point to the first point of the line sequence.

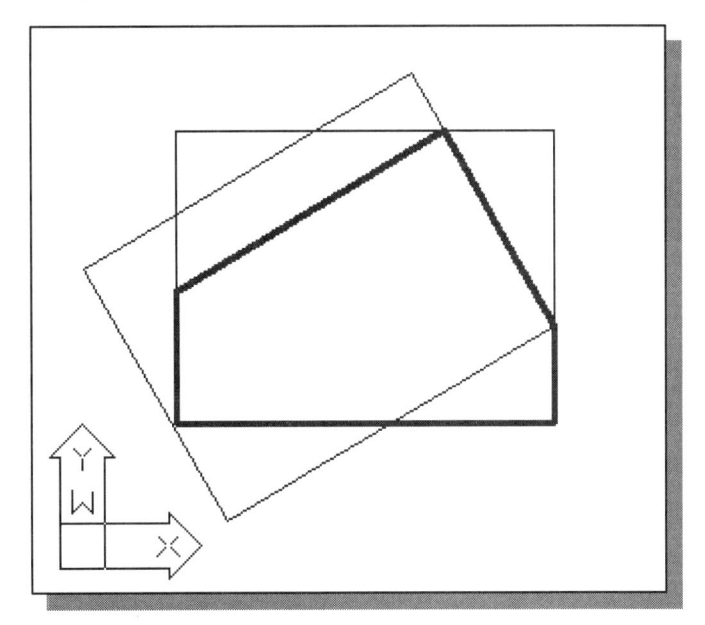

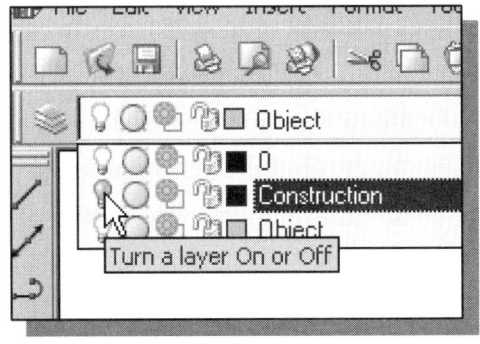

19. On your own, turn *OFF* the display of the *Construction Layer* by using the left-mouse-button.

20. Move the cursor into the graphics window and **left-mouse-click once** to accept the layer control settings.

Creating a Region

- In AutoCAD, **regions** are 2D enclosed areas, which can be created from one or more closed shapes called **loops**. A *loop* is a curve or a sequence of connected curves that defines an area on a plane with a boundary that does not intersect itself. *Loops* can be combinations of lines, polylines, circles, arcs, ellipses, elliptical arcs, splines, 3D faces, traces, and solids. The objects that make up the loops must either be closed or form closed areas by sharing endpoints with other objects. The objects must also be coplanar, which means they must be on the same plane. We can create regions out of multiple loops and out of open curves whose endpoints are connected. Objects that intersect with themselves or with other objects cannot form regions; for example, intersecting lines, intersecting arcs or self-intersecting curves. Once a region is created, we can analyze properties such as its area and moments of inertia.

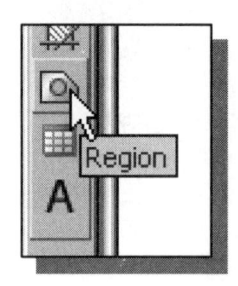

1. Select the **Region** command icon in the *Draw* toolbar.

2. In the command prompt area, the message "*Select objects:*" is displayed. Select displayed entities by creating a selection window enclosing all objects.

3. Inside the graphics window, **right-mouse-click** to accept the selection and create a region.

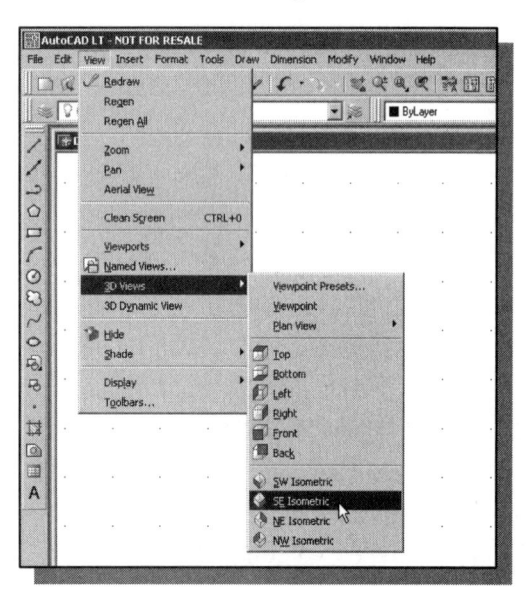

4. In the pull-down menus, select:

   [View] → [3D Views] → [SE Isometric]

- Notice the orientation of the 2D region in relation to the displayed AutoCAD user coordinate system. By default, the 2D sketch-plane is aligned to the XY plane of the world coordinate system.

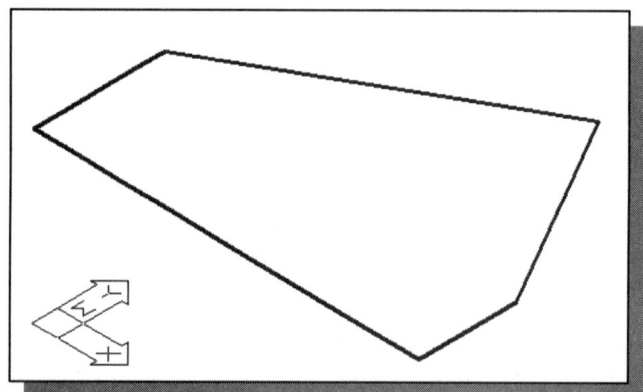

Solids Toolbar

1. Move the cursor to the *Standard* toolbar area and right-mouse-click once on any icon in the toolbar area to display a list of toolbar menu groups.

2. Select **Solids**, with the left-mouse-button, to display the *Solids* toolbar on the screen.

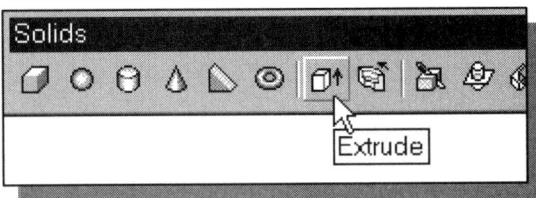

3. Select **Extrude** in the *Solids* toolbar as shown.

4. In the command prompt area, the message "*Select objects:*" is displayed. Select the region by clicking on any of the displayed segments.

5. Inside the graphics window, right-mouse-click to accept the selection and proceed with the extrude command.

6. In the command prompt area, the message "*Specify height of extrusion or [path]:*" is displayed. Enter: **2.0 [ENTER]**.

7. In the command prompt area, the message "*Specify angle of taper for extrusion <0>:*" is displayed. Enter: **0.0 [ENTER]**.

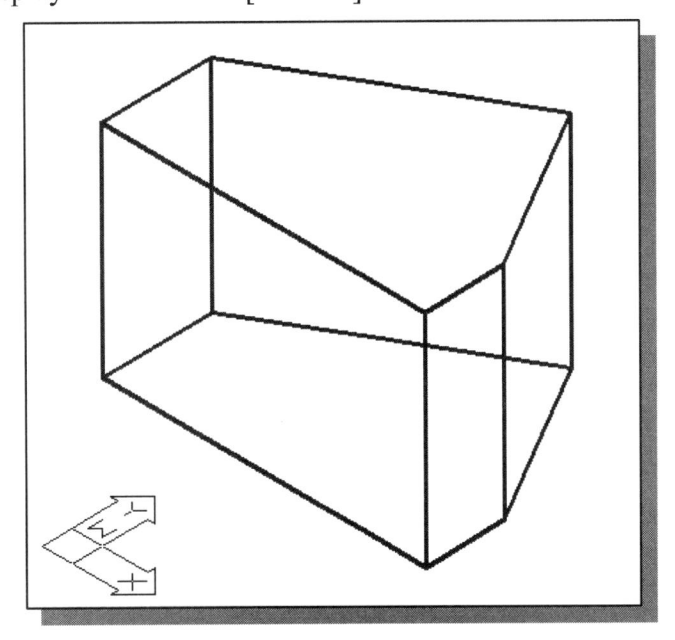

- Note that the orientation and position of solid models, in relation to the world coordinate system, can be adjusted using the **Move** and **Rotate 3D** commands.

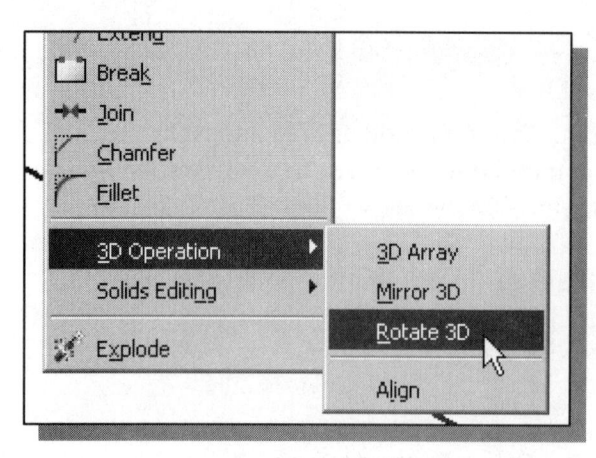

8. In the pull-down menus, select:

   [Modify] → **[3D Operation]** → **[Rotate 3D]**

9. In the command prompt area, the message "*Select objects:*" is displayed. Select the solid model by clicking on any of the displayed edges of the model.

10. Inside the graphics window, right-mouse-click to accept the selection and proceed with the **Rotate 3D** command.

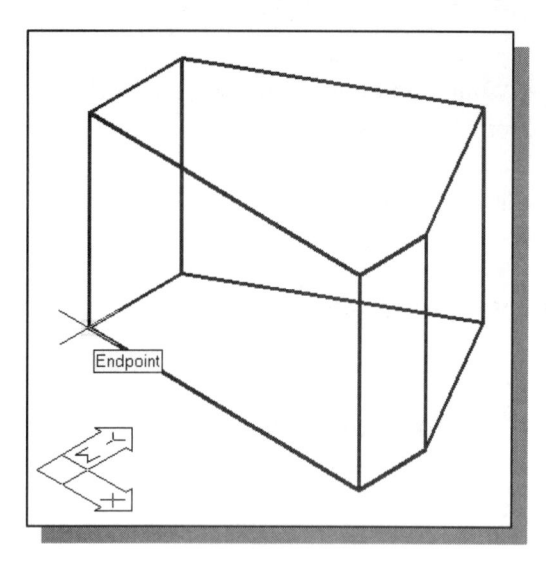

11. In the command prompt area, the message "*Specify first point on axis or define axis by [Object/Last/View/Xaxis/Yaxis/Zaxis/2points]:*" is displayed. Select the lower left corner of the model as shown

    - Note that the **Rotate 3D** command requires the definition of a *direction vector*, which is used as the axis of rotation. The rotation follows the conventional engineering *right-hand rule*. The first point defines the base of the direction vector.

12. In the command prompt area, the message "*Specify second point on axis:*" is displayed. Select the lower right corner of the model as shown

13. In the command prompt area, the message "*Specify rotation angle or [Reference]:*" is displayed. Enter: **90** [**ENTER**]

14. On your own, use the available **Shade** commands to examine the created solid object.

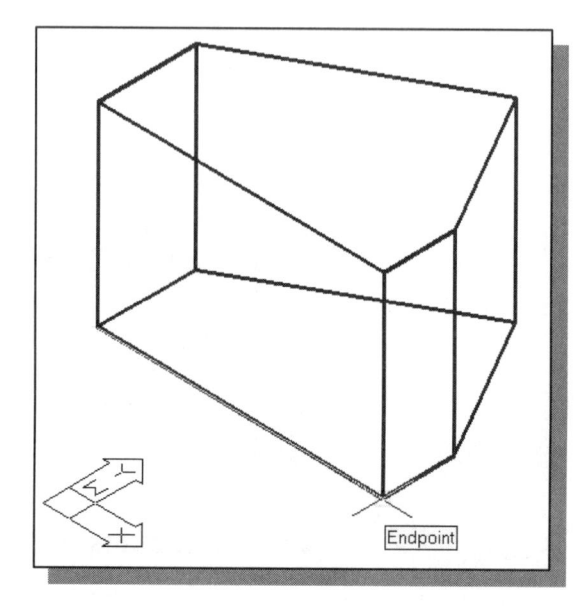

Creating a 2D Sketch at the Base of the Model

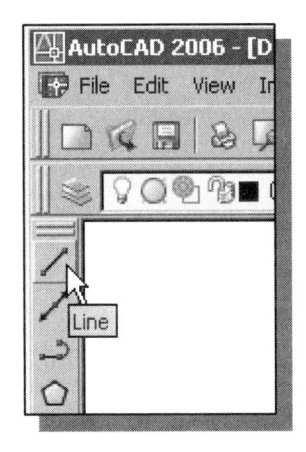

1. Select the **Line** command icon in the *Draw* toolbar. In the command prompt area, the message "*_line Specify first point:*" is displayed.

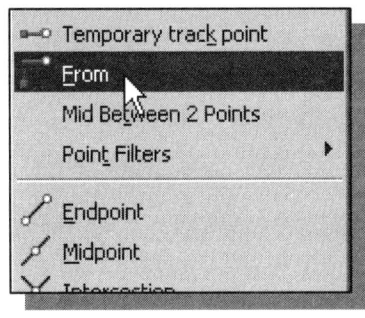

2. Inside the graphics window, hold down the **[SHIFT]** key and **right-mouse-click** once to bring up the *Object Snap* shortcut menu.

3. Select the **From** option in popup window.

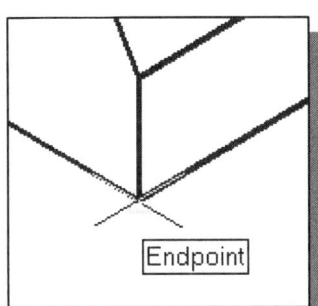

4. Pick the lower right corner of the solid model as the base point as shown in the figure.

5. We will place the first point above the selected corner. Enter: **@0,0.75 [ENTER]**

- Note the orientation of the world XY plane.

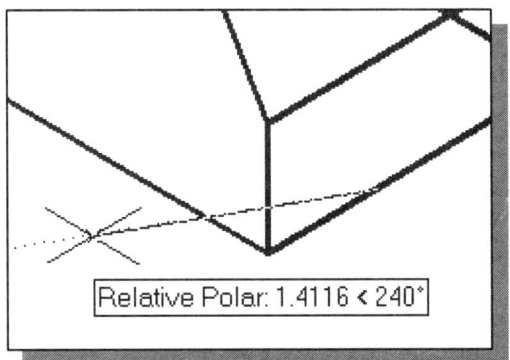

6. In the command prompt area, the message "*Specify next point or [Undo]:*" is displayed. Using the *POLAR Tracking* option (30 degrees increment), create a line that is about 1.4 inches long, as shown in the figure.

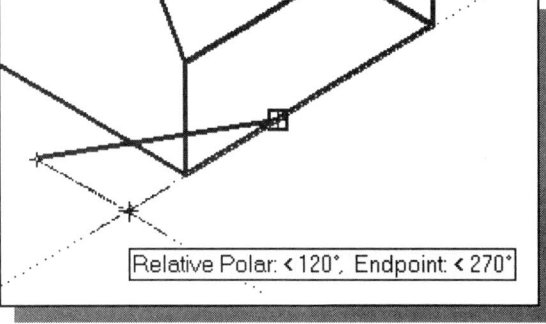

7. Move the cursor on top of the starting point of the last line to activate the *OTRACK* option and select the location that forms a 90-degrees angle as shown.

8. Inside the graphics window, right-mouse-click to activate the option menu and select **Close** with the left-mouse-button to create a line connecting to the starting point of the line-segments and end the Line command.

Creating a Copy of the 2D Sketch

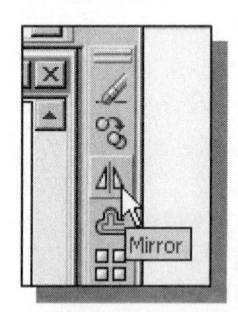

1. Click on the **Mirror** icon in the *Modify* toolbar.

2. In the command prompt area, the message "*Select objects:*" is displayed. Select the three line segments we just created.

3. Inside the graphics window, right-mouse-click to accept the selection and proceed with the **Extrude** command.

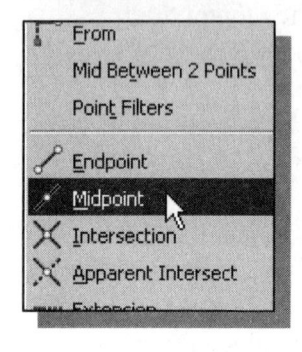

4. Inside the graphics window, hold down the [**SHIFT**] key and **right-mouse-click** once to bring up the *Object Snap* shortcut menu.

5. Select the **Midpoint** option in popup window.

6. Click on the bottom left edge of the model. With the *Midpoint SNAP* option, the midpoint of the edge is selected.

7. Move the cursor toward the right vertical line in the active viewport, and left-click once when the **Perpendicular** tooltip is displayed as shown in the below figure.

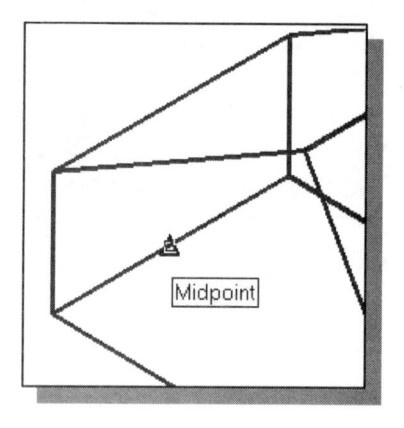

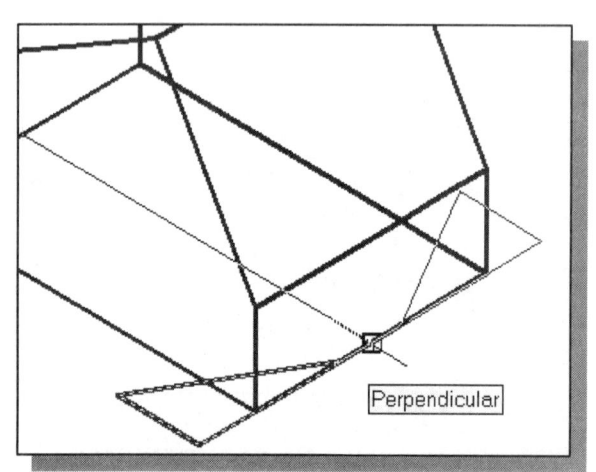

8. In the command prompt area, the message "*Delete Source Objects? [Yes/No] <N>:*" is displayed. Inside the graphics window, right-mouse-click once to bring up the option menu and select **No** to keep both sets of objects.

Creating the Cutter Solids

* The two triangular shapes will be the extruded in the Z-direction to form two solids that can be used to create the two vertical cuts.

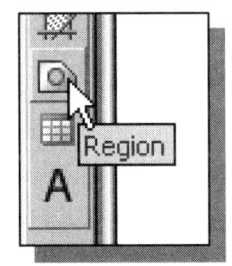

    1. Select the **Region** command icon in the *Draw* toolbar.

    2. In the command prompt area, the message "*Select objects:*" is displayed. Select the **six line segments** we just created.

    3. Inside the graphics window, right-mouse-click to accept the selection and create two regions.

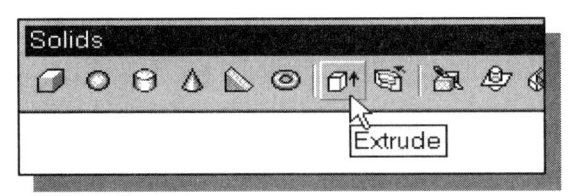

    4. Select **Extrude** in the *Solids* toolbar as shown.

    5. In the command prompt area, the message "*Select objects:*" is displayed. Select the **two regions** by clicking on the two triangular regions.

    6. Inside the graphics window, right-mouse-click to accept the selection and proceed with the **Extrude** command.

    7. In the command prompt area, the message "*Specify height of extrusion or [path]:*" is displayed. Enter: **2.5** [**ENTER**].

    8. In the command prompt area, the message "*Specify angle of taper for extrusion <0>:*" is displayed. Enter: **0.0** [**ENTER**].

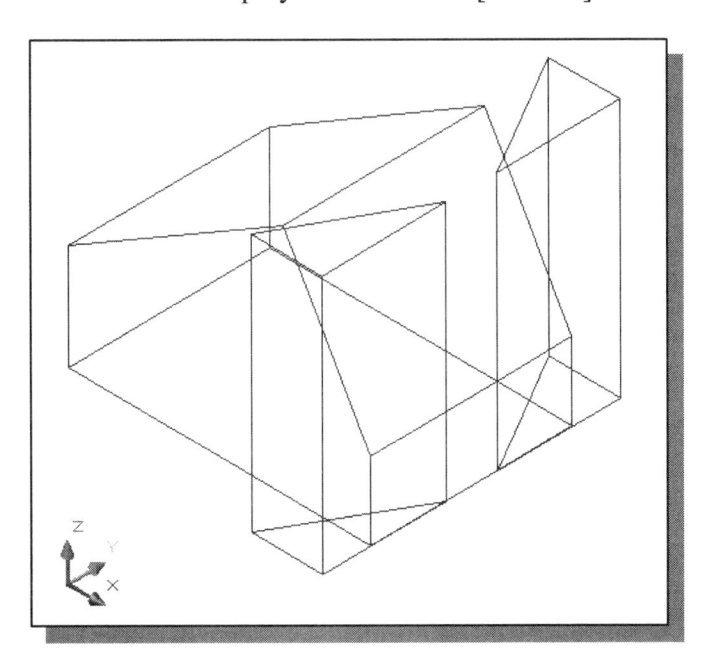

* Note that the two cutters were intentionally made larger than the dimensions shown on the design. In fact, we could create a single cutter, instead of the two we have created, provided the two triangular shapes are included and positioned at the proper locations.

Boolean Operation – *Subtract*

1. Move the cursor to the *Standard* toolbar area and **right-click** on any icon of the *Standard* toolbar to display a list of toolbar menu groups.

2. Select **Solids Editing**, with the left-mouse-button, to display the *Solids Editing* toolbar on the screen.

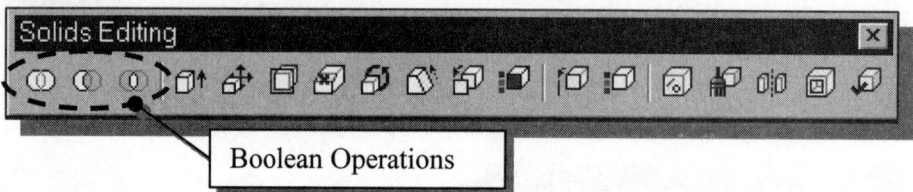

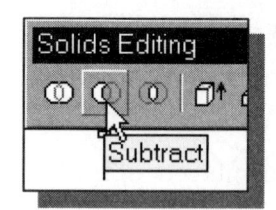

   Boolean Operations

3. In the *Solids Editing* toolbar, click on the **Subtract** icon. In the command prompt area, the message "*_subtract Select solids and regions to subtract from .. Select Objects:*" is displayed.

4. Pick the base block solid by clicking on one of the edges.

5. Inside the graphics window, right-mouse-click to accept the selection and proceed with the **Subtract** command.

6. In the command prompt area, the message "*Select solids and regions to subtract .. Select Objects:*" is displayed. Pick the two triangular blocks by clicking on the edges of the two cutters.

7. Inside the graphics window, right-mouse-click to accept the selection and proceed with the **Subtract** command.

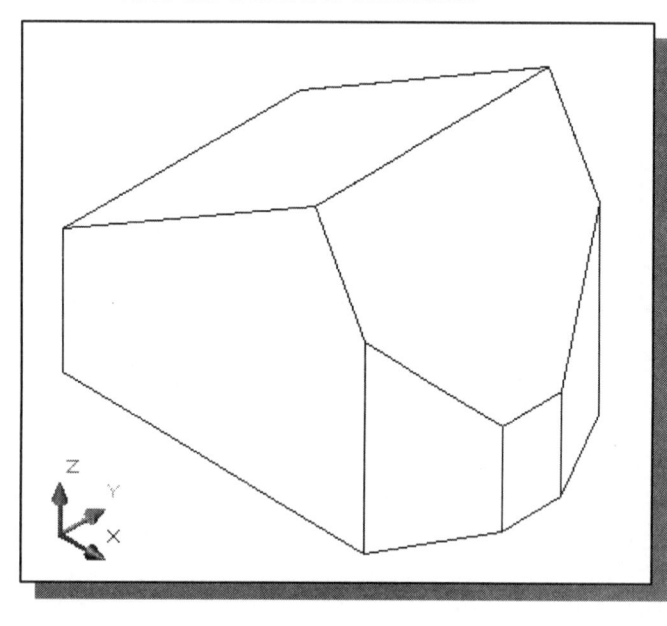

- Note that the two cut features are created fairly quickly. In comparison to the wireframe or surface models created in the previous chapters, the solid modeling approach is much more straightforward and parallel to the manufacturing processes.

Mass Properties of the Solid Model

1. In the pull-down menus, select:
   [Tools] → [Inquiry] → [Region/Mass Properties]

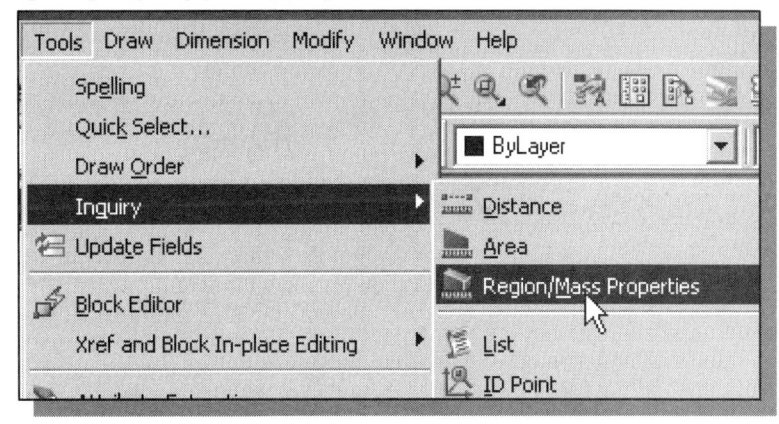

2. In the command prompt area, the message "*Select objects:*" is displayed. Select the solid model by clicking on any of the displayed edges of the model.

3. Inside the graphics window, right-mouse-click to accept the selection and proceed with the **Region/Mass Properties** command.

```
---------------- SOLIDS ----------------

Mass:                    9.2470
Volume:                  9.2470
Bounding box:       X:   2.3685  --  5.3685
                    Y:  -0.3266  --  1.6734
                    Z:   0.0000  --  2.2500
Centroid:           X:   3.8756
                    Y:   0.6734
                    Z:   0.8555
Moments of inertia: X:  16.5894
                    Y: 153.5781
                    Z: 151.3567
Products of inertia:XY: 24.1336
                    YZ:  5.3273
                    ZX: 31.2975
Radii of gyration:  X:   1.3394
                    Y:   4.0753
                    Z:   4.0458
Principal moments and X-Y-Z directions about centroid:
                    I:   5.4823 along [0.9748 0.0000 0.2231]
                    J:   7.9168 along [0.0000 1.0000 0.0000]
                    K:   8.4158 along [-0.2231 0.0000 0.9748]

Press ENTER to continue:
```

- In the *AutoCAD Text Window*, a list of properties related to the created solid model is displayed. The *Mass* and *Volume* properties are listed as the same since a mass density of one is used for the calculation.

4. Press the **ENTER** key twice to close the *AutoCAD Text Window*.

Align the UCS to the Inclined Face

➢ We will create an UCS plane aligning to the inclined face of the design.

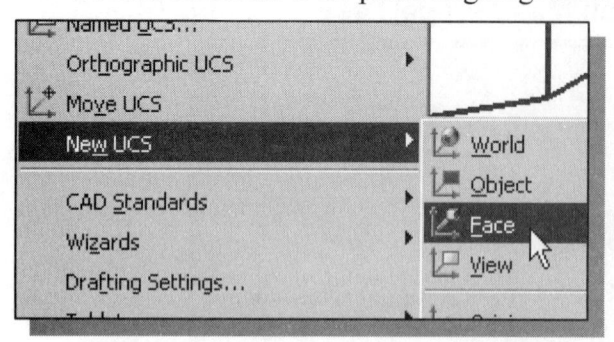

1. In the pull-down menus, select:

   [Tools] → [New UCS] → [Face]

   - Note that other options are also available to aid the orientation of new UCS planes.

2. In the command prompt area, the message "*Select face of solid object:*" is displayed. Pick the **inclined face** by clicking on the surface near the upper left corner.

3. In the command prompt area, the message "*Enter an option [Next/Xflip/Yflip] <accept>:*" is displayed. Press the **[ENTER]** key to accept the new UCS positioned as shown.

4. On your own, bring up the *UCS* toolbar.

5. In the *UCS* toolbar, select the **Z Axis Rotate UCS** command icon as shown.

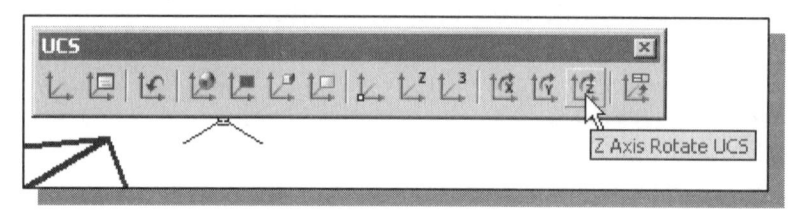

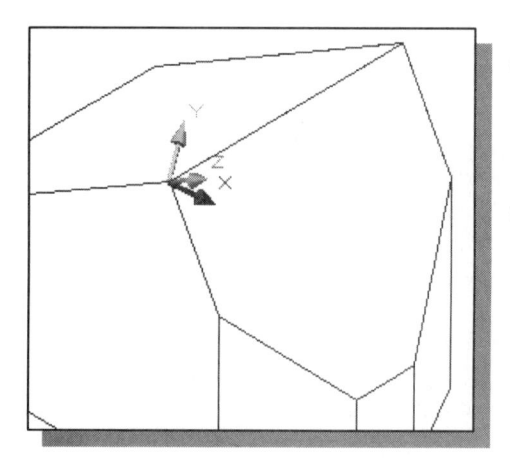

6. In the command prompt area, the message "*Specify rotation angle about Z-axis <90>:*" is displayed. Enter: **-45 [ENTER]**.

   - The negative value entered rotates the UCS clockwise about the current UCS Z-axis.

Creating the *V-cut*

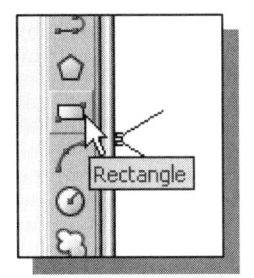

1. Select the **Rectangle** icon in the *Draw* toolbar. In the command prompt area, the message "*Specify first corner point or [Chamfer/Elevation/Fillet/Thickness/Width]:*" is displayed.

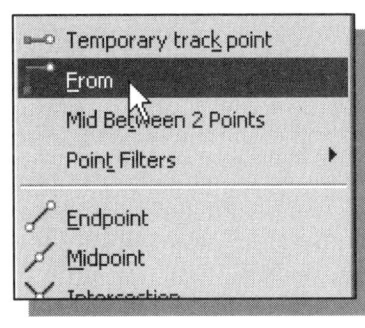

2. Inside the graphics window, hold down the [**SHIFT**] key and **right-mouse-click** once to bring up the *Object Snap* shortcut menu.

3. Select the **From** option in popup window.

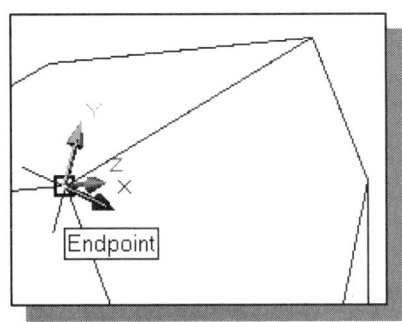

4. Pick the upper left corner of the inclined surface as the base point as shown in the figure.

5. We will place the first point above the selected corner. Enter: **@0.2<45 [ENTER]**

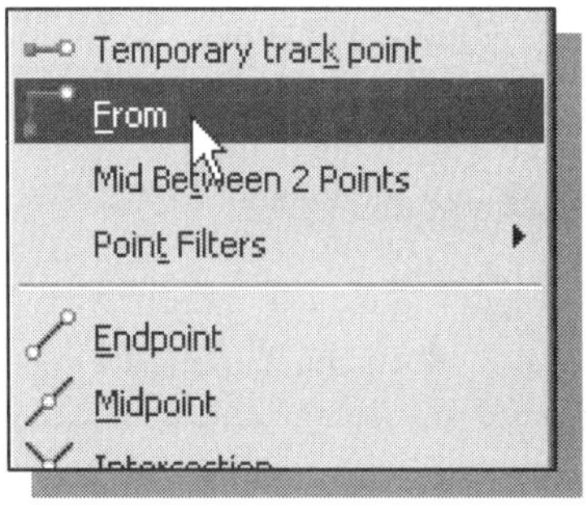

6. Inside the graphics window, hold down the [**SHIFT**] key and right-mouse-click once to bring up the *Object Snap* shortcut menu.

7. Select the **From** option in popup window.

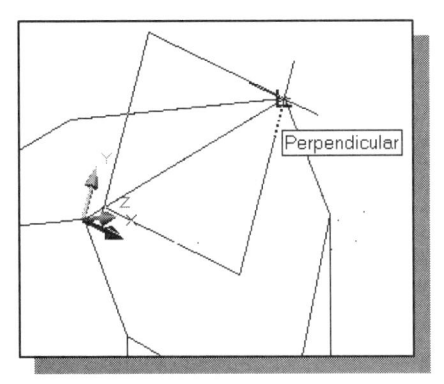

8. Pick the upper right corner of the inclined surface as the base point as shown in the figure.

9. We will place the first point above the selected corner. Enter: **@0.2<-135 [ENTER]**

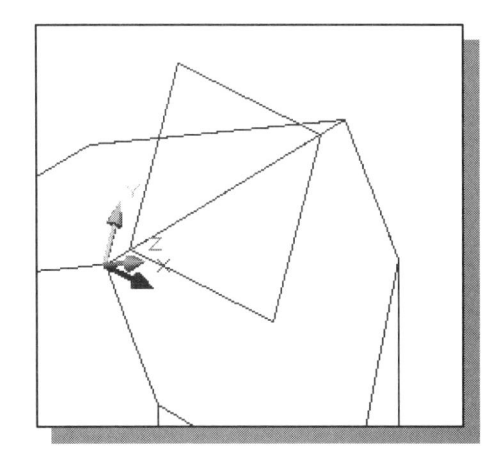

- The above procedure of creating the rectangle at the proper location is simplified by first orienting the UCS.

❖ Note that a rectangle is a valid filled polygon and therefore can be extruded into a solid object.

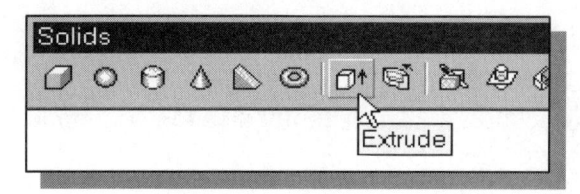

10. Select **Extrude** in the *Solids* toolbar as shown.

11. In the command prompt area, the message "*Select objects:*" is displayed. Select the rectangle by clicking on the any edges of the created rectangle.

12. Inside the graphics window, right-mouse-click to accept the selection and proceed with the **Extrude** command.

13. In the command prompt area, the message "*Specify height of extrusion or [path]:*" is displayed. Enter: **-4.5** [**ENTER**].

- The negative number indicates the extrusion direction, which needs to be in the negative Z-direction of the current UCS.

14. In the command prompt area, the message "*Specify angle of taper for extrusion <0>:*" is displayed. Enter: **0.0** [**ENTER**].

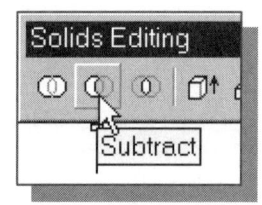

15. In the *Solids Editing* toolbar, click on the **Subtract** icon. In the command prompt area, the message "*_subtract Select solids and regions to subtract from .. Select Objects:*" is displayed.

16. Pick the base solid block by clicking on one of the edges.

17. Inside the graphics window, right-mouse-click to accept the selection and proceed with the **Subtract** command.

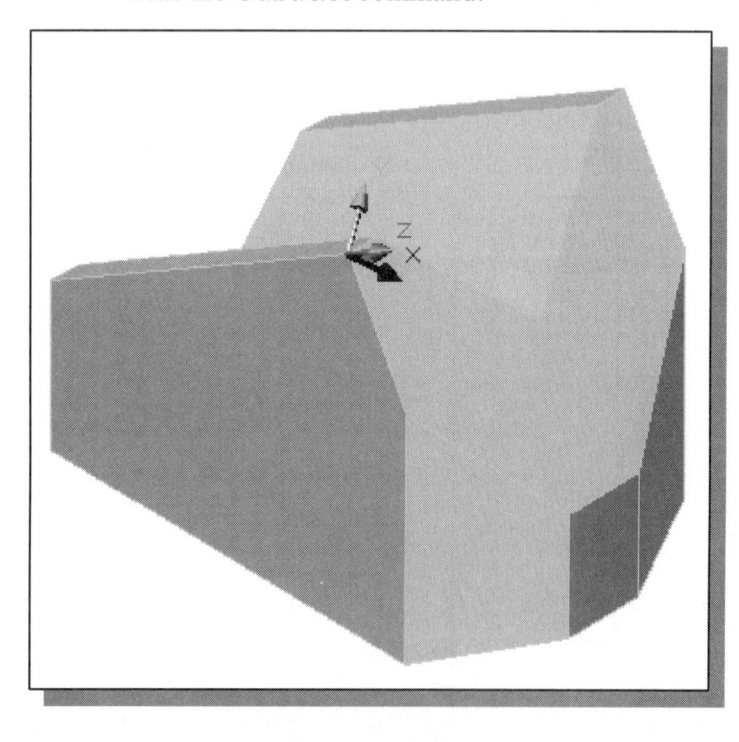

18. In the command prompt area, the message "*Select solids and regions to subtract .. Select Objects:*" is displayed. Pick the rectangular block.

19. Inside the graphics window, right-mouse-click to accept the selection and proceed with the **Subtract** command.

- On your own, use the dynamic viewing commands to examine the completed solid model.

20. In the *UCS* toolbar, select the **World UCS** command icon as shown.

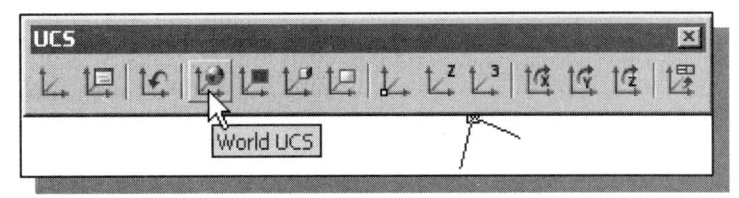

- The **World UCS** command aligns the UCS to the world coordinate system.

21. On your own, reset the display to 2D wireframe.

22. In the pull-down menus, select:
    [Tools] → **[Inquiry]** → **[Region/Mass Properties]**

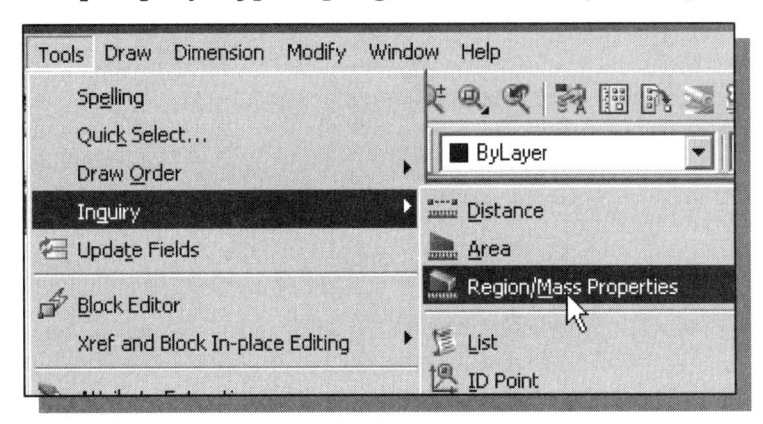

23. In the command prompt area, the message "*Select objects:*" is displayed. Select the solid model by clicking on any of the displayed edges of the model.

24. Inside the graphics window, right-mouse-click to accept the selection. On your own, compare the displayed values to those shown on page 7-17.

```
---------------- SOLIDS ----------------

Mass:                    7.5715
Volume:                  7.5715
Bounding box:      X: 2.3685  --  5.3685
                   Y: -0.3266 --  1.6734
                   Z: 0.0000  --  2.2500
Centroid:          X: 3.9578
                   Y: 0.6734
                   Z: 0.7446
Moments of inertia: X: 12.2687
                   Y: 128.8923
                   Z: 129.1229
Products of inertia: XY: 20.1798
                   YZ: 3.7965
                   ZX: 22.9310
Radii of gyration: X: 1.2729
                   Y: 4.1259
                   Z: 4.1296
Principal moments and X-Y-Z directions about centroid:
                   I: 4.4903 along [0.9729 0.0000 0.2314]
                   J: 6.0925 along [0.0000 1.0000 0.0000]
                   K: 7.2345 along [-0.2314 0.0000 0.9729]
Press ENTER to continue:
```

Questions:

1. What is an auxiliary view and why would it be important?

2. When is a line viewed as a point? When is a surface viewed as true size and shape?

3. How do we define a **Region** in AutoCAD? What are the advantages of using the Regions?

4. How is a **Solid Model** different from a **Surface Model**?

5. List some of the different mass properties that are available with 3D models in AutoCAD.

6. Identify the following commands:

   (a)

   (b)

   (c)

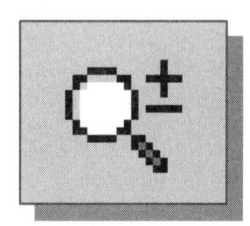

   (d)

Exercises: All dimensions are in inches.

1.

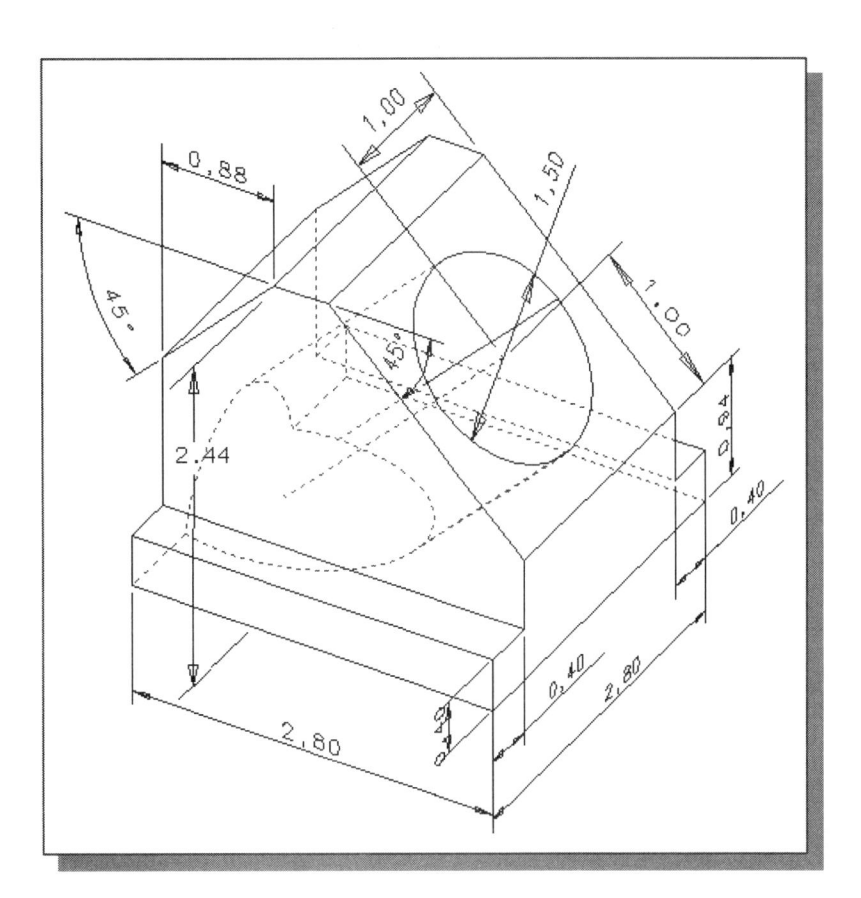

2.

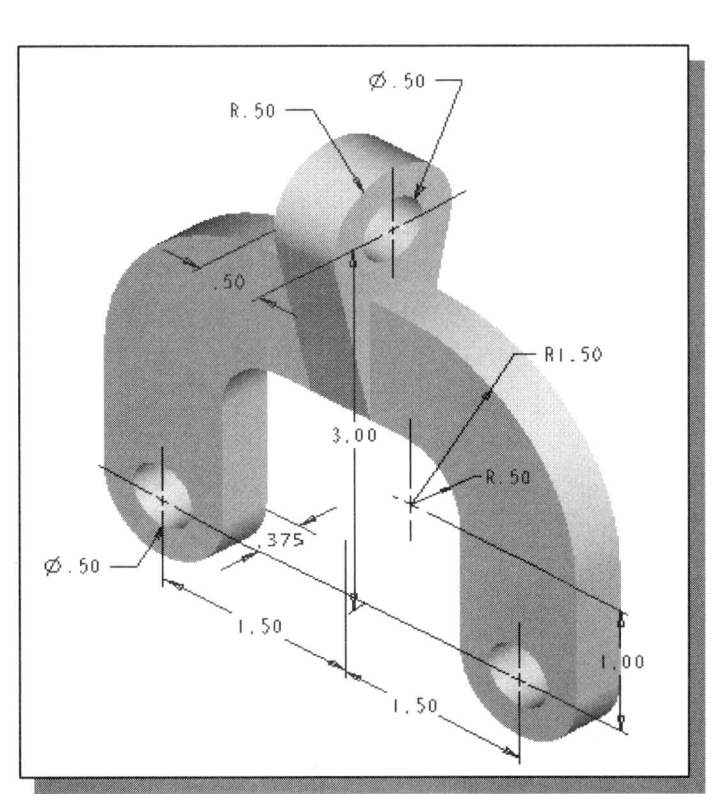

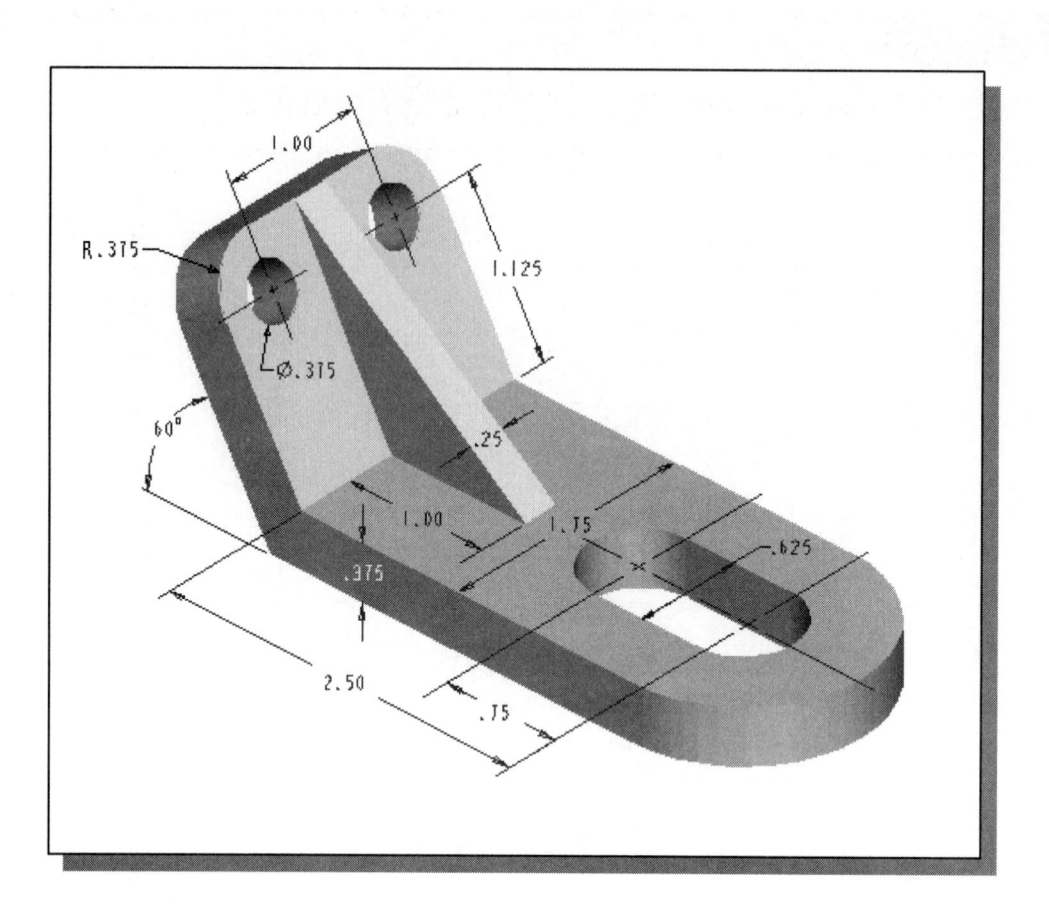

3.

Chapter 8
Multiview Drawings from 3D Models

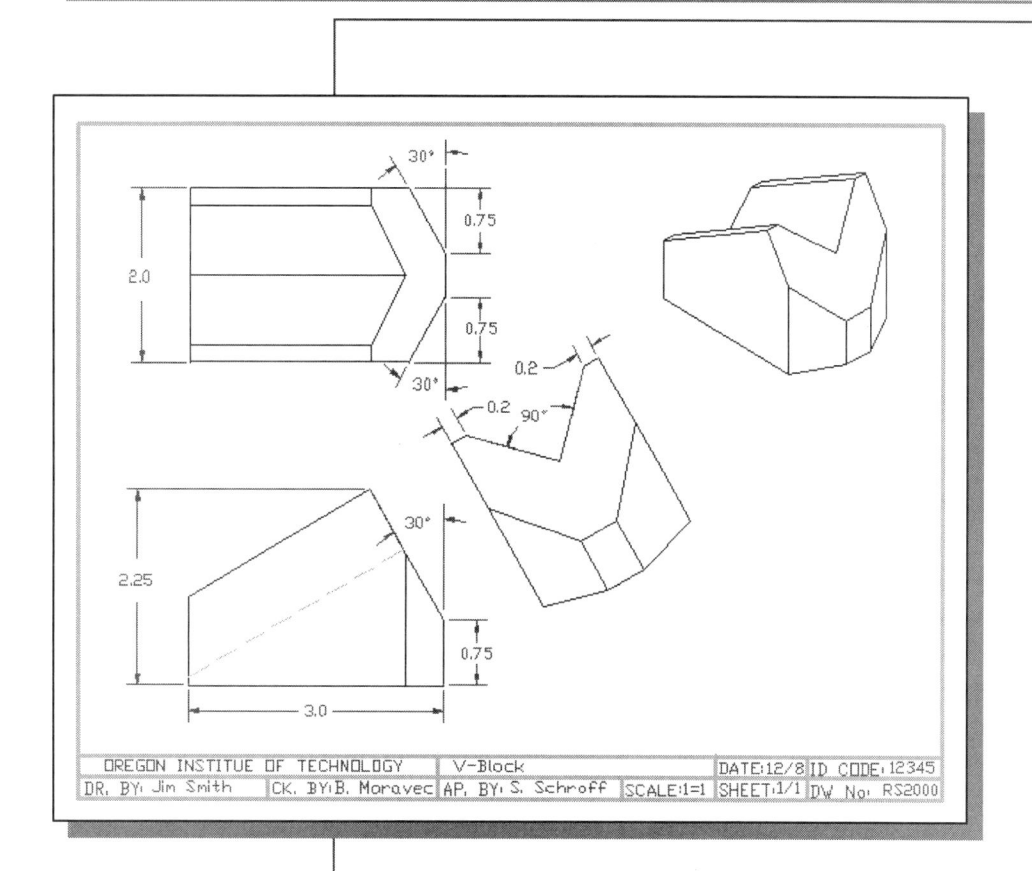

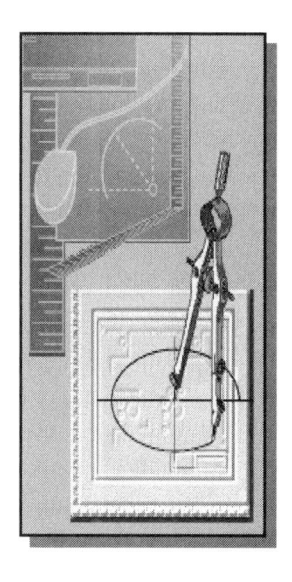

Learning Objectives

- ♦ Create Multiview Drawings from 3D Models
- ♦ Create Borders and Title Block in the Layout Mode
- ♦ Create 2D Projections from 3D Models
- ♦ Create Multiple Viewports in Paper Space
- ♦ Align and Position 2D Projected Views in Layout Mode

Introduction

With the capabilities of solid modeling software, the importance of two-dimensional drafting is decreasing. Drafting is now considered as one of the downstream applications of using solid modeling. In many production facilities, solid models are being used to generate machine tool paths for *computer numerical control* (CNC) machine tools. Solid models can also used in *rapid prototyping* to create 3D physical models out of plastic resins, powdered metal, etc. Ideally, the solid model database should be used directly to generate the final product. However, many organizations still require the use of two-dimensional drawings for the applications in their production facilities. In AutoCAD, model views can be used to automatically create 2D views in AutoCAD *paper space*. Using the 3D model as the starting point for a design, 3D modeling tools can easily create all the necessary two-dimensional views. In this sense, 3D modeling tools are making the process of creating two-dimensional drawings more efficient and effective.

An important rule concerning multiview drawings is to create enough views to accurately describe the design. This usually requires two or three of the regular views, such as a front view, a top view and/or a side view. Many designs have features located on inclined surfaces that are not parallel to the regular planes of projection. To truly describe the feature, the true shape of the feature must be shown using an **auxiliary view**. An *auxiliary view* has a line of sight that is perpendicular to the inclined surface, as viewed looking directly at the inclined surface. An *auxiliary view* is a supplementary view that can be constructed from any of the regular views.

In this chapter, the procedure of creating 2D drawings from 3D models is presented. We will create orthographic views, including an auxiliary view, of the *V-block* design.

The *V-Block* Design

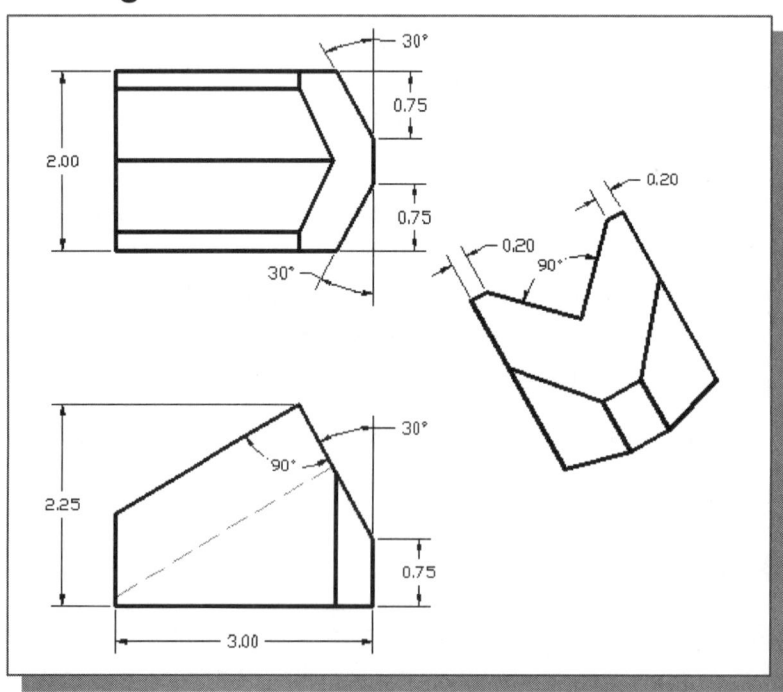

Starting Up AutoCAD® 2006

1. Select the **AutoCAD 2006** option on the *Program* menu or select the **AutoCAD 2006** icon on the *Desktop*.

2. In the *AutoCAD* startup dialog box, select the **Open a Drawing** option with a single click of the left-mouse-button.

3. Open the *V-blockS* solid model created in the previous chapter by clicking on the corresponding filename. (Use the Browse option to locate the solid model file if it is not displayed.)

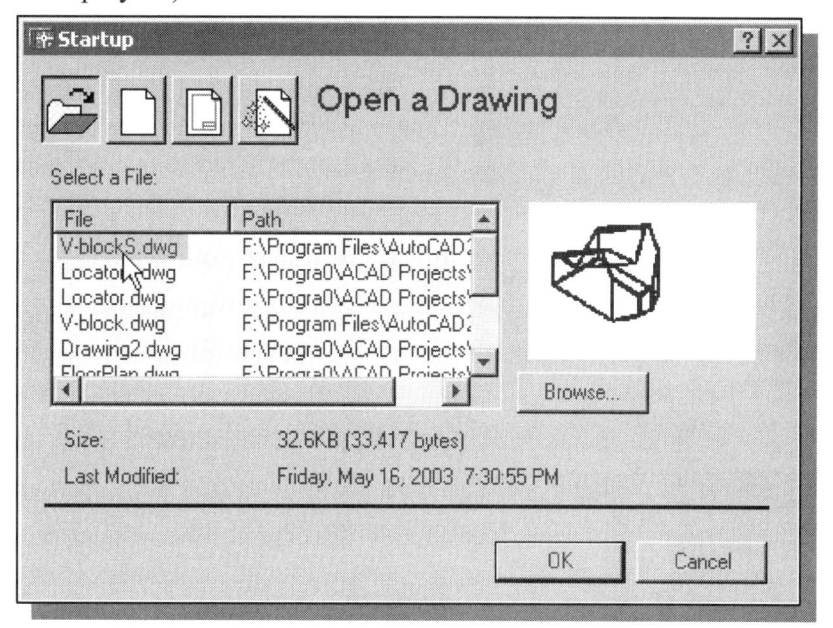

- The *V-blockS* solid model is retrieved and displayed in the graphics window.

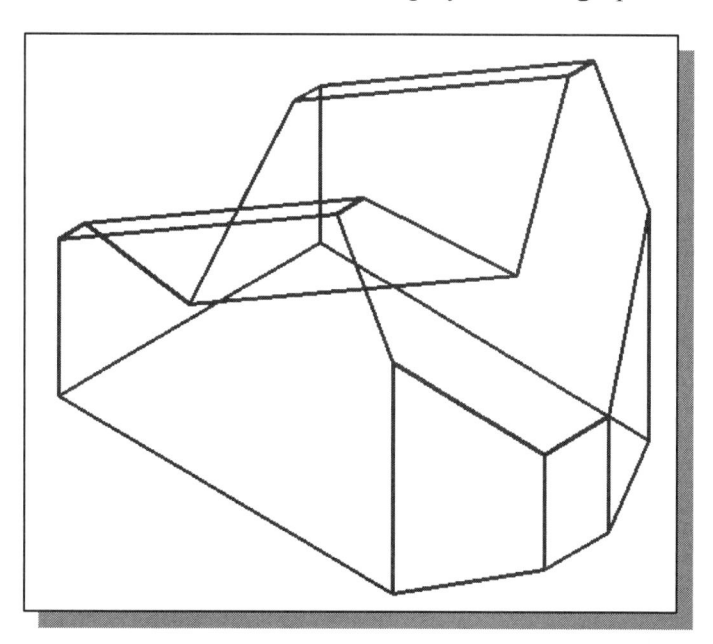

AutoCAD Paper Space

- Until now, we have been working in ***model space*** to create our design in **full size**. Once the 3D model is constructed, we can arrange our design on a two-dimensional sheet of paper so that the plotted hardcopy is exactly what we want. This two-dimensional sheet of paper is known as ***paper space*** in AutoCAD. We can place borders and title blocks, the objects that are less critical to our design, on *paper space*.

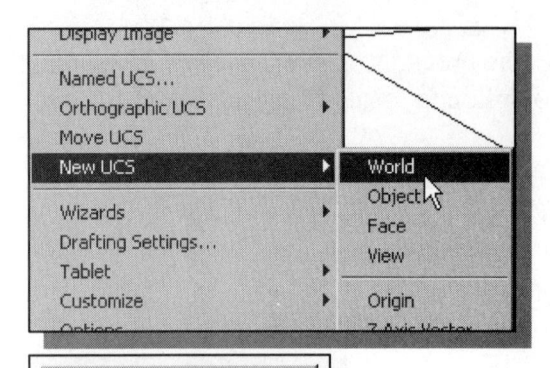

1. Set the UCS to the ***World UCS*** by selecting:
   [Tools] → [New UCS] → [World]

- The default 2D views generated in *paper space* are oriented based on the *World Coordinate System*. It is always a good idea to align the UCS to the WCS prior to entering the AutoCAD *paper space*.

2. On your own, reset the display to **2D Wireframe**.

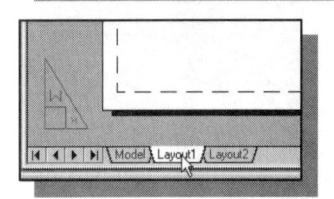

3. Click the **Layout1** tab to switch to the two-dimensional paper space.

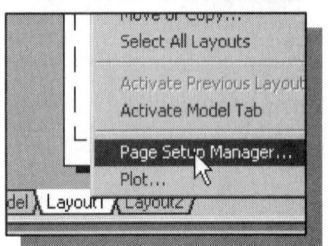

4. To adjust any *Page Setup* options, right-mouse-click once on the tab and select **Page Setup Manager**.

5. Choose the default layout in the *Page Setup Manager*.

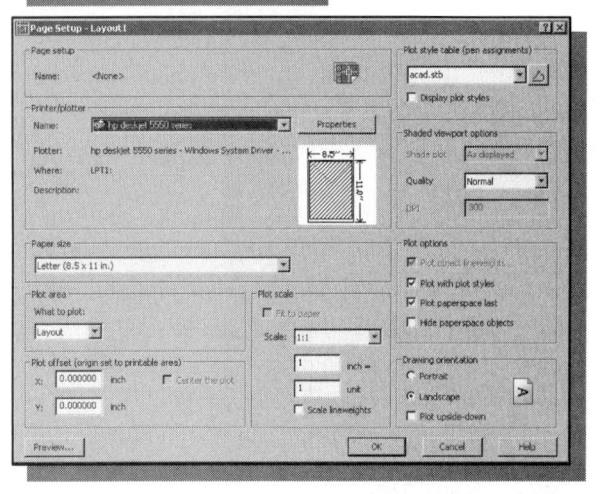

6. In the *Page Setup* dialog box, select a plotter/printer that is available to your system. Consult with your instructor or technical support personnel if you have difficulty identifying the hardware.

7. Confirm the *Paper size* is set to **Letter** or equivalent (8.5" by 11") and the *Drawing orientation* is set to **Landscape**.

8. Click **OK** button to accept the settings.

Multiview Drawings from 3-D Models   8-5

❖ In the graphics window, a rectangular outline on a gray background indicates the paper size. The dashed lines displayed within the paper indicate the *printable area*. The 3D model appears inside a solid rectangle, which is known as a **viewport**. Note that the orientation of the model is identical to that of the model space.

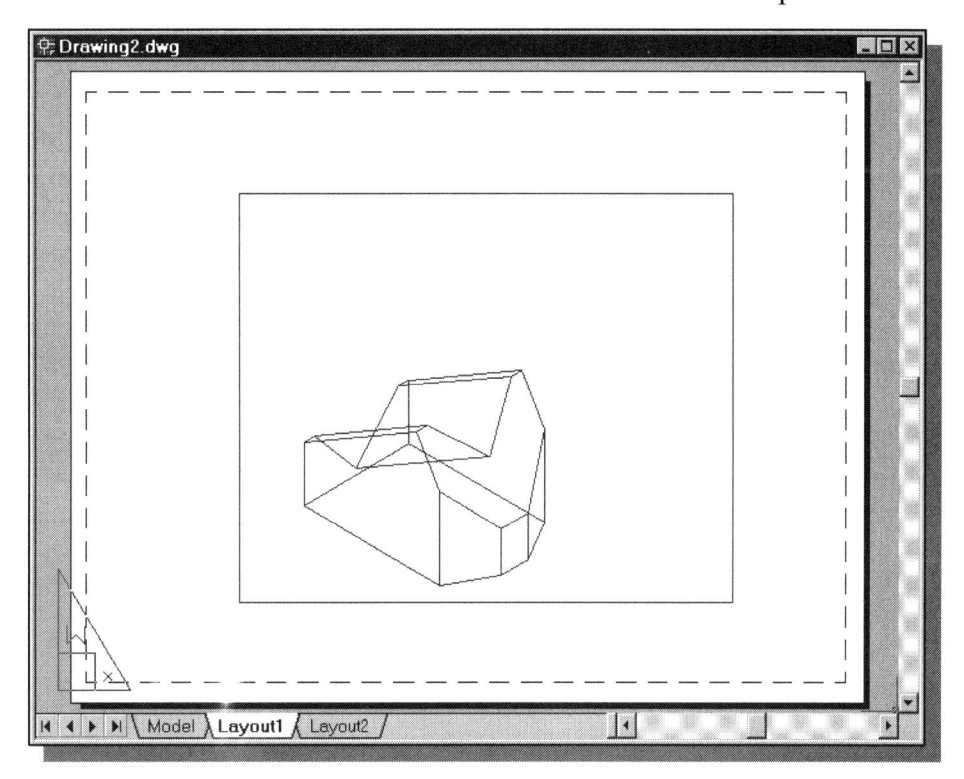

Deleting the Displayed Viewport

- Viewports can be considered as objects with a view into model space that we can move and resize. Floating viewports can be overlapping or separated from one another. We can also create new viewports, as well as delete them.

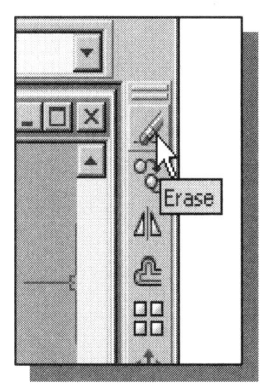

1. Pick **Erase** in the *Modify* toolbar.

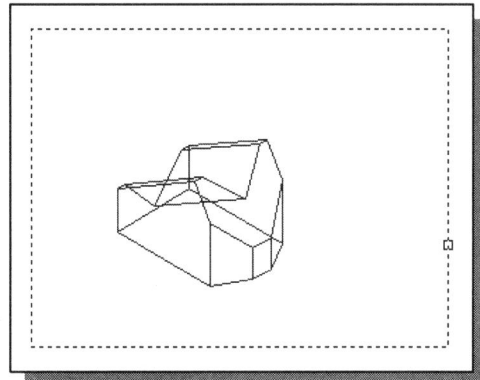

2. The message "*Select objects*" is displayed in the command prompt area and AutoCAD awaits us to select the objects to erase. Pick the viewport by clicking on one of the edges of the viewport border.

3. Inside the graphics window, **right-mouse-click** to accept the selection and delete the viewport.

Adding Borders and Title Block in the Layout

> **AutoCAD® 2006** allows us to create plots/prints to any exact scale on the paper. We can place borders and title blocks, the objects that are less critical to our design, on *paper space*.

1. On your own, create two new layers and set *Title_Block* as the *Current Layer*.

Layer	Color	LineType	Lineweight	PlotStyle
Title_Block	**Green**	**Continuous**	**1.2mm**	**Normal**
Dimensions	**Magenta**	**Continuous**	**Default**	**Normal**
Viewport	**White**	**Continuous**	**Default**	**Normal**

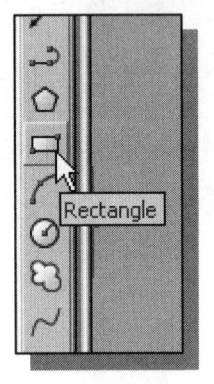

2. Select the **Rectangle** icon in the *Draw* toolbar. In the command prompt area, the message "*Specify first corner point or [Chamfer/Elevation/Fillet/Thickness/Width]:*" is displayed.

3. Pick a location that is on the inside and near the lower left corner of the dashed rectangle.

4. In the command prompt area, use the *relative coordinate entry method* and create a **10.25″ × 7.75″** rectangle.

5. Complete the title block as shown in the below figure. (Place the *text* in the *Dimensions* layer.)

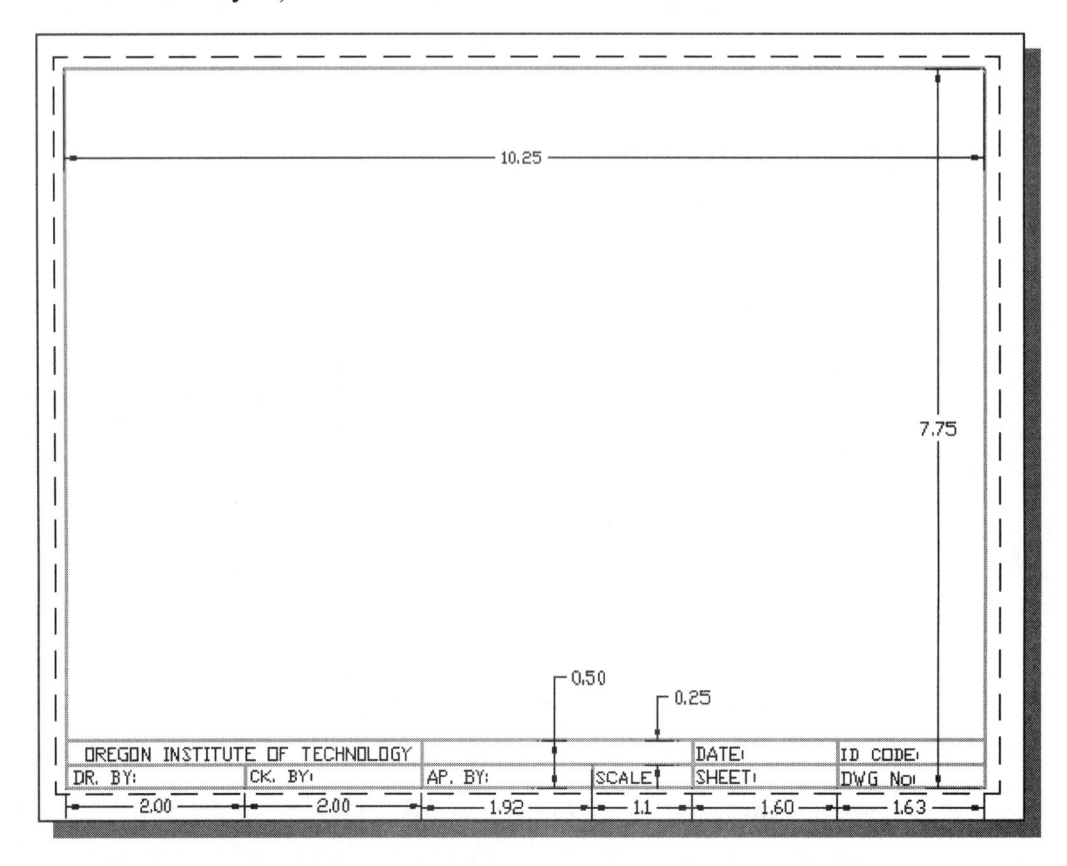

Setting Up Viewports inside the Titleblock

- Viewports can be used in both model mode and in layout mode. In *model mode*, only *tiled viewports* (viewports cannot overlap each other) can be created. In *layout mode*, *floating viewports* (viewports can overlap each other) can be created to set up standard 2D views from 3D models.

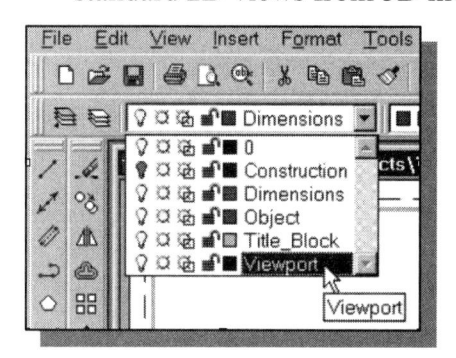

1. In the *Object Properties* toolbar area, select the *Layer Control* box and set layer *Viewport* layer as the *Current Layer*.

2. In the pull-down menus, select:

   [View] → [Viewports] → [4 viewports]

3. In the command prompt area, the message "*Specify first corner of viewport or [Fit] <Fit>:*" is displayed. Pick a corner near the lower left corner of the created border as shown in the figure below.

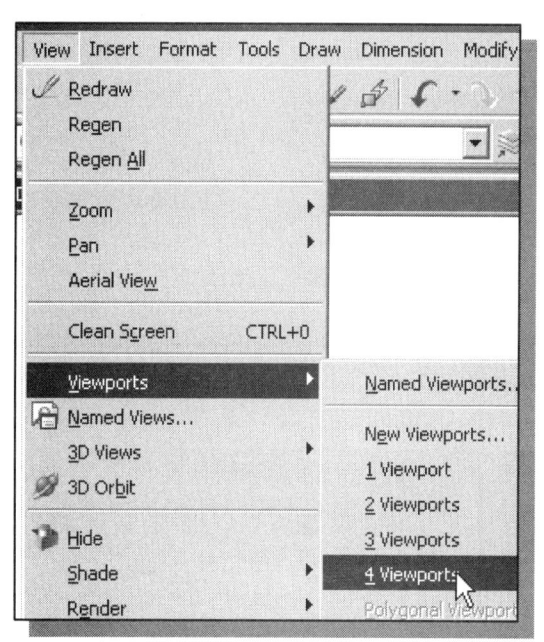

4. In the command prompt area, the message "*Specify opposite corner:*" is displayed. Pick a corner near the upper right corner of the border as shown.

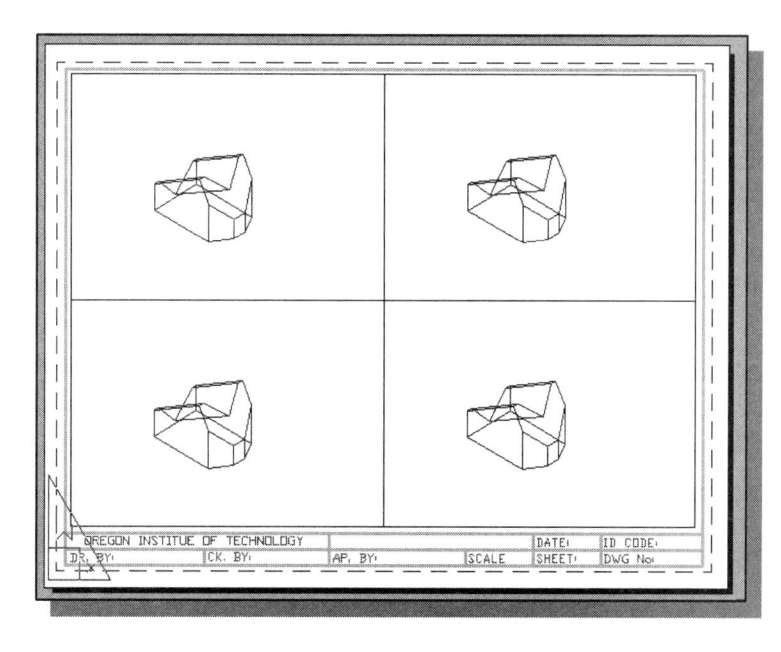

- Four viewports, with rectangular viewport borders, displaying the same isometric view are created.

Setting Up the Standard Views

➢ AutoCAD allows us to control the display of the model inside the created viewports. This is accomplished by switching to *model mode* within each viewport. In *layout mode*, either the *paper space* is activated or one of the created viewports is activated at one time. When a viewport is activated, the *model space* of the viewport is opened, and the model mode commands become available within the viewport. The viewport with the darker border is the active viewport.

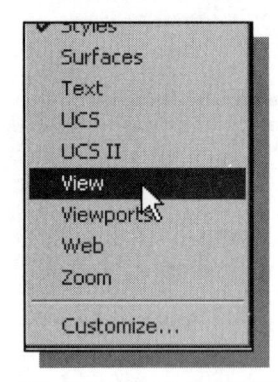

1. Move the cursor on top of any icon and right-click on any icon of the *Standard* toolbar to display a list of toolbar menu groups.

2. Select **View**, with the left-mouse-button, to display the *View* toolbar on the screen.

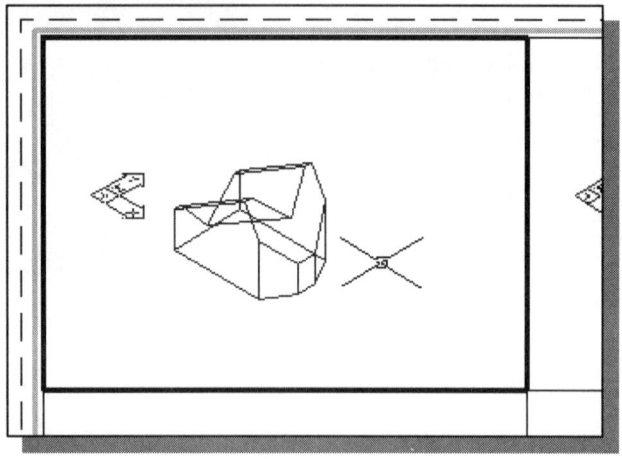

3. Move the cursor inside the upper left viewport.

4. Double-click with the left-mouse-button to activate the viewport.

 • Note the darker border and the cursor display of the active viewport.

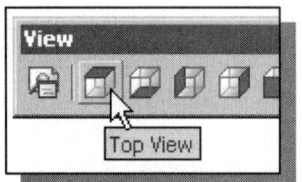

5. Pick **Top View** in the *View* toolbar as shown.

 • The display is changed to the top view based on the current User Coordinate System, which is aligned to the WCS.

6. Pick **Pan Realtime** in the *Standard* toolbar as shown.

7. On your own, reposition the model so that it is roughly near the center of the viewport.

 • Note that the **Pan Realtime** command only adjusts the display within the active viewport. Model space commands are only available in the active viewport.

- On your own, experiment with other display related commands, such as **3D Orbit** and **Zoom**.

8. Move the cursor inside the lower right viewport. Left-click once inside the viewport to activate it.

9. On your own, repeat the above steps and set the four viewports to display the standard four engineering views, as shown in the figure below. (The standard four engineering views are: **top**, **front**, **right**, and **isometric**)

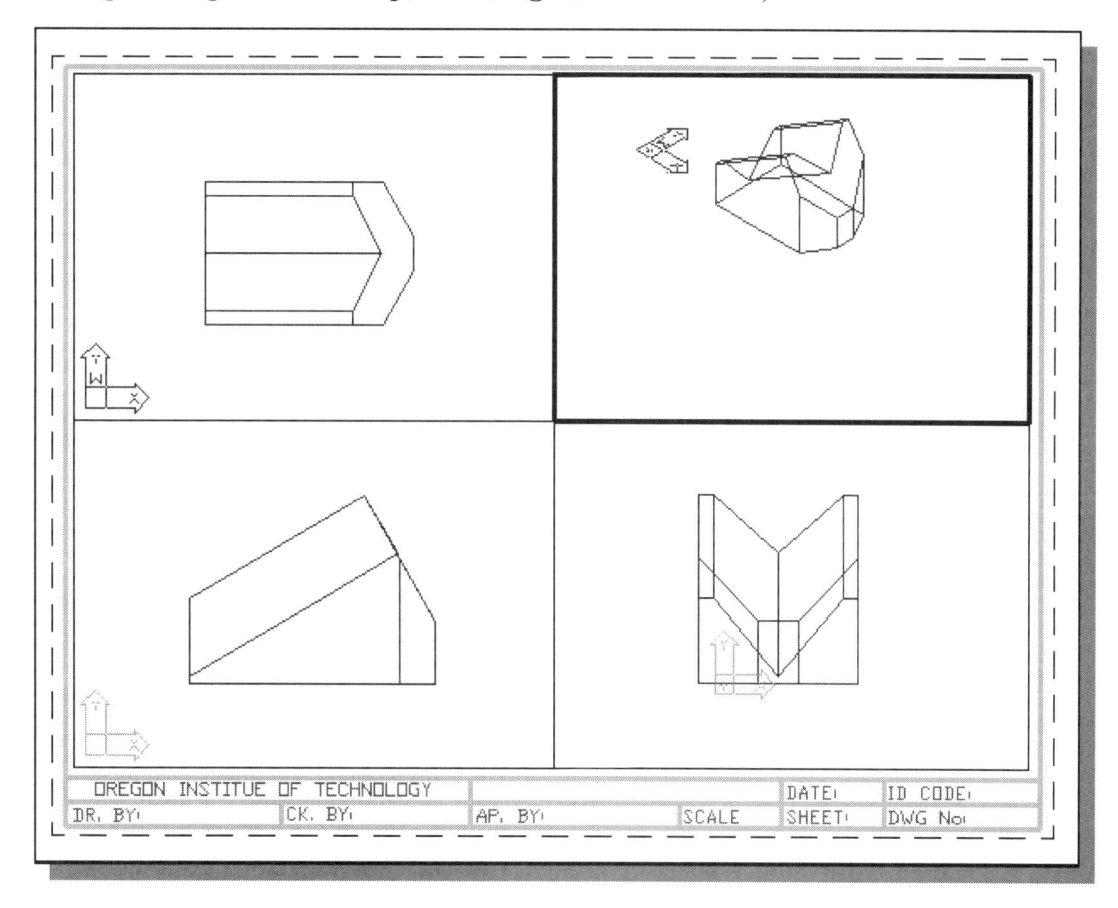

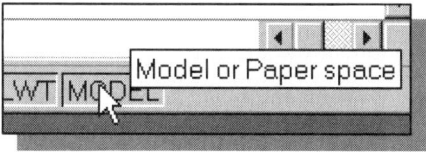

10. In the status bar area, switch back to the *paper space* by left-clicking once on *MODEL*.

11. Click on *PAPER* to switch back to *model space* and note the last active viewport is activated again.

- The 3D model is displayed in all viewports; note that there is only one 3D model. The multiple viewports simply allow us to view the same model from different angles simultaneously. This approach assures the correctness and accuracy of the 2D views of the model since the views are generated from the database of the same 3D model.

Determining the Necessary 2D Views

- Multiview drawings usually require two or three of the regular views, such as a front view, a top view and/or a side view. Many designs have features located on inclined surfaces that are not parallel to the regular planes of projection. For the *V-block* design, to truly describe the V-cut feature, the true shape of the feature must be shown using an **auxiliary view**. An *auxiliary view* has a line of sight that is perpendicular to the inclined surface, as viewed looking directly at the inclined surface. An *auxiliary view* is a supplementary view that can be constructed from any of the regular views.

➢ Examining the current four standard views, we can conclude that the right side view provides no crucial information on the design. Instead, an *auxiliary view* is needed to truly describe the V-cut feature.

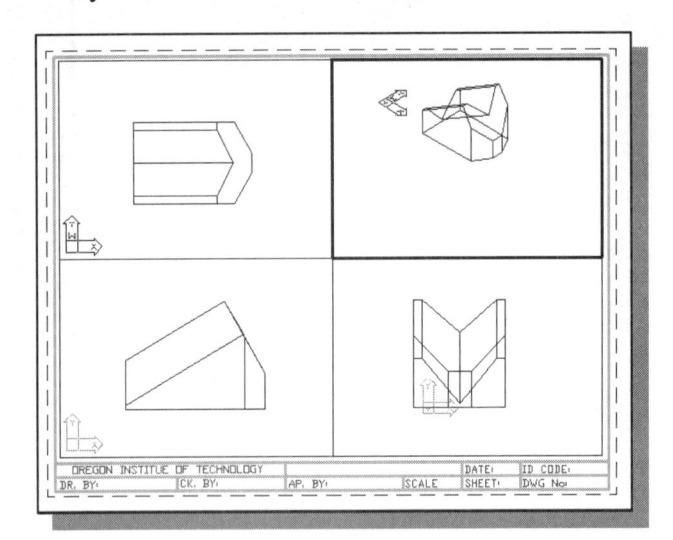

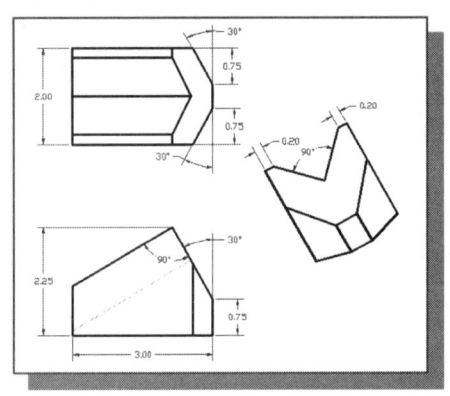

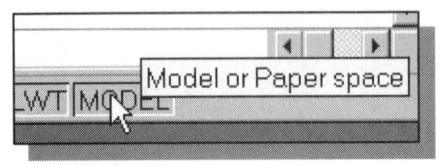

1. In the *Status Bar* area, switch back to the *paper space* by left-clicking once on *MODEL*.

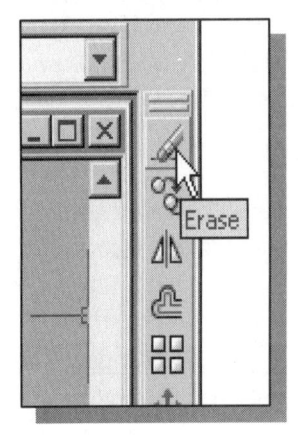

2. Pick **Erase** in the *Modify* toolbar.

3. The message "*Select objects*" is displayed in the command prompt area and AutoCAD awaits us to select the objects to erase. Pick the lower right viewport by clicking on one of the edges of the viewport border.

4. Inside the graphics window, right-mouse-click to accept the selection and delete the viewport.

Establishing an Auxiliary View in Model Mode

➢ AutoCAD provides us with many options to establish new viewpoints. The easiest approach is to create a new UCS aligned to the inclined plane of the design and then adjust the display view to the UCS.

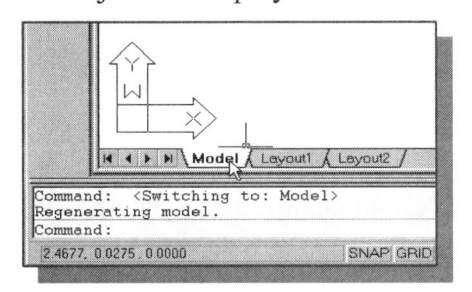

1. Pick the **Model** tab to switch back to model space.

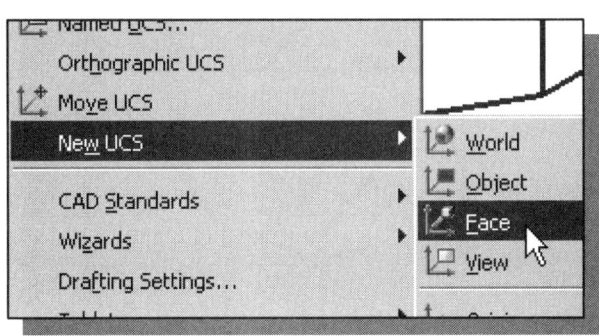

2. In the pull-down menus, select:

   [Tools] → **[New UCS]** → **[Face]**

3. In the command prompt area, the message "*Select face of solid object:*" is displayed. Pick the **inclined face** by clicking on the surface near the bottom horizontal edge of the surface.

4. In the command prompt area, the message "*Enter an option [Next/Xflip/Yflip] <accept>:*" is displayed. Press the **[ENTER]** key once to accept the new UCS positioned as shown.

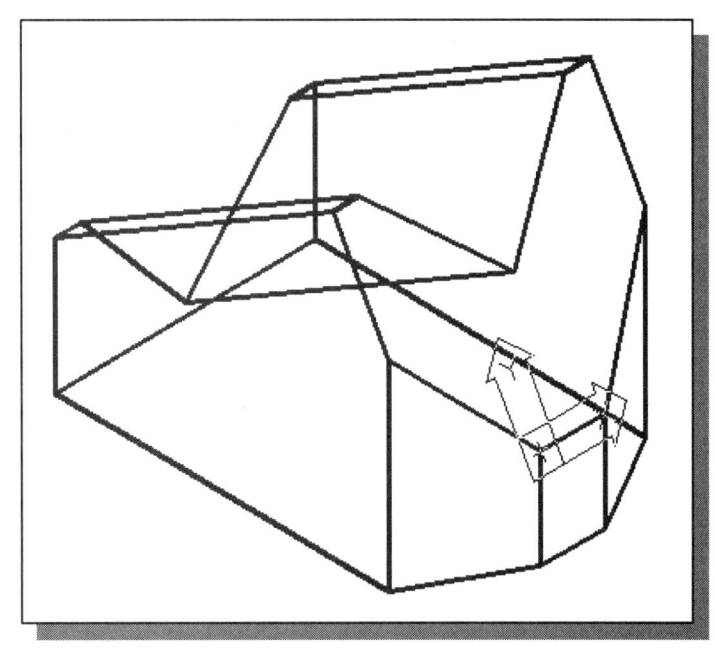

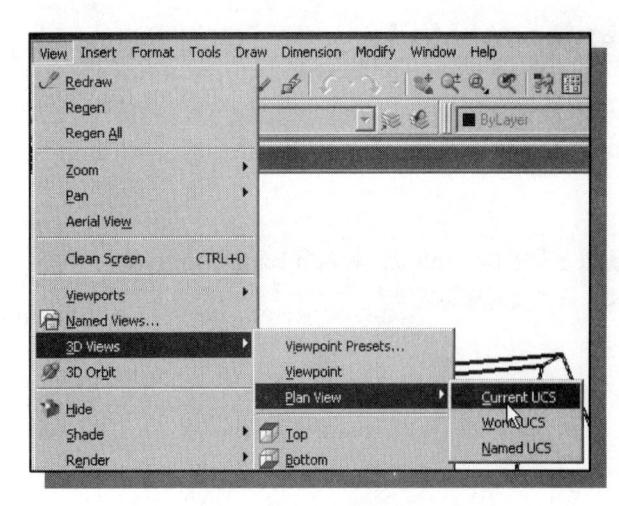

5. In the pull-down menus, select

   [View] → [3-D Views] → [Plan View] → [Current UCS]

   - The graphics window is now adjusted to display the current UCS plane.

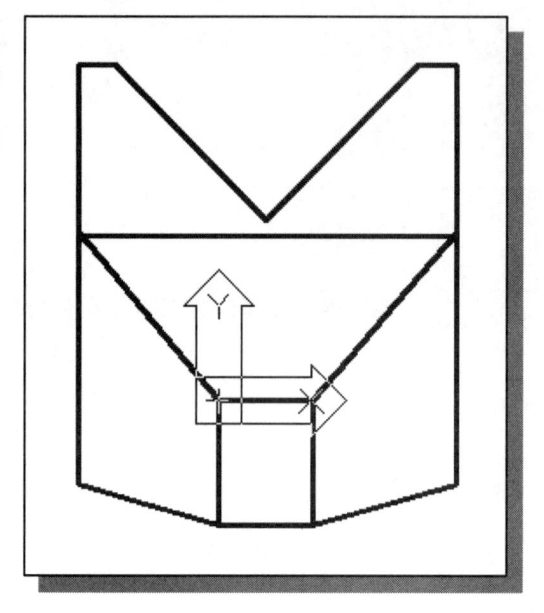

- Note the orientation of the UCS in the display view. This is the auxiliary view that is needed in the multiview drawing.

6. Click on the **Named Views** icon in the *View* toolbar.

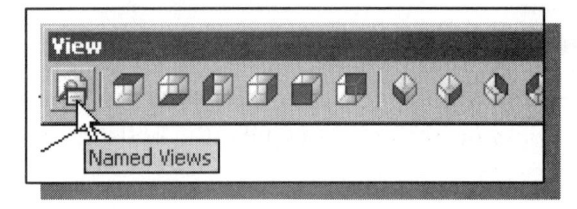

7. In the *View* dialog box, click on the **New** button to create a new named view.

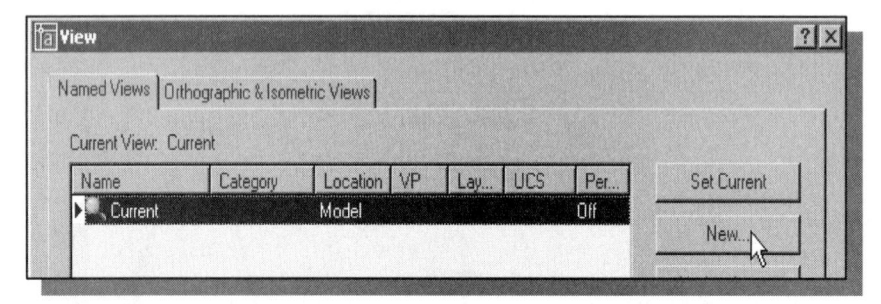

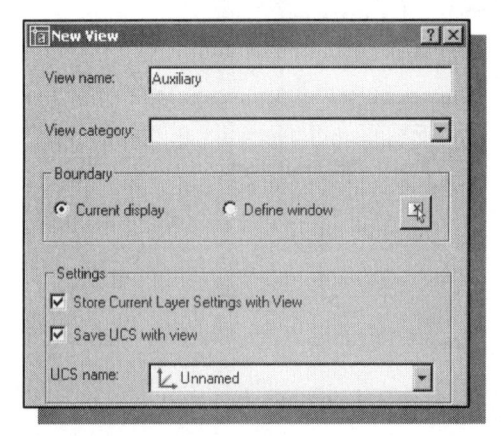

8. In the *New View* dialog box, enter **Auxiliary** as the new *View name* and confirm the *Current display* option and the *Save UCS with view* option are switched *ON* as shown.

9. Click on the **OK** button to accept the settings and create the named view.

10. In the *View* dialog box, the new named view (*Auxiliary*) is added to the *Named Views* list. We can also switch to this view by using the **Set Current** option.

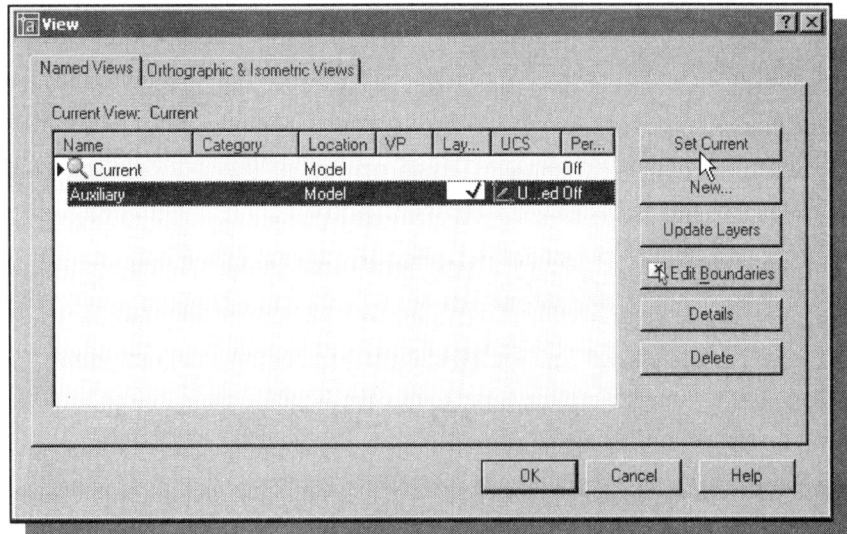

11. Click on the **OK** button to accept the settings.

Adding a Viewport for an Auxiliary View

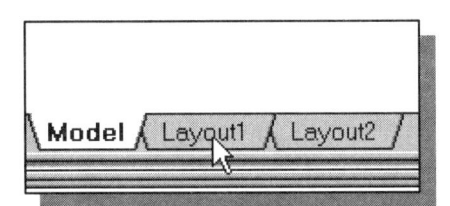

1. Pick the **Layout1** tab to switch back to the two-dimensional paper space.

2. In the status bar area, reset the option buttons so that only *LWT* and *PAPER* are switched *ON*.

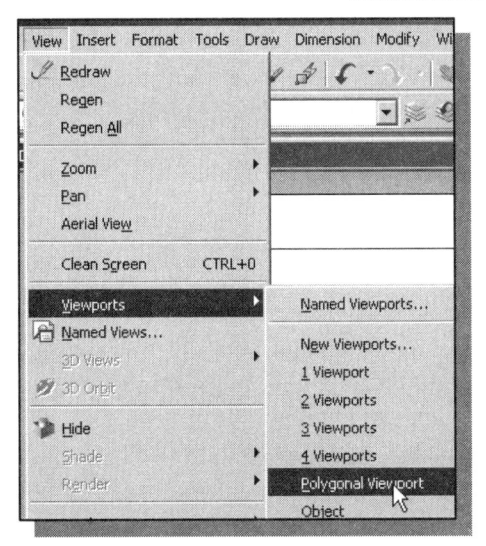

3. In the pull-down menus, select

   [View] → [Viewports] → [Polygonal Viewport]

   - In **AutoCAD® 2006**, viewports can be of arbitrary polygonal shapes. In the layout mode, the viewports can also overlap each other. We will create a four-sided viewport overlapping the other viewports.

4. In the command prompt area, the message "*Specify start point:*" is displayed. Create a four-sided shape as shown. (Hint: use the **Close** option to assure the last segment is connected to the start point.)

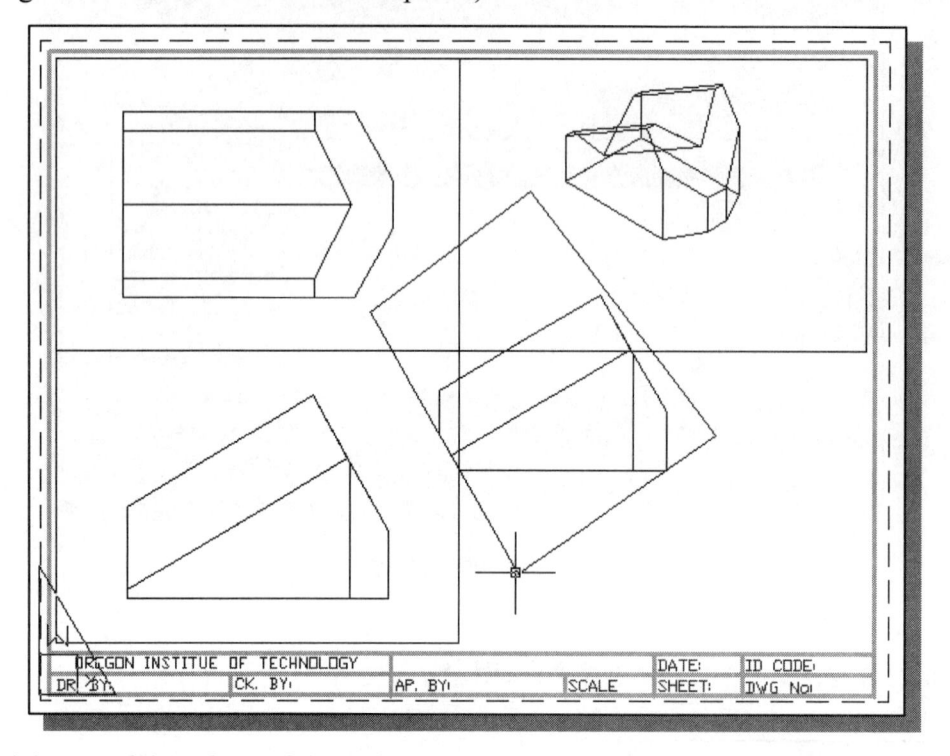

5. Pick one of the edges of the viewport border and notice that four grip points are displayed.

6. Adjust the size and shape of the viewport by dragging the grip points.

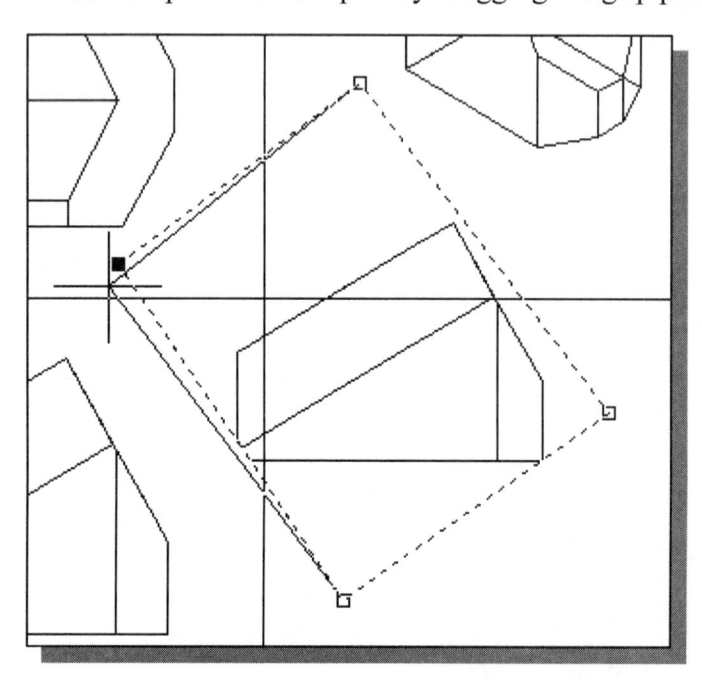

7. Press the **[Esc]** key once to deselect the highlighted objects.

8. Double-click inside the viewport we just created to activate the viewport.

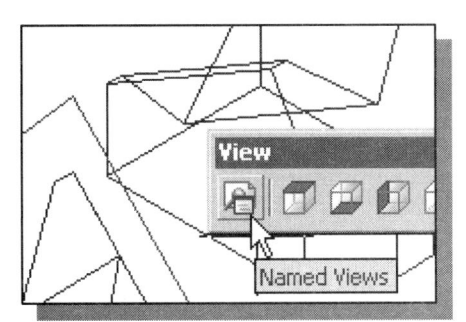

9. Pick **Named Views** in the *View* toolbar.

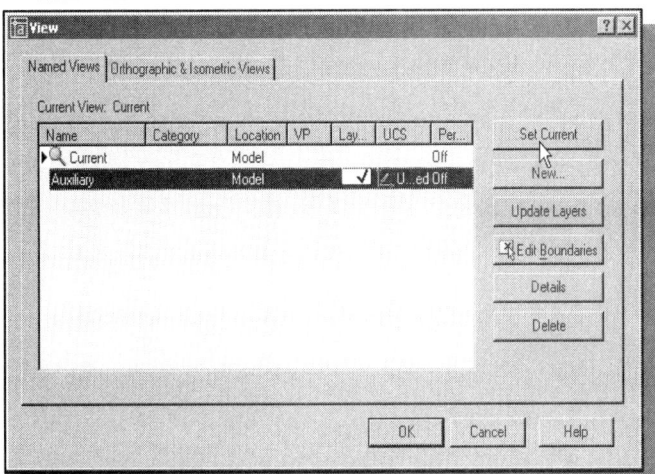

10. Pick the *Auxiliary* view in the *Named Views* list as shown.

11. Click on the **Set Current** to change the current display to the saved *Auxiliary* view.

12. Click **OK** to accept the settings.

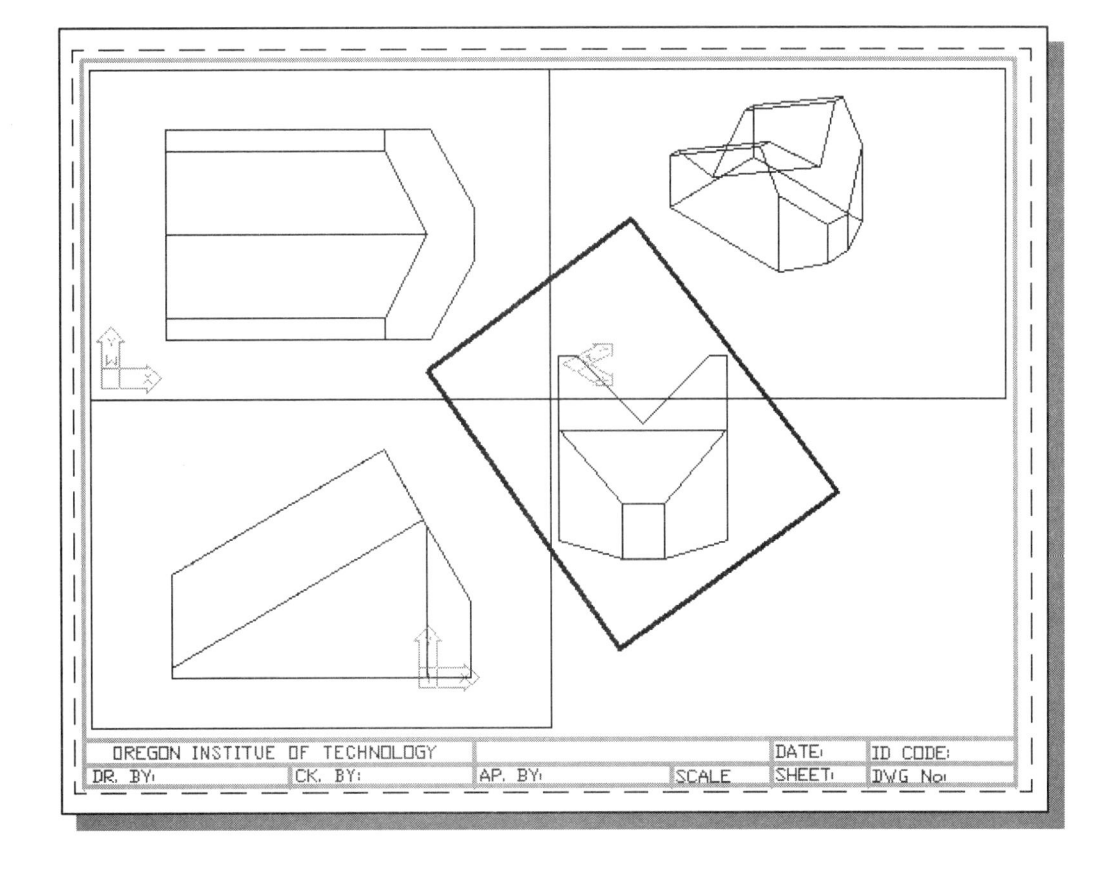

Using the DVIEW Command

- We can use the DVIEW command to zoom, pan, and twist views. We can also use DVIEW to remove objects from in front of and behind a clipping plane and to remove hidden lines during a dynamic viewing operation.

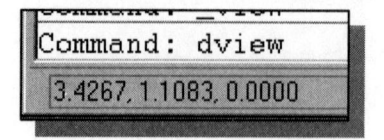

  ```
  Command: dview
  3.4267, 1.1083, 0.0000
  ```

  1. At the command prompt, enter: **dview** [ENTER]

  2. In the command prompt area, the message "*Select objects or <use DVIEWBLOCK>:*" is displayed. Pick the solid model inside the active viewport.

  3. Inside the active viewport, right-mouse-click once to accept the selection.

  4. In the command prompt area, the message "*Enter option:*" is displayed. Choose the **Twist** option by entering: **TW** [ENTER]

  5. In the command prompt area, the message "*Specify the view twist angle <0.00>:*" is displayed. Enter: **30** [ENTER]

  6. Press the [**ENTER**] key once to end the DVIEW command.

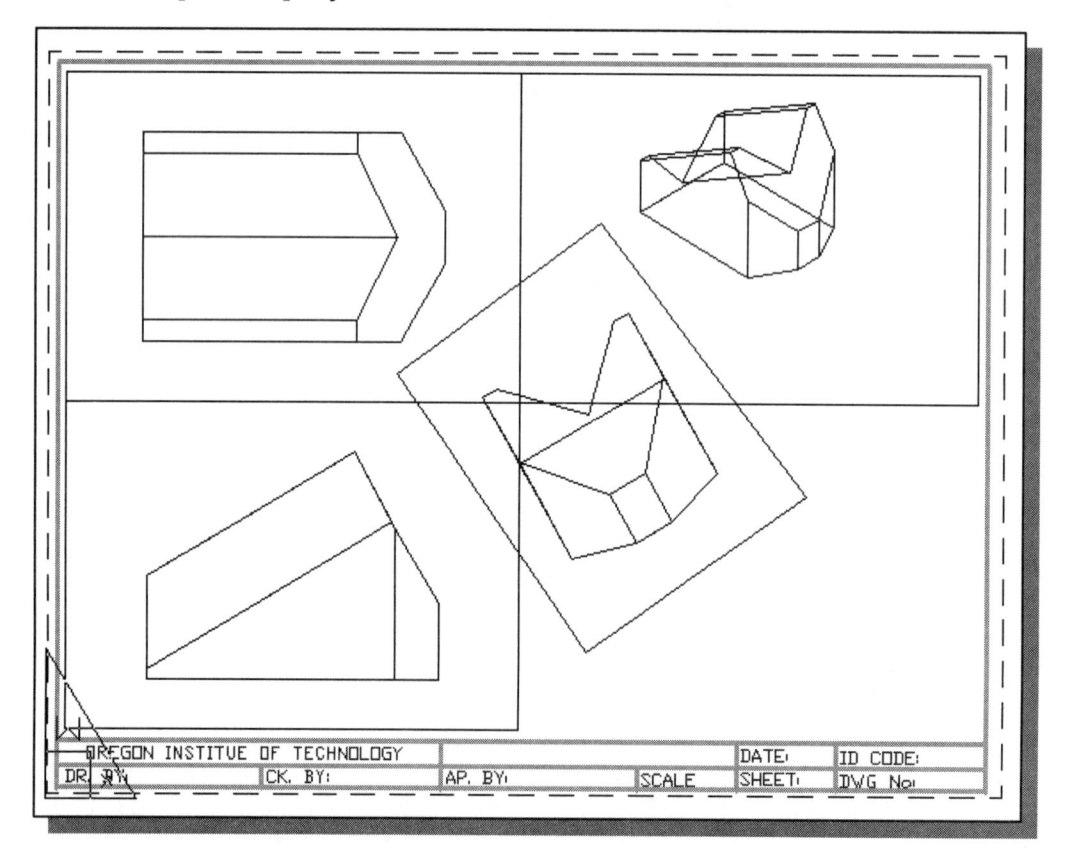

Adjusting the Viewport Scale

- A uniform scale for the 2D views is necessary for view alignments and dimensioning the features.

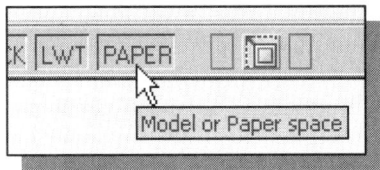

1. Left-mouse-click in the status bar area, switch back to the *paper space* mode.

2. Pick **Properties** in the *Standard* toolbar as shown.

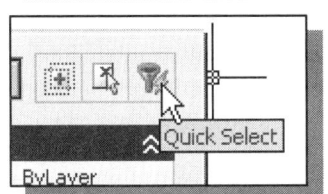

3. In the *Properties* window, click on the **Quick Select** icon.

- The **Quick Select** option allows us to select items based on the object properties.

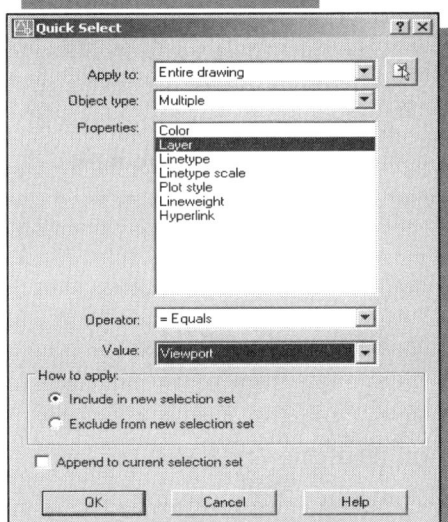

4. In the *Quick Select* window, choose **Layer** in the *Properties* list and set the selection operation to **Equal Viewport** as shown in the figure.

5. Click **OK** to accept the setting and proceed to select objects matching the quick select settings.

6. Choose the 4 viewports in the selection list as shown.

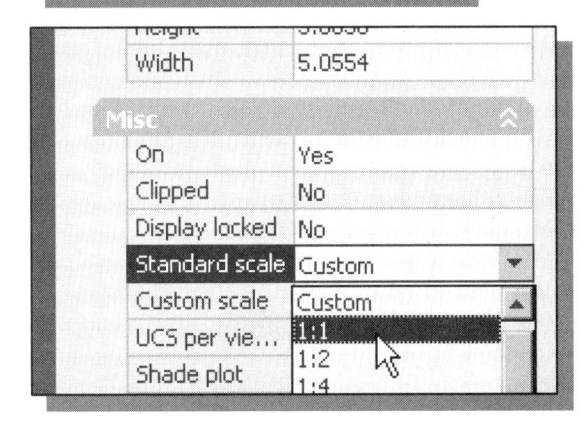

7. Pick the *Standard scale* property in the *Misc* list as shown.

8. Select **1:1** to set the scale to one to one. This will set all displayed views to *full scale*.

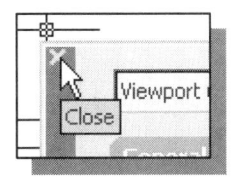

9. Click on the [**X**] button to close the dialog box.

Locking the Base View

1. Pre-select the lower left viewport by clicking on one of the viewport borders.

2. Pick **Properties** in the *Standard* toolbar as shown.

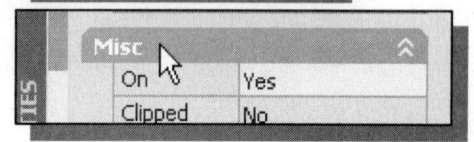

3. Find the **Misc** tab in the Properties Palette as shown.

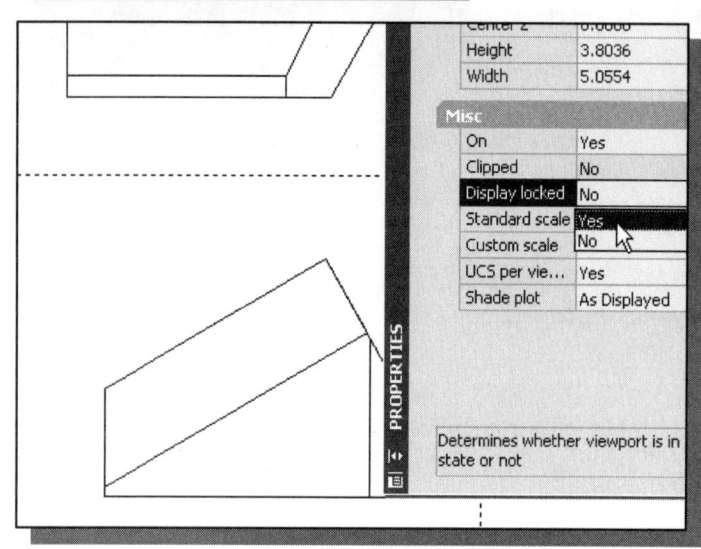

4. Pick *Display Locked* option in the *Property* lists.

5. Set the option to **Yes** as shown.

6. Close the *Properties* window by clicking on the [**X**].

- By setting this option, the displayed view is locked at its current location. We will use this view as the base view.

Aligning the 2D Views

- We will create two construction lines to assist the alignments of the 2D views.

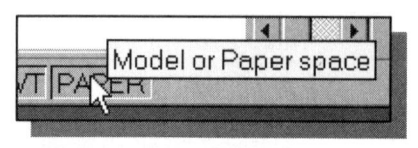

1. Confirm that the *paper space* is active. Click on the *MODEL* button if it is not set to the *paper space* as shown.

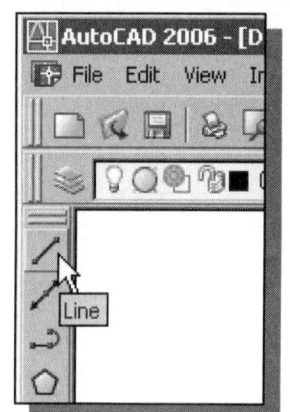

2. Click **Line** in the *Draw* toolbar as shown.

3. Switch *ON* and set the **Construction** layer as the *Current Layer* as shown.

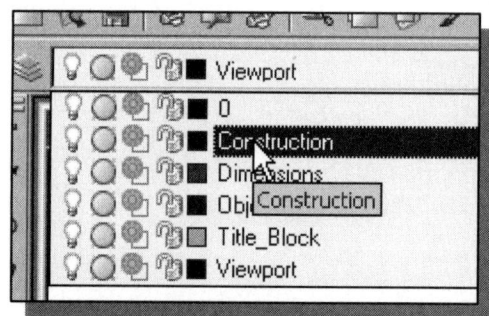

4. On your own, create two lines aligned to the upper left corner of the front view as shown. (Hint: turn *ON* the *OSNAP* and *OTRACK* options to help the constructions.)

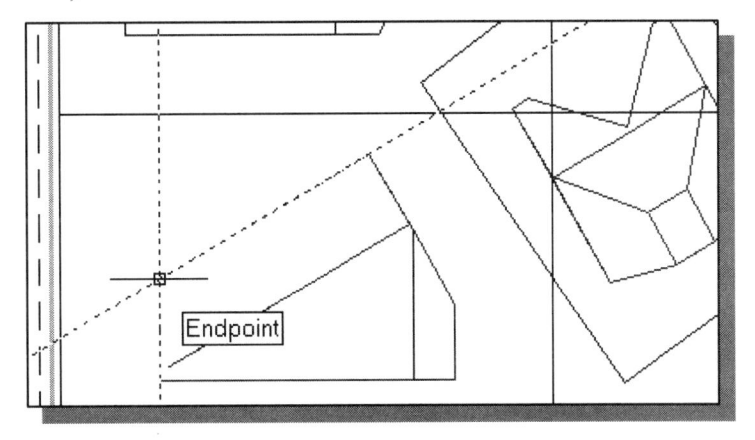

5. Double-click inside the upper left viewport to activate the viewport.

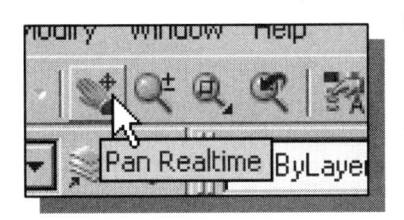

6. Pick **Pan Realtime** in the *Standard* toolbar as shown.

7. On your own, reposition the displayed top view so that it is aligned vertically to the front view. Use the vertical construction line to assist the alignment.

8. Repeat the above steps and reposition the *Auxiliary* view so that it is aligned to the front view as shown in the below figure.

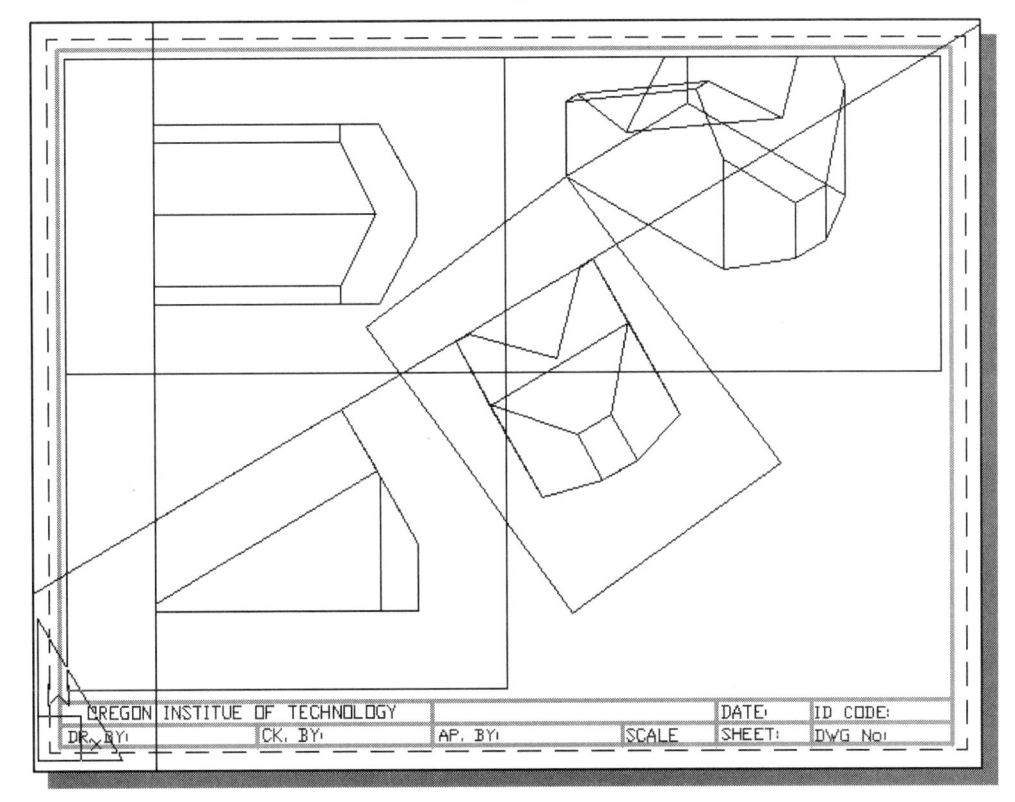

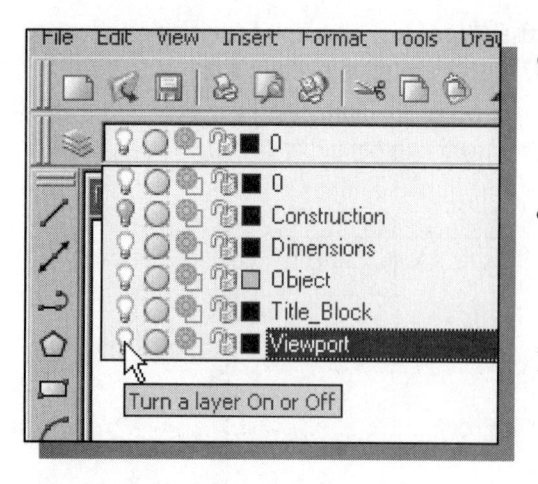

9. Turn **OFF** layer *Viewport* and layer *Construction* in the *Layer Control* box as shown.

- Note that the viewport borders and the two construction lines are switched off since they are placed in the layer *Viewport*.

10. Double-click on the inside of the isometric view and note the upper right viewport is activated.

11. Click **Zoom Realtime** in the *Standard* toolbar as shown.

12. On your own, resize and reposition the isometric view so that it does not overlap with the *Auxiliary* view.

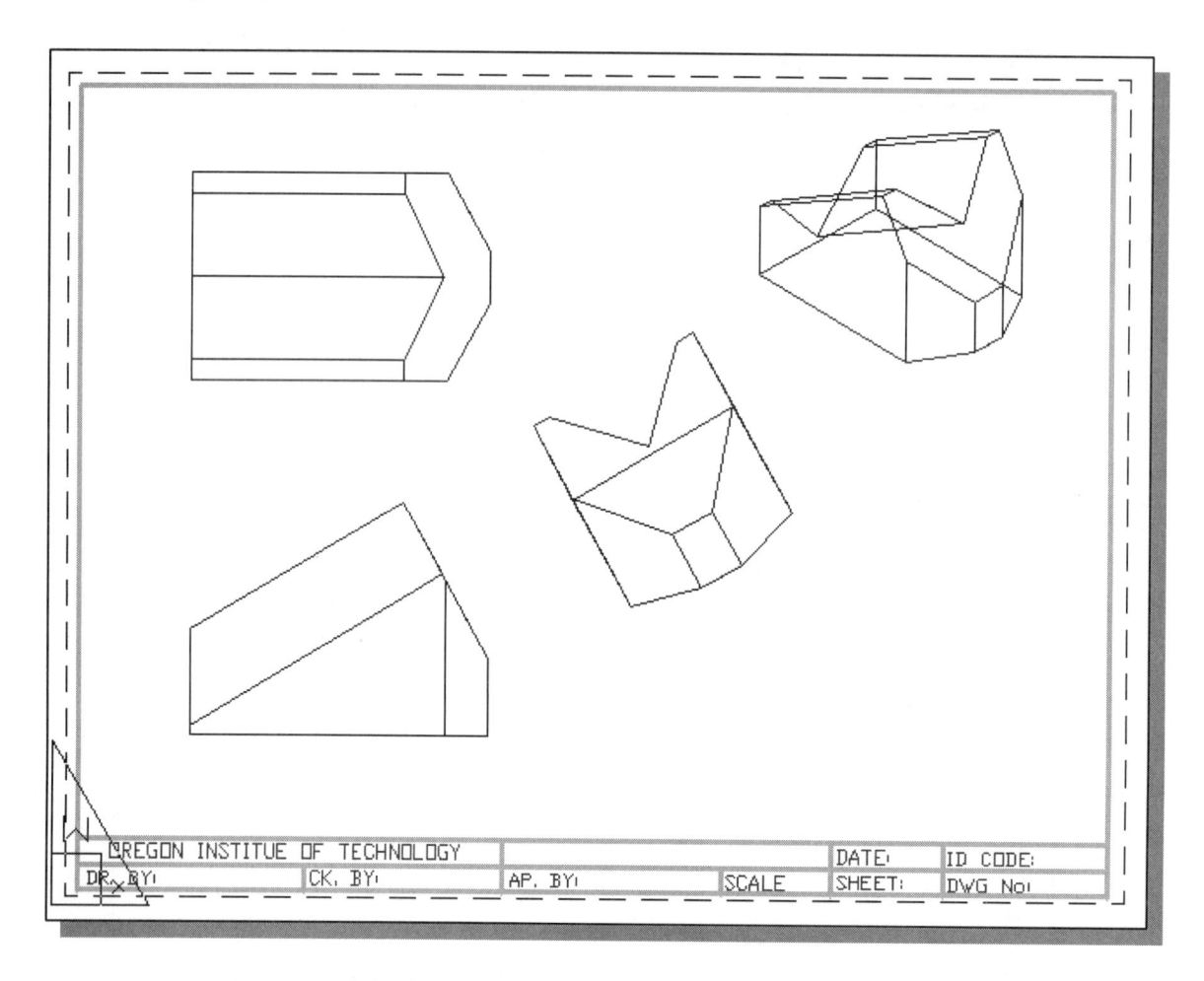

Creating 2D Projected Entities – SOLPROF

- The views created are still 3D in nature. This prevents us from setting color or linetype for individual edges, so that hidden features can be described correctly based on engineering drafting standards. To overcome that problem, we will generate 2D entities from the established viewports. We can then use layer controls to set color and linetype options, which allows us to distinguish visible and hidden objects in individual views.

    1. Double-click inside the lower left viewport to activate the viewport.

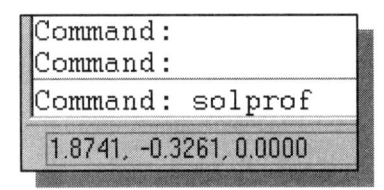

    2. At the *command prompt*, enter: **solprof** [ENTER]

    3. In the command prompt area, the message "*Select objects:*" is displayed. Pick the solid model inside the active viewport.

    4. Inside the active viewport, right-mouse-click once to accept the selection.

    5. In the command prompt area, the message "*Display hidden profile lines on separate layer? [Yes/No] <Y>:*" is displayed. Enter: **Y** [ENTER]

    6. In the command prompt area, the message "*Project profile lines onto a plane? [Yes/No] <Y>:*" is displayed. Enter: **Y** [ENTER]

    7. In the command prompt area, the message "*Delete tangential edges? [Yes/No] <Y>:*" is displayed. Enter: **Y** [ENTER]

- The display in the layout appears to be the same before the SOLPROF command was executed. Projected 2D entities (profile lines) are created in the view direction of the viewport.

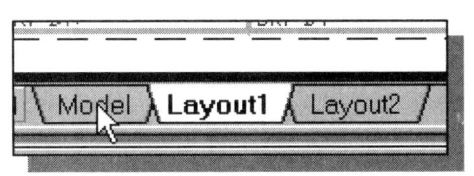

    8. Pick the **Model** tab to switch back to the 3D *model space*.

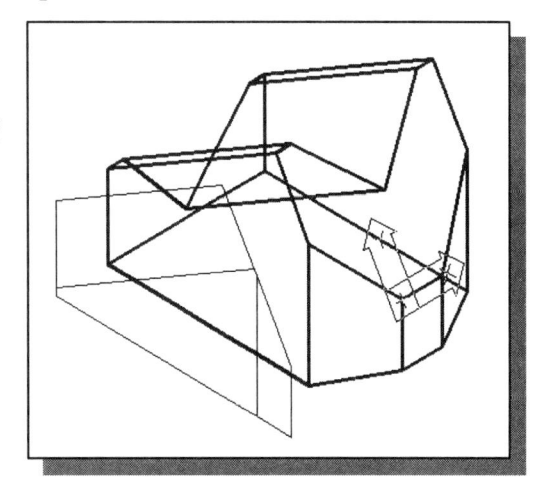

- The projected 2D entities appear in *model space*. Note that the locations of the projected entities might be at a different location on your screen than is displayed in the figure.

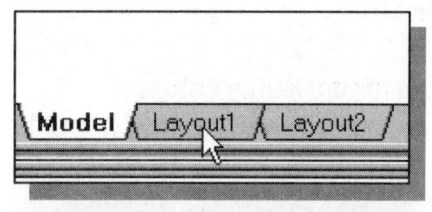

9. Pick the **Layout1** tab to switch back to the two-dimensional *paper space*.

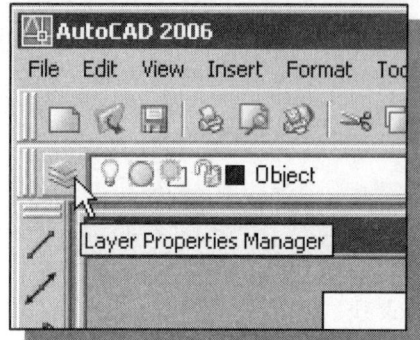

10. Click on the **Layers** icon in the *Object Properties* toolbar.

- Note that two additional layers are created by the **SOLPROF** command. The *PV-xx* layer is for the projected visible lines and *PH-xx* layer is for the projected hidden lines.

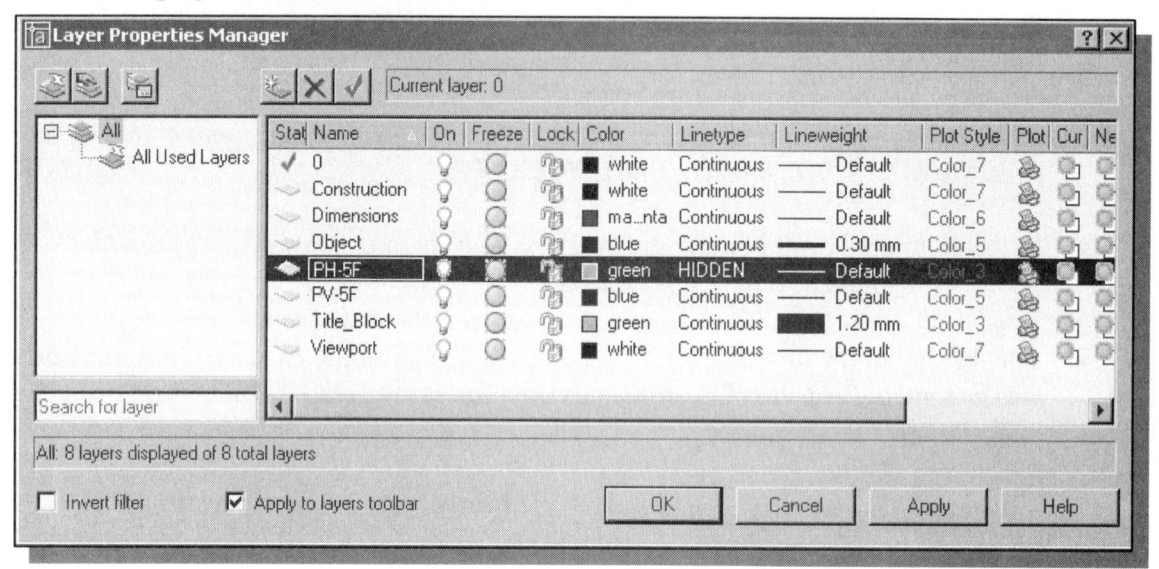

11. On your own, change the two layers to the following settings:

Layer	*Color*	*LineType*	*Lineweight*	*PlotStyle*
PH-xx	**Green**	**HIDDEN**	**Default**	**Normal**
PV-xx	**Blue**	**Continuous**	**0.3mm**	**Normal**

12. On your own, *Freeze* the layer *Object* to remove the solid object from the display.

13. Click the **OK** button to accept the settings. (Note that only the projected entities are displayed.)

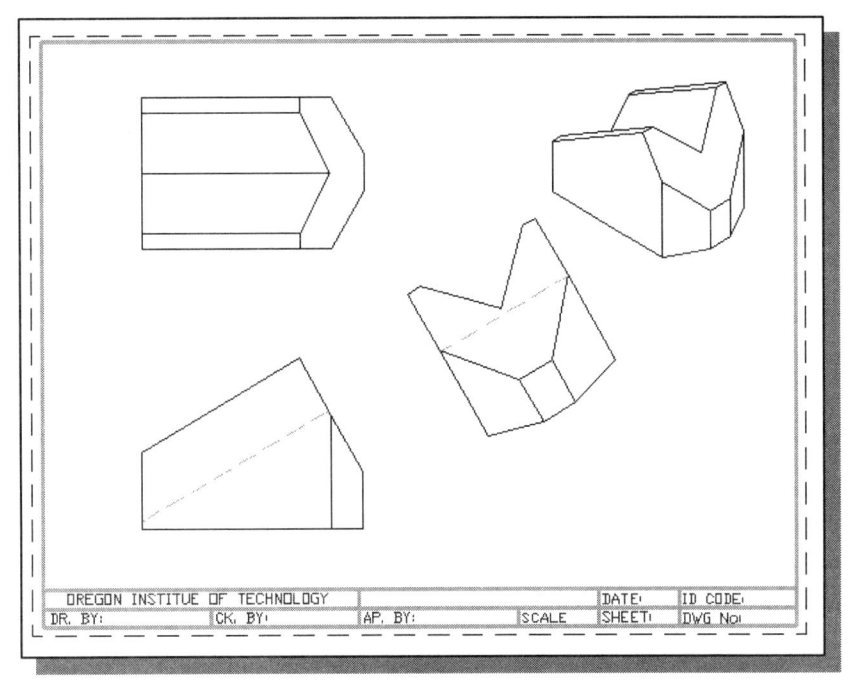

14. On your own, ***Thaw*** the *Object* layer and repeat the above steps to generate 2D projected entities for the other three views.

Completing the 2D Drawing

- Dimensions and notes can be placed in the *model space* or the *paper space*. From a design/modeling perspective, it is generally agreed that the dimensions should be placed in the 2D *paper space*.

1. Confirm the active mode is set to ***paper space*** as shown in the *Status Bar* area.

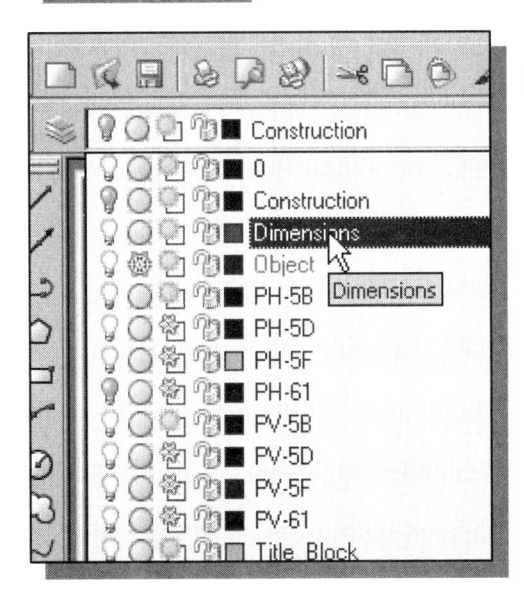

2. On your own, set layer **Dimensions** as the *Current Layer* in the *Layer Properties* box as shown.

3. Move the cursor to the *Standard* toolbar area and right-mouse-click on any icon in the *Standard* toolbar to display a list of toolbar menu groups.

4. Select **Dimension**, with the left-mouse-button, to display the *Dimension* toolbar on the screen.

5. On your own, create the necessary dimensions and complete the 2D drawing as shown.

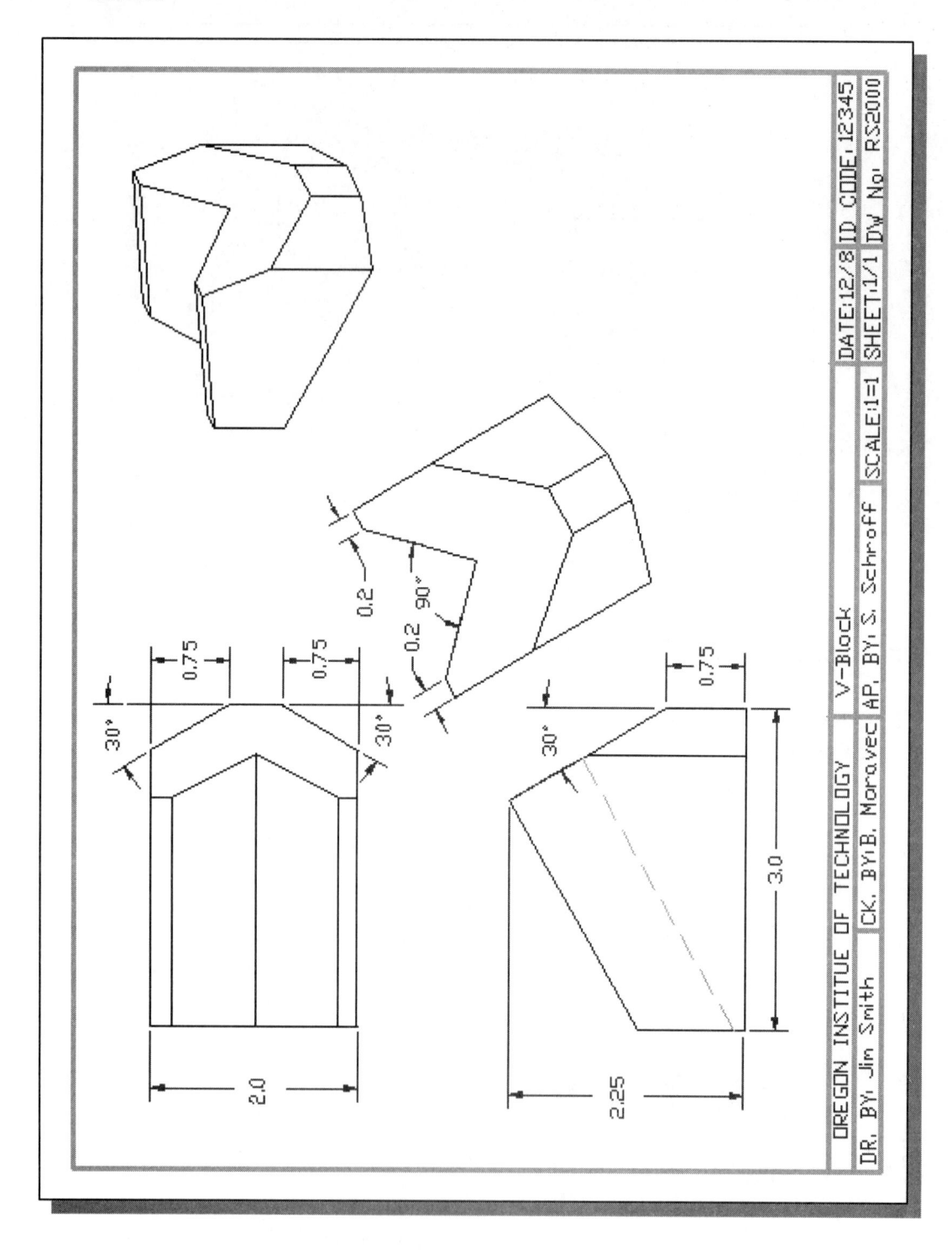

Questions:

1. What is an auxiliary view and why would it be important?

2. List and describe the types 2D views used in the lesson.

3. Why and when should you use the **SOLPROF** command?

4. What is the difference between *model space* and *paper space* in AutoCAD?

5. Why and when should you use the **DVIEW** command?

6. Identify the following commands:

   (a)

   (b)

   (c)

   (d)

Exercises: Create the following solid models and generate multiview drawings from the 3D models. (All dimensions are in inches.)

1.

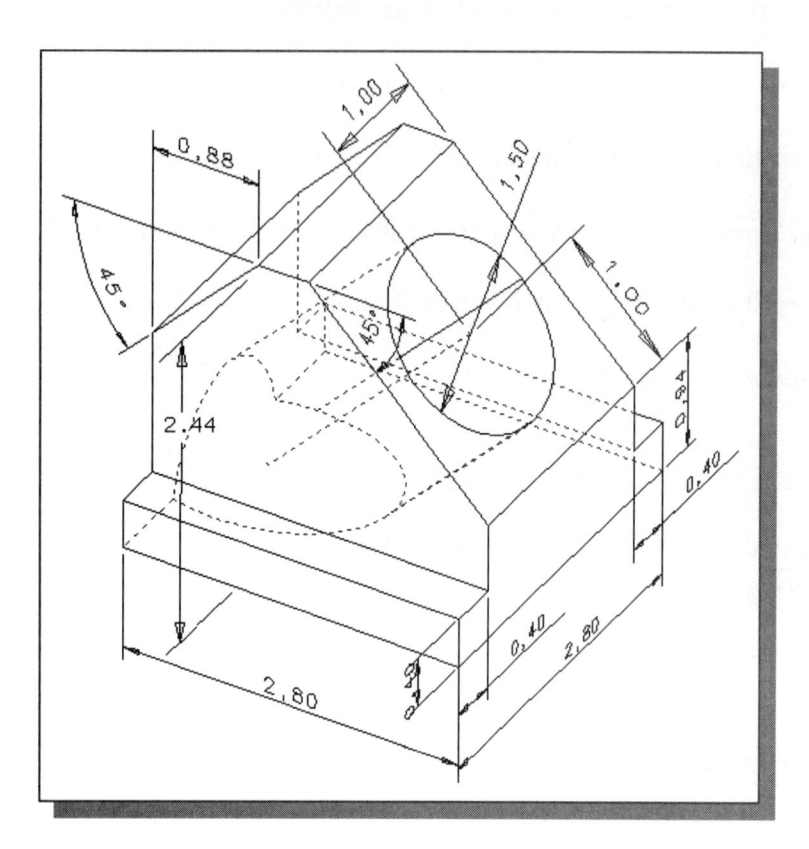

2.

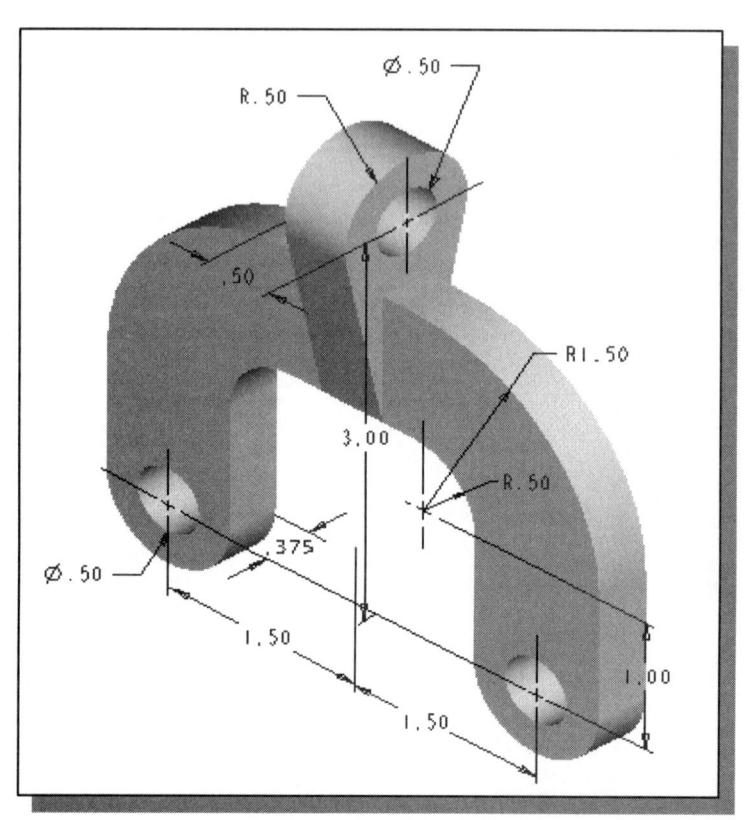

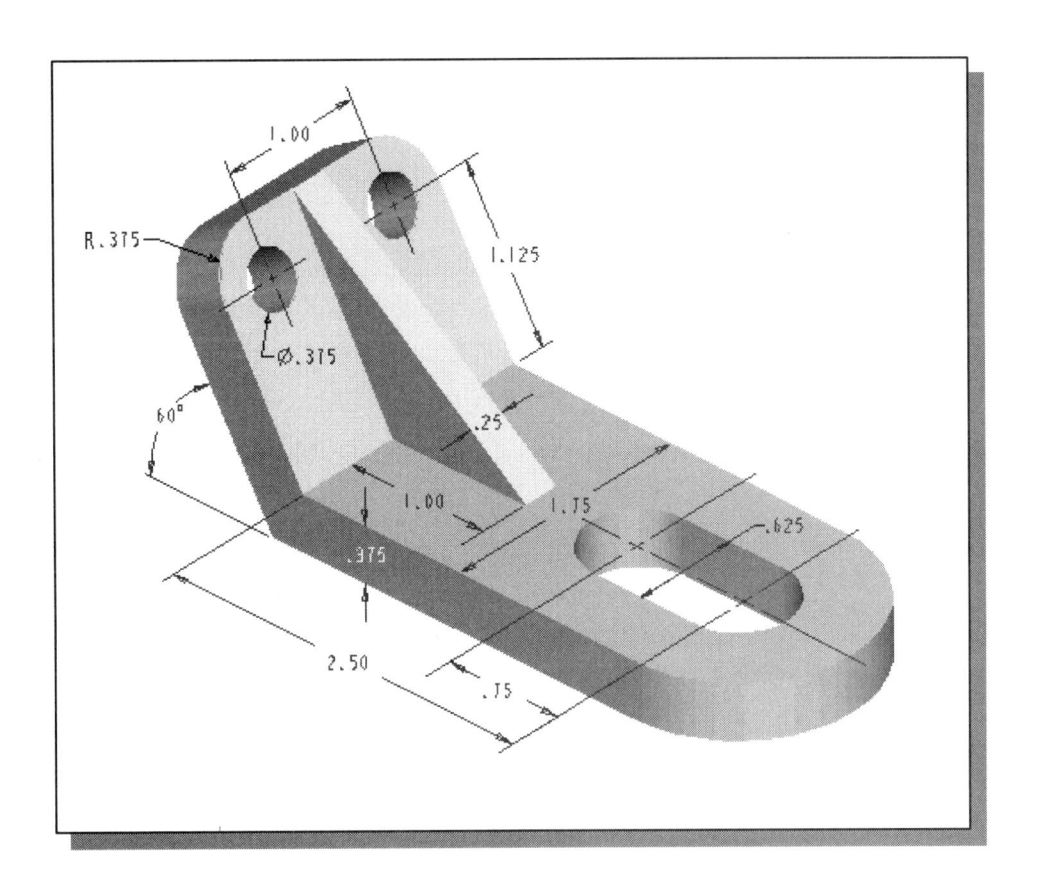

3.

NOTES:

Chapter 9
Symmetrical Features in Designs

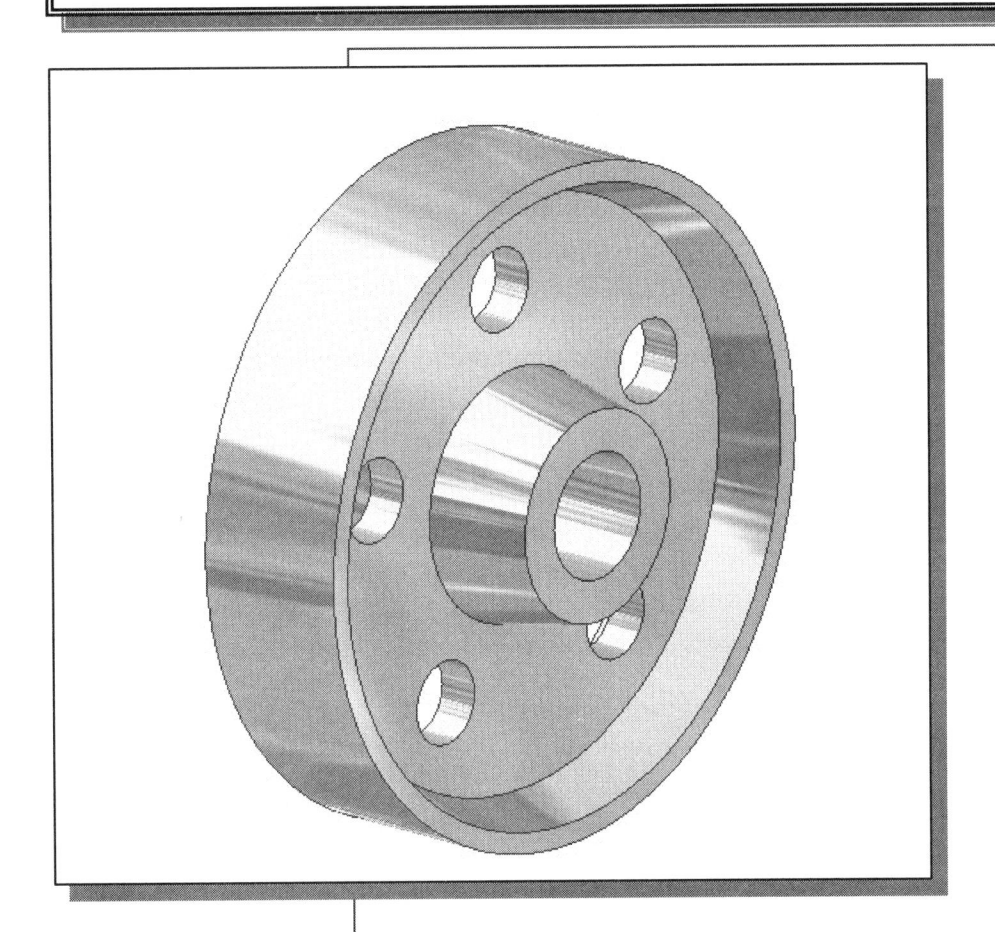

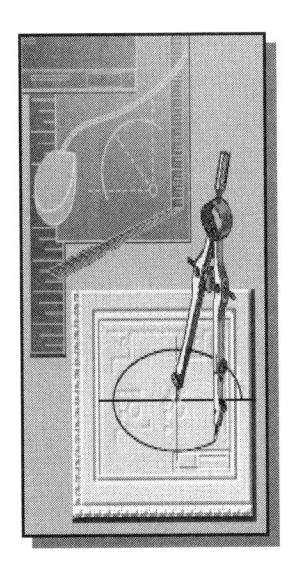

Learning Objectives

- **Create Revolved Features**
- **Use the Mirror Part Command**
- **Understand and Create Construction Geometry**
- **Create Combined Parts**
- **Create and Modify Feature Arrays**
- **Understand the Importance of Identifying Symmetrical Features in Designs**

Introduction

In solid modeling, it is important to identify and determine the features that exist in the design. *Modern solid modeling systems* enable us to build complex designs by working on smaller and simpler units. This approach simplifies the modeling process and allows us to concentrate on the characteristics of the design. Symmetry is an important characteristic that is often seen in designs. Symmetrical features can be easily accomplished by the assortments of tools that are available in feature-based modeling systems, such as AutoCAD.

The modeling technique of extruding two-dimensional sketches along a straight line to form three-dimensional features, as illustrated in the previous chapters, is an effective way to construct solid models. For designs that involve cylindrical shapes, shapes that are symmetrical about an axis, revolving two-dimensional sketches about an axis can form the needed three-dimensional features. In solid modeling, this type of feature is called a **revolved feature**.

In AutoCAD, besides using the **Revolve** command to create revolved features, several options are also available to handle symmetrical features. For example, we can create multiple identical copies of symmetrical features with the **Array** command, or create mirror images of models using the **Mirror** command. In this chapter, the construction and modeling techniques of these more advanced features are illustrated.

A Revolved Design: *Pulley*

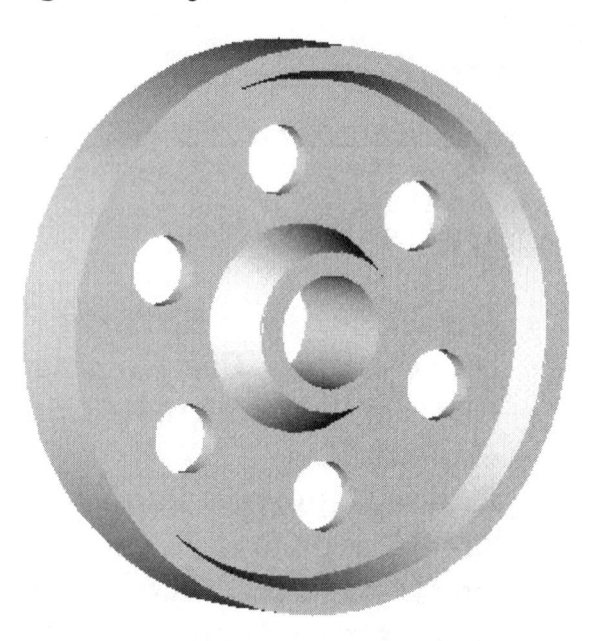

❖ Based on your knowledge of AutoCAD, how many features would you use to create the design? Which feature would you choose as the starting point for creating the model? Identify the symmetrical features in the design and consider other possibilities in creating the design. You are encouraged to create the model on your own prior to following through the tutorial.

Modeling Strategy – A Revolved Design

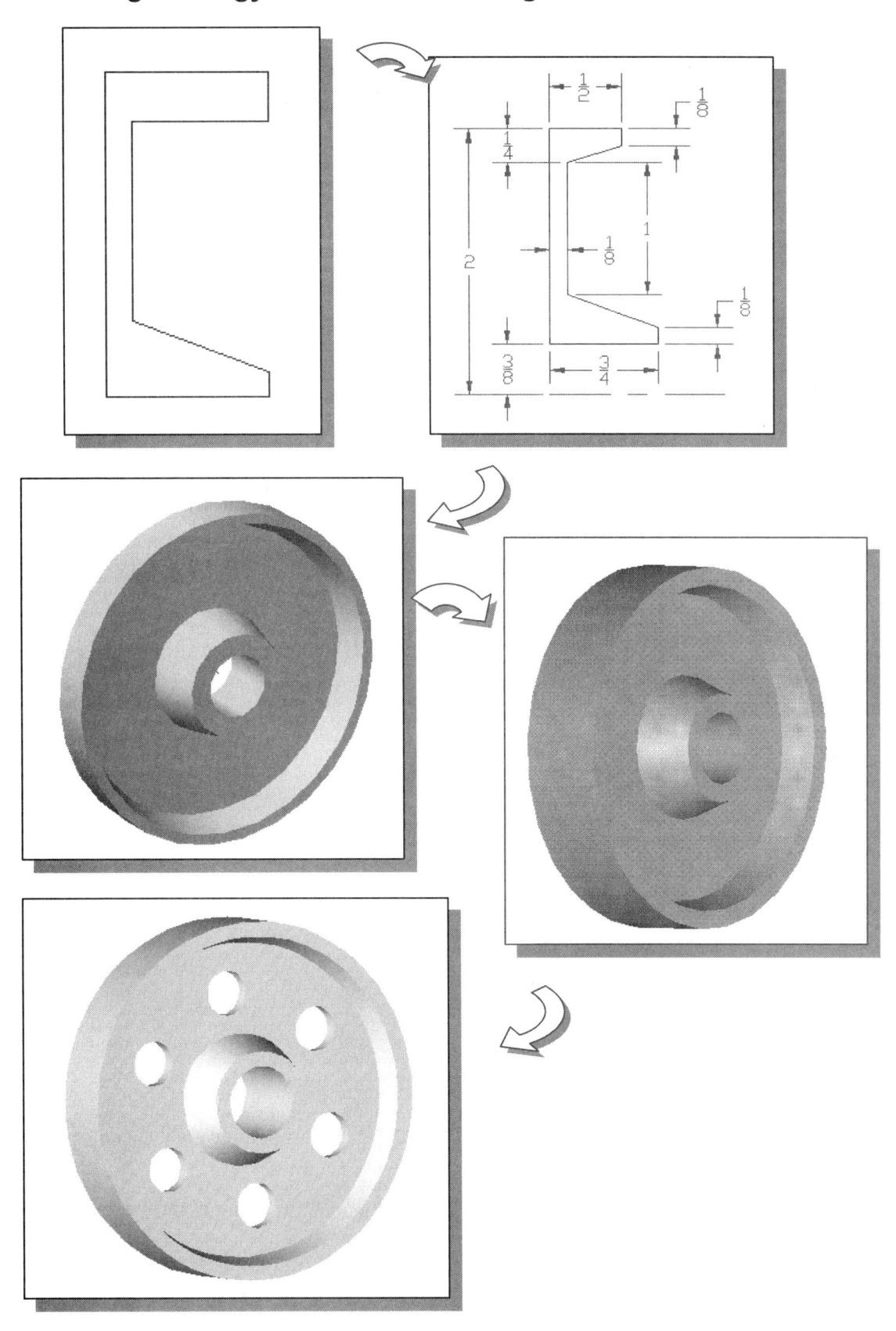

Starting Up AutoCAD® 2006

1. Select the **AutoCAD 2006** option on the *Program* menu or select the **AutoCAD 2006** icon on the *Desktop*.

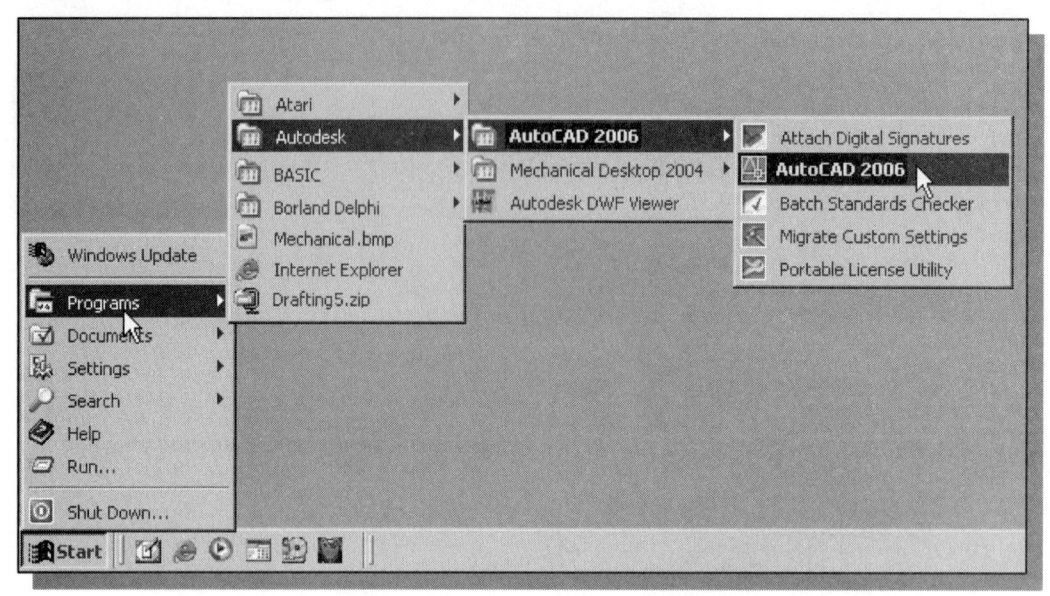

2. In the startup window, select **Start from Scratch**, as shown in the figure below.

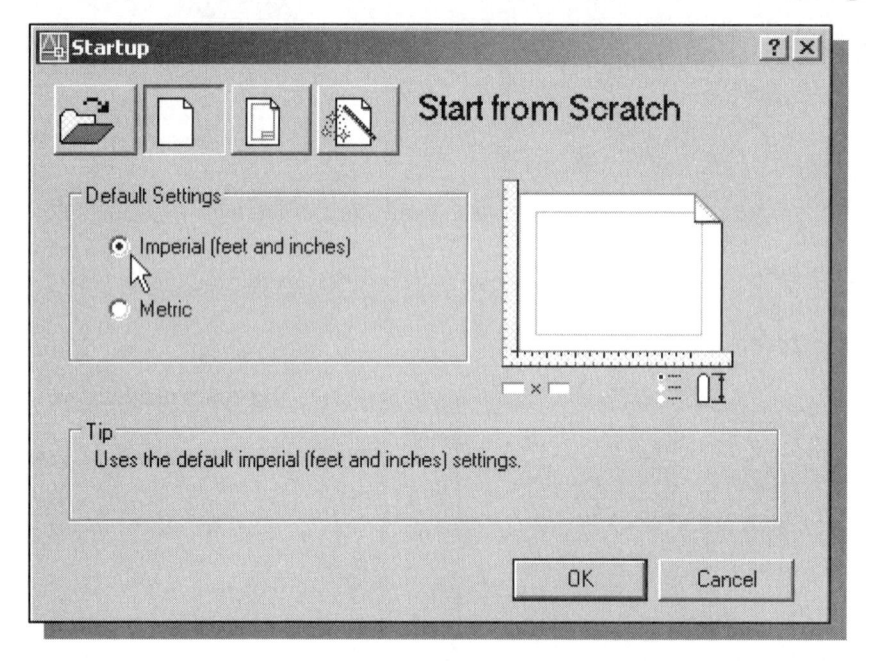

3. In the *Default Settings* section, pick **Imperial (feet and inches)** as the drawing units.

4. Pick **OK** in the startup dialog box to accept the selected settings.

Layers Setup

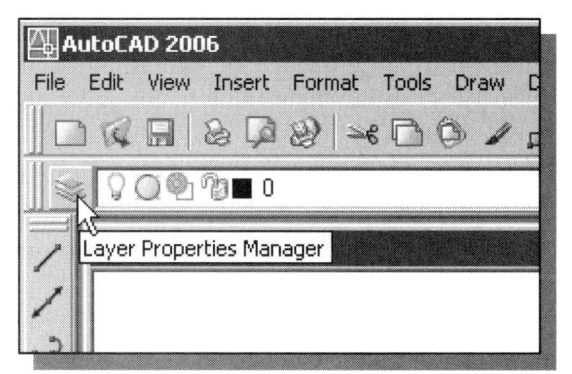

1. Pick **Layer Properties Manager** in the *Object Properties* toolbar.

2. Click on the **New** button to create new layers.

3. Create **three new layers** with the following settings:

Layer	Color	LineType	LineWeight
Center_lines	Red	Center	Default
Dimensions	Megenta	Continuous	Default
Object_lines	Cyan	Continuous	0.30mm

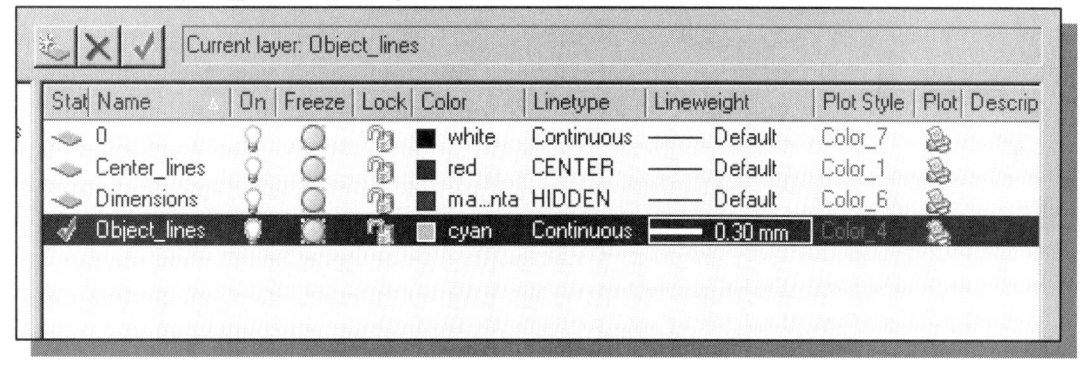

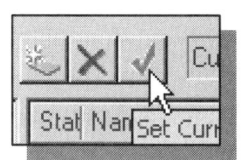

4. Highlight the layer *Object_lines* in the list of layers.

5. Click on the **Current** button to set layer *Object_lines* as the *Current Layer*.

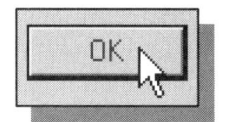

6. Click on the **OK** button to accept the settings and exit the *Layer Properties Manager* dialog box.

7. In the *Status Bar* area, reset the options and turn *ON* the *POLAR, OSNAP, OTRACK, DYN, LWT* and *MODEL* options.

Setting Up a 2D Sketch for the Revolved Feature

- Note that the *Pulley* design is symmetrical about a horizontal axis as well as a vertical axis, which allows us to simplify the 2D sketch as shown below. Instead of creating the 2D sketch by the **Line** command, we will use some of the more advanced editing options available with the **Region** command.

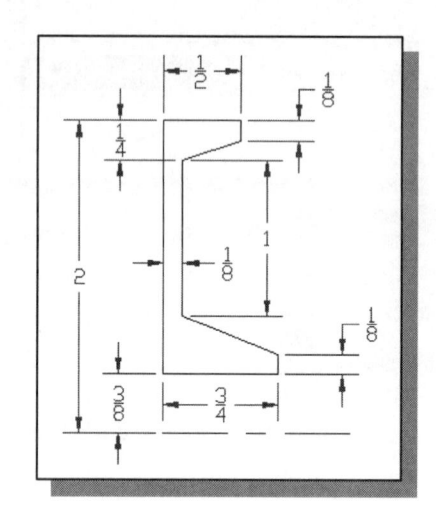

1. Select the **Rectangle** icon in the *Draw* toolbar. In the command prompt area, the message "*Specify first corner point or [Chamfer/Elevation/Fillet/Thickness/Width]:*" is displayed.

2. Place the first corner point of the rectangle near the center of the screen. Do not be overly concerned about the actual coordinates of the selected location; the CAD drawing space is as big as you can imagine.

3. We will create a 0.75" × 1.625" rectangle for the main body of the sketch. Enter **@0.75,1.625 [ENTER]**.

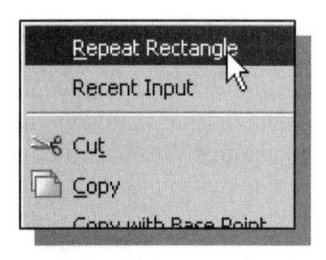

4. Inside the graphics window, click once with the right-mouse-button to bring up the popup option menu.

5. Pick **Repeat Rectangle**, with the left-mouse-button, in the popup menu to repeat the last command.

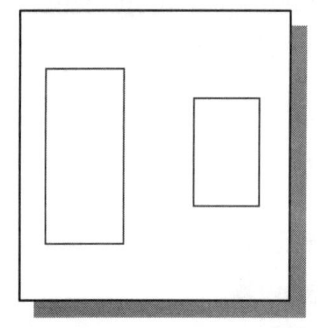

6. Place the first corner point of the rectangle toward the right side of the previous rectangle.

7. We will create a 0.625" × 1.0" rectangle as a cutter sketch. Enter **@0.625,1.0 [ENTER]**.

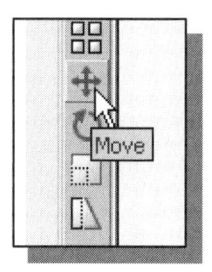

8. Select the **Move** icon in the *Draw* toolbar.

9. In the command prompt area, the message "*Select Objects:*" is displayed. Pick the cutter rectangle, the smaller rectangle, to move.

10. Inside the graphics window, right-mouse-click to accept the selection.

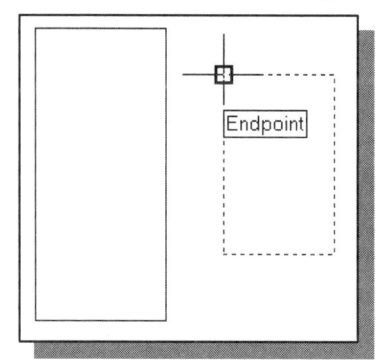

11. In the command prompt area, the message "*Specify base point or displacement:*" is displayed. Pick the upper left corner of the smaller rectangle as shown.

12. Inside the graphics window, hold down the [**SHIFT**] key and **right-mouse-click** once to bring up the *Object Snap* shortcut menu.

13. Select the **From** option in popup window.

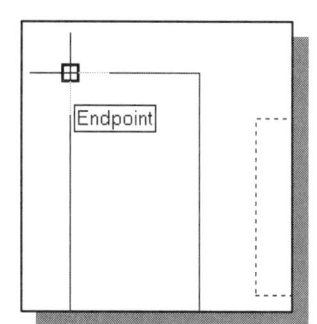

14. Pick the upper left corner of the larger rectangle as shown.

15. In the command prompt area, the message "*Specify base point or displacement: _from Base point: <Offset>:*" is displayed. Enter: **@0.125,-0.25** [**ENTER**].

16. Pre-select the smaller rectangle to activate the grip-editing function.

17. Pick the lower right grip point displayed along the right vertical edge.

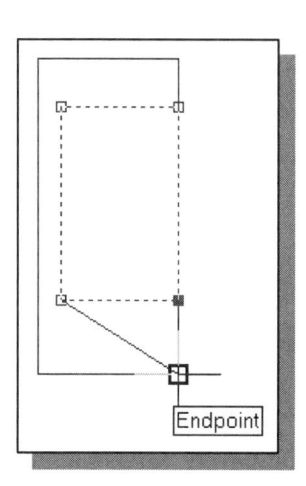

18. Inside the graphics window, hold down the [**SHIFT**] key and **right-mouse-click** once to bring up the *Object Snap* shortcut menu.

19. Select the **From** option in popup window.

20. Pick the lower right corner of the larger rectangle.

21. At the command prompt: **@0,0.125** [**ENTER**]

22. Press the [**Esc**] key once to deselect any entities.

Perform 2D Boolean Operations

- AutoCAD's 2D regions are special types of filled polygons. We can create composite 2D regions by applying *Boolean operations* to subtract, find the intersection, or combine two or more regions.

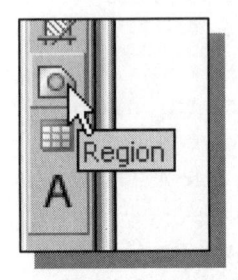

1. Select the **Region** command icon in the *Draw* toolbar.

2. In the command prompt area, the message "*Select objects:*" is displayed. Select the polygon we just edited.

3. Inside the graphics window, right-mouse-click to accept the selection and create a region.

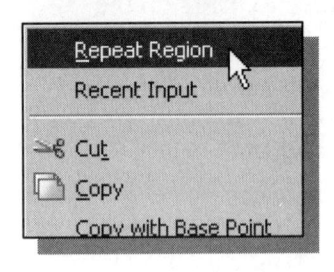

4. Inside the graphics window, click once with the right-mouse-button to bring up the popup option menu.

5. Pick **Repeat Region**, with the left-mouse-button, in the popup menu to repeat the last command.

6. In the command prompt area, the message "*Select objects:*" is displayed. Select the rectangle by clicking on one of the edges.

7. Inside the graphics window, right-mouse-click to accept the selection and create a region.

8. Move the cursor to the *Standard* toolbar area and right-click on any icon of the *Standard* toolbar to display a list of toolbar menu groups.

9. Select **Solids Editing**, with the left-mouse-button, to display the *Solids Editing* toolbar on the screen.

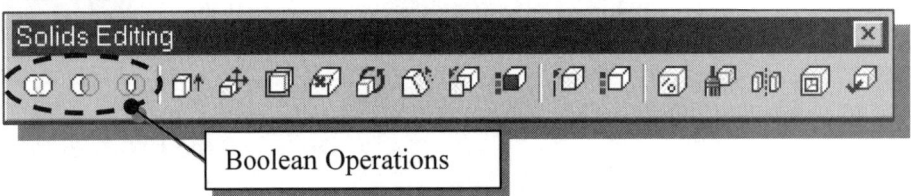

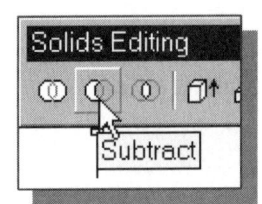

10. In the *Solids Editing* toolbar, click on the **Subtract** icon. In the command prompt area, the message "*_subtract Select solids and regions to subtract from .. Select Objects:*" is displayed.

11. Pick the **rectangle** by clicking on one of the edges.

12. Inside the graphics window, *right-mouse-click* to accept the selection and proceed with the **Subtract** command.

13. In the command prompt area, the message "*Select solids and regions to subtract .. Select Objects:*" is displayed. Pick the four-sided **polygon** by clicking on one of the edges.

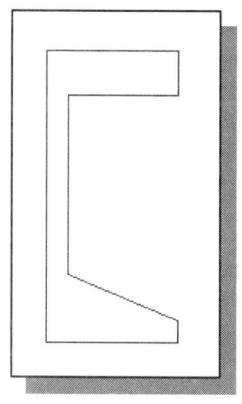

14. Inside the graphics window, right-mouse-click to accept the selection and proceed with the **Subtract** command. The completed 2D sketch is as shown.

   - AutoCAD's region is a special type of geometric entity. It was developed based on the CSG solid modeling approach and is valuable in the sense that it provides a different method to construct complex geometric shapes. In many cases, the Boolean operations are more efficient than creating geometry by entering point-to-point coordinates.

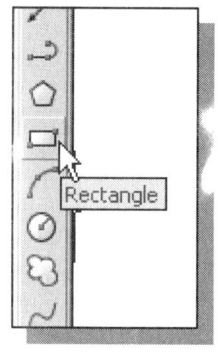

15. Select the **Rectangle** icon in the *Draw* toolbar.

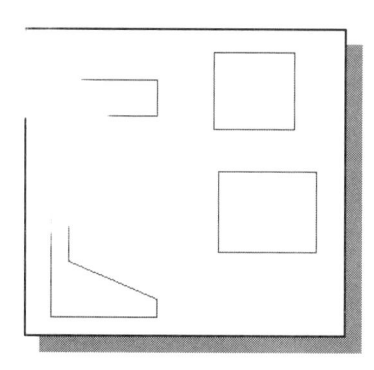

16. On your own, create two rectangles of arbitrary sizes as shown. We will use these two rectangles to shape the upper regions of the design.

17. On your own, convert the two rectangles into two **regions**.

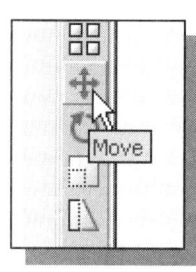

18. Select the **Move** icon in the *Draw* toolbar.

19. In the command prompt area, the message "*Select Objects:*" is displayed. Pick one of the rectangles to move.

20. Inside the graphics window, right-mouse-click to accept the selection.

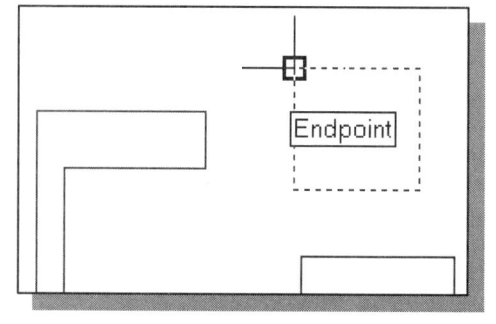

21. In the command prompt area, the message "*Specify base point or displacement:*" is displayed. Pick the upper left corner of the selected rectangle as shown.

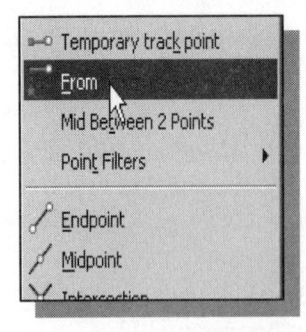

22. Inside the graphics window, hold down the [SHIFT] key and **right-mouse-click** once to bring up the *Object Snap* shortcut menu.

23. Select the **From** option in popup window.

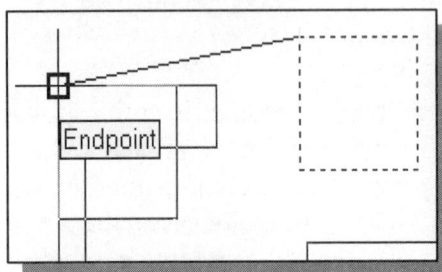

24. Pick the upper left corner of the larger rectangle as shown.

25. In the command prompt area, the message "*Specify base point or displacement: _from Base point: <Offset>:*" is displayed.
    Enter: **@0.5,0.0 [ENTER]**

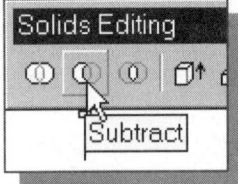

26. On your own, perform the **Boolean Subtract** operation and create the 2D sketch as shown.

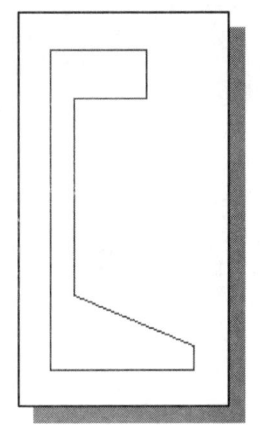

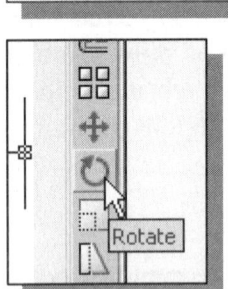

27. Select the **Rotate** icon in the *Modify* toolbar.

28. In the command prompt area, the message "*Select Objects:*" is displayed. Pick the rectangle to rotate.

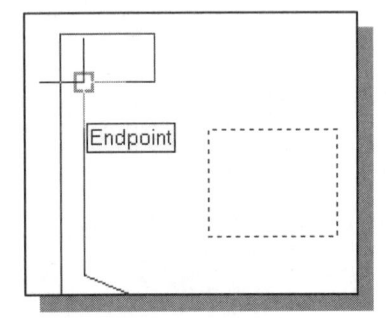

29. Inside the graphics window, right-mouse-click to accept the selection.

30. In the command prompt area, the message "*Specify base point or displacement:*" is displayed. Pick the inside corner of the 2D sketch as shown.

31. Inside the graphics window, hold down the [SHIFT] key and **right-mouse-click** once to bring up the *Object Snap* shortcut menu.

32. Select the **From** option in popup window.

Symmetrical Features in Designs 9-11

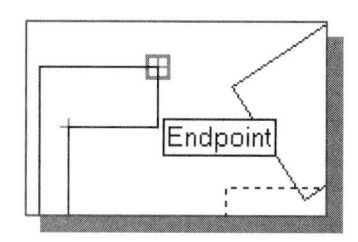

33. Pick the upper right corner of the 2D sketch as shown.

34. At the command prompt: **@0,-0.125** [ENTER]

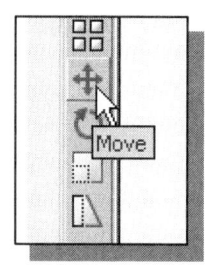

35. Select the **Move** icon in the *Draw* toolbar.

36. In the command prompt area, the message "*Select Objects:*" is displayed. Pick the rotated rectangle to move.

37. Inside the graphics window, right-mouse-click to accept the selection.

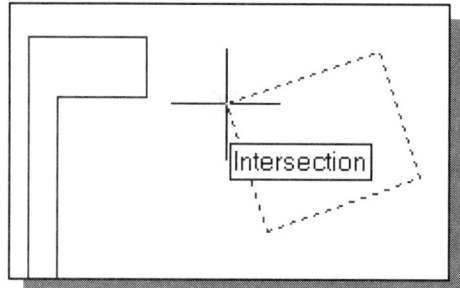

38. In the command prompt area, the message "*Specify base point or displacement:*" is displayed. Pick the upper left corner of the selected rectangle as shown.

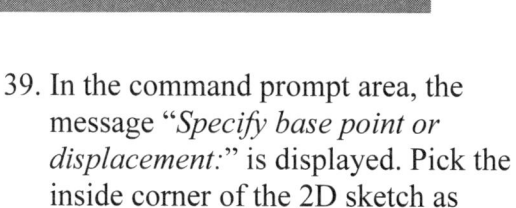

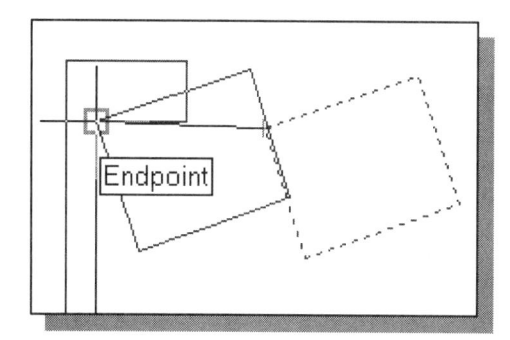

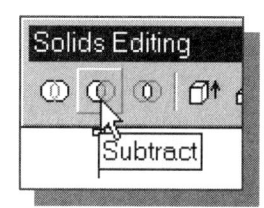

39. In the command prompt area, the message "*Specify base point or displacement:*" is displayed. Pick the inside corner of the 2D sketch as shown.

40. On your own, perform the **Boolean Subtract** operation and complete the 2D sketch as shown.

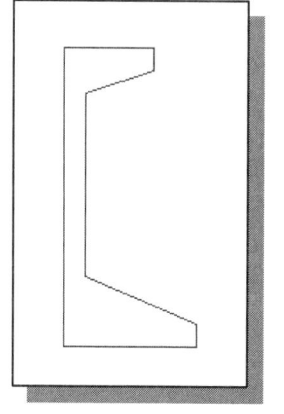

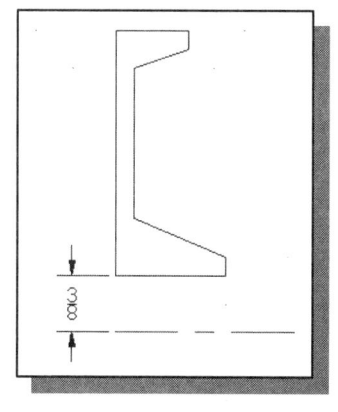

41. On your own, create a centerline that is 3/8 below the 2D sketch as shown. This line will be used as the axis of rotation for the revolved feature. (Trim the line so that the left endpoint is aligned to the left edge of the 2D sketch.)

Creating the Revolved Feature

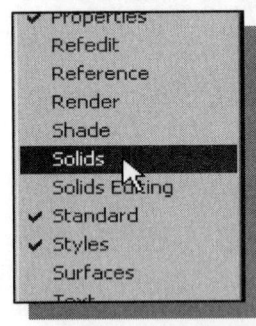

1. Move the cursor to the *Standard* toolbar area and right-mouse-click once on any icon in the toolbar area to display a list of toolbar menu groups.

2. Select **Solids**, with the left-mouse-button, to display the *Solids* toolbar on the screen.

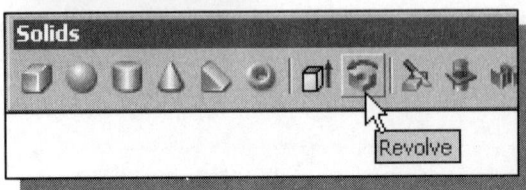

3. Select **Revolve** in the *Solids* toolbar as shown.

4. In the command prompt area, the message "*Select objects:*" is displayed. Select the 2D region by clicking on any of the displayed segments.

5. Inside the graphics window, right-mouse-click to accept the selection and proceed with the **Revolve** command.

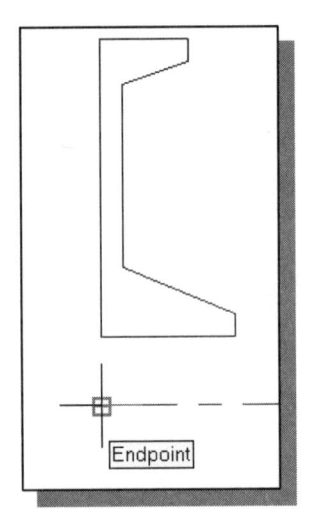

6. In the command prompt area, the message "*Define axis by [Object/X (axis)/Y (axis)}]:*" is displayed. Pick the **left endpoint of the centerline** as shown.

7. In the command prompt area, the message "*Specify endpoint of axis:*" is displayed. Pick the **right endpoint of the centerline**.

8. In the command prompt area, the message "*Specify angle of revolution <360>:*" is displayed. Enter: **360.0 [ENTER]**

- The revolved feature is created as shown.

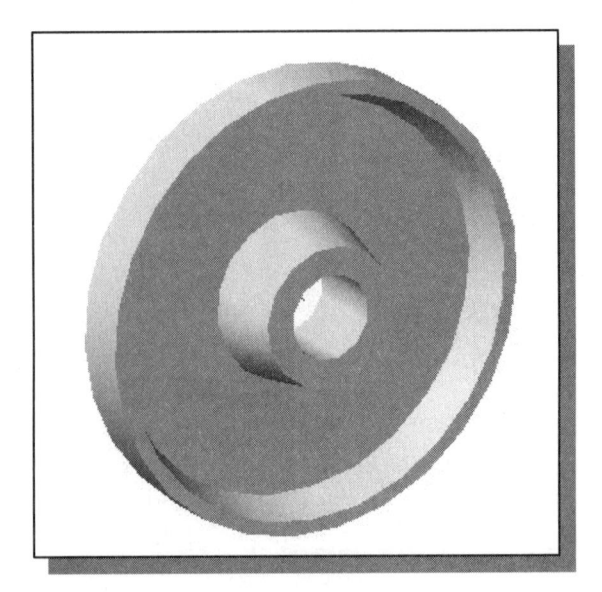

Mirroring Part

- In AutoCAD, we can mirror a 2D sketch or a 3D part about a line or a specified surface. We can define the line by selecting existing points or by picking locations on the screen. We can physically flip the part about a reference plane or create a new mirrored part.

1. On your own, change the display to **SW Isometric View** or use the **3D Orbit** command to display the flat face of the solid model.

2. In the pull down menus, select
   [Modify] → [3D Operation] → [Mirror 3D]

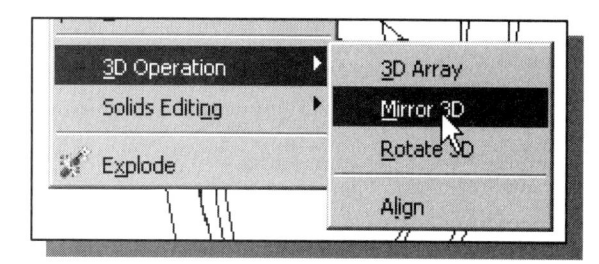

3. In the command prompt area, the message "*Select objects:*" is displayed. Select any edge of the 3D model.

4. Inside the graphics window, right-mouse-click to accept the selection.

5. In the command prompt area, the message "*Specify first point of mirror plane (3 points) or [Object/Last/Zaxis/View/XY/YZ/ZX/3points] <3points>:*" is displayed. Choose the YZ plane by entering: **YZ [ENTER]**

6. In the command prompt area, the message "*Select point on YZ plane <0,0,0>:*" is displayed. Pick the **left endpoint of the centerline**.

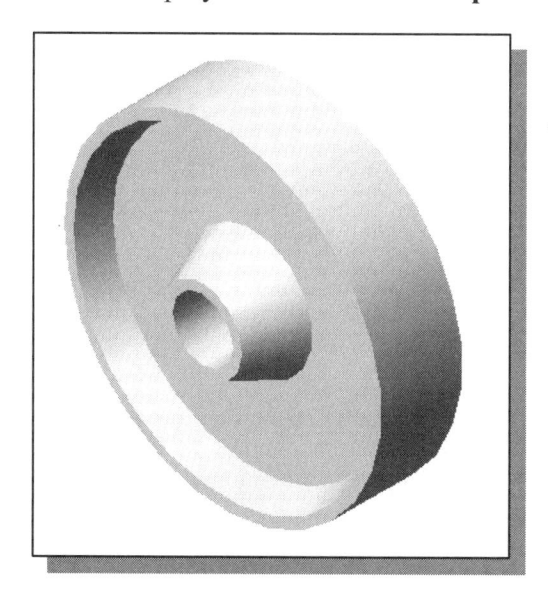

7. In the command prompt area, the message "*Delete source objects? [Yes/No] <N>:*" is displayed. Inside the graphics window, right-mouse-click to bring up the option menu and select **No** to create a new part.

Combining Parts

- We have created two separate parts with the **Mirror 3D** command. We will combine the two parts into a single part using the CSG *Boolean operations*.

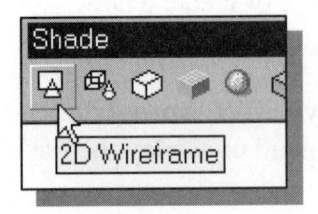

1. Select **2D Wireframe** by left-clicking once on the icon in the *Shade* toolbar.

- Note the edges separating the two parts displayed on the screen.

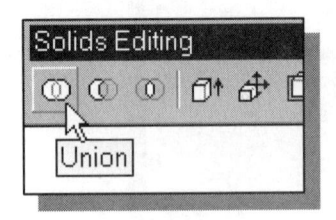

2. Pick **Union** in the *Solids Editing* toolbar as shown.

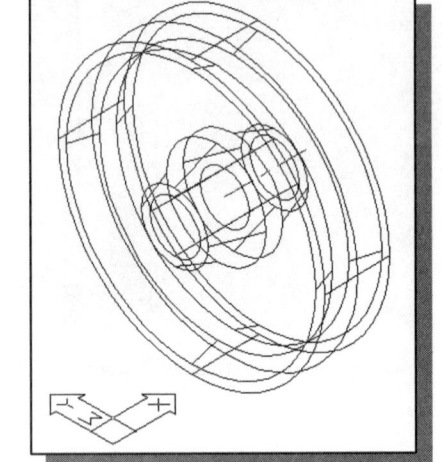

3. In the command prompt area, the message "*Select Objects:*" is displayed. Pick the two parts.

4. Inside the graphics window, right-mouse-click to accept the selection and proceed with the Union command to combine the two parts into a single part.

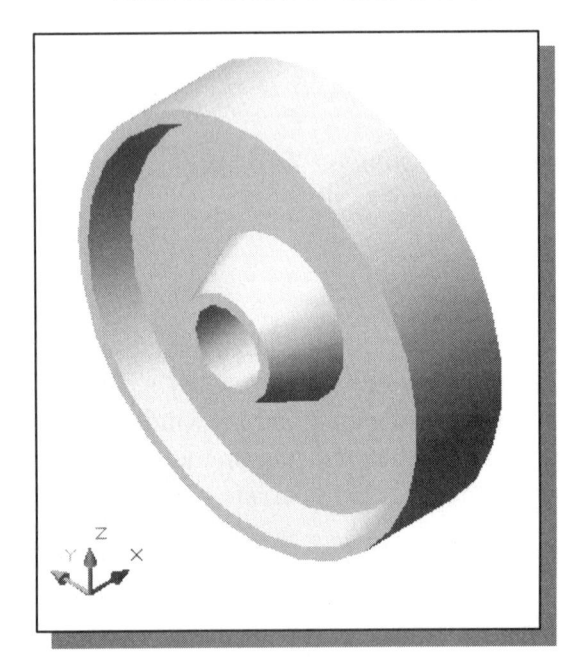

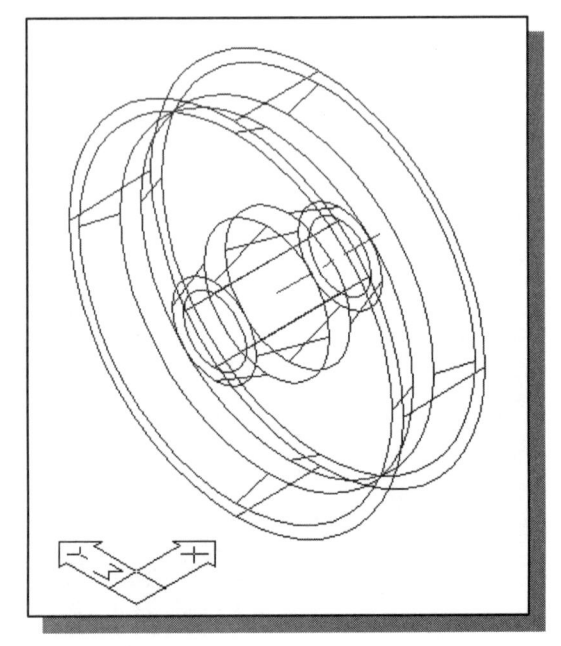

- Reset the display to **2D Wireframe** mode before proceeding to the next section.

3D Array

- In AutoCAD, existing 2D and/or 3D features can be easily duplicated. The **Array** command allows us to create both rectangular and polar arrays of features.

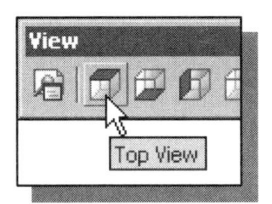

  1. In the *View* toolbar, select the **Top View** option to reset the display.

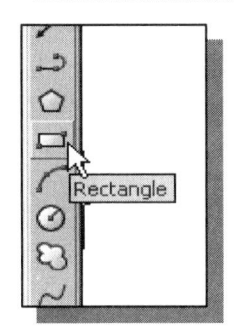

  2. Select the **Rectangle** icon in the *Draw* toolbar.

  3. On your own, create a rectangle that is **1.0″ × .25″** toward the right side of the solid model.

  4. On your own, convert the rectangle into a *region*.

- The rectangle will be used to create a cylindrical solid model. The cylinder will be used to generate a polar array using the Array command.

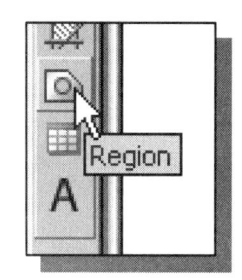

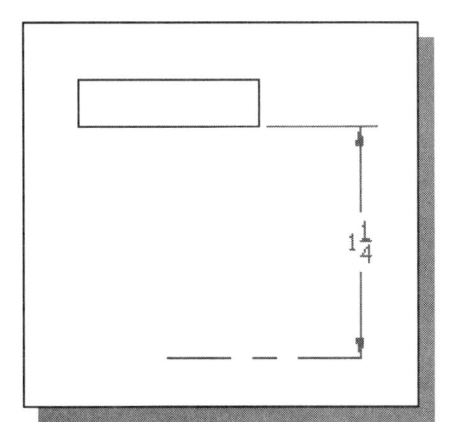

  5. On your own, create a centerline that is **1¼″** below the rectangle with the <u>left endpoint aligned to the center of the rectangle</u> as shown.

- The centerline will be used as the axis of rotation to create the *polar array*.

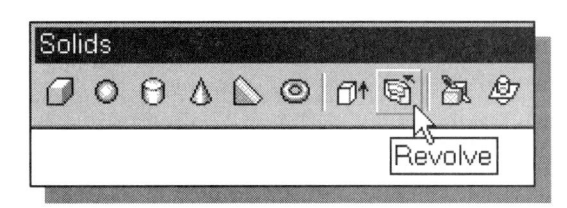

  6. Select **Revolve** in the *Solids* toolbar as shown.

  7. In the command prompt area, the message "*Select objects:*" is displayed. Select the rectangular region by clicking on any of the displayed segments.

  8. Inside the graphics window, right-mouse-click to accept the selection and proceed with the **Revolve** command.

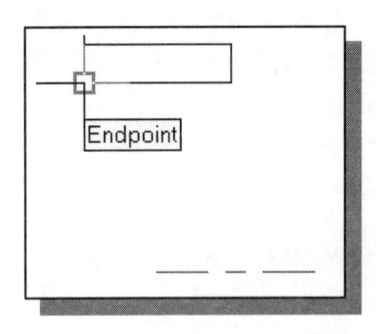

9. In the command prompt area, the message "*Define axis by [Object/X (axis)/Y (axis)}]:*" is displayed. Pick the **lower left corner** of the region as shown.

10. In the command prompt area, the message "*Specify endpoint of axis:*" is displayed. Pick the **lower right corner** of the region.

11. In the command prompt area, the message "*Specify angle of revolution <360>:*" is displayed.
    Enter: **360.0 [ENTER]**

12. In the *View* toolbar, click on the **SE Isometric View** icon.

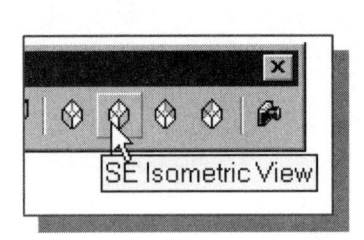

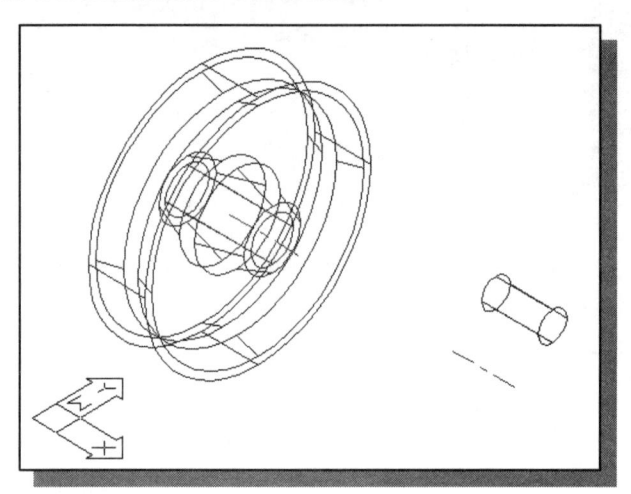

13. In the pull down menus, select
    [Modify] → [3D Operation] → [3D Array]

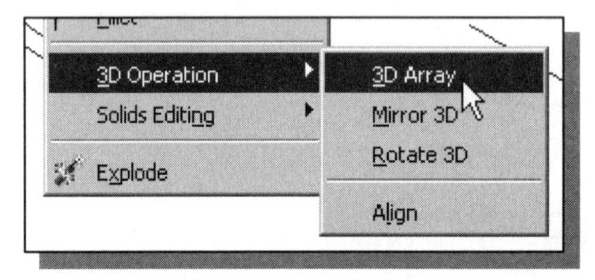

14. In the command prompt area, the message "*Select objects:*" is displayed. Select any edge of the small cylinder.

15. Inside the graphics window, right-mouse-click to accept the selection.

16. In the command prompt area, the message "*Enter the type of array [Rectangular/Polar] <R>:*" is displayed.
    Choose the *Polar* option by entering: **P [ENTER]**

17. In the command prompt area, the message "*Enter the number of items in the array:*" is displayed. We will create 6 items by entering: **6 [ENTER]**

18. In the command prompt area, the message "*Specify the angle to fill (+=ccw, -=cw) <360>:*" is displayed. Enter: **360 [ENTER]**

19. In the command prompt area, the message "*Rotate arrayed objects?[Yes/No] <Y>:*" is displayed. Enter: **N [ENTER]**

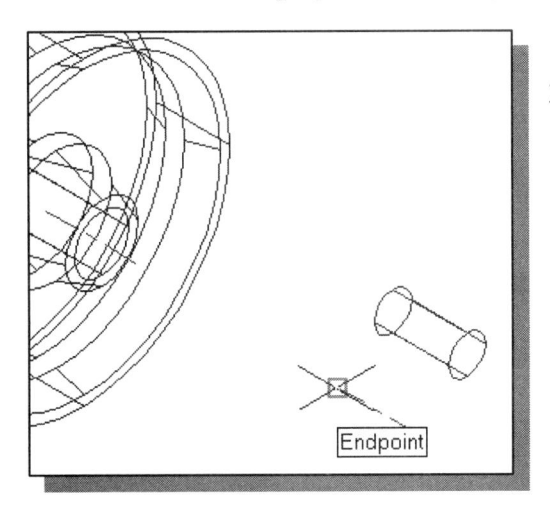

20. In the command prompt area, the message "*Specify center point of array:*" is displayed. Pick the **left endpoint of the centerline** as shown.

21. In the command prompt area, the message "*Specify second point on axis of rotation:*" is displayed. Pick the right endpoint of the centerline as shown.

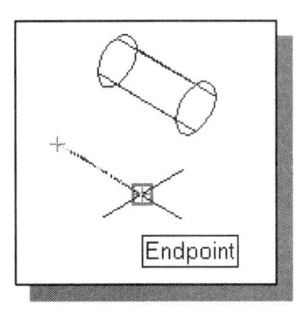

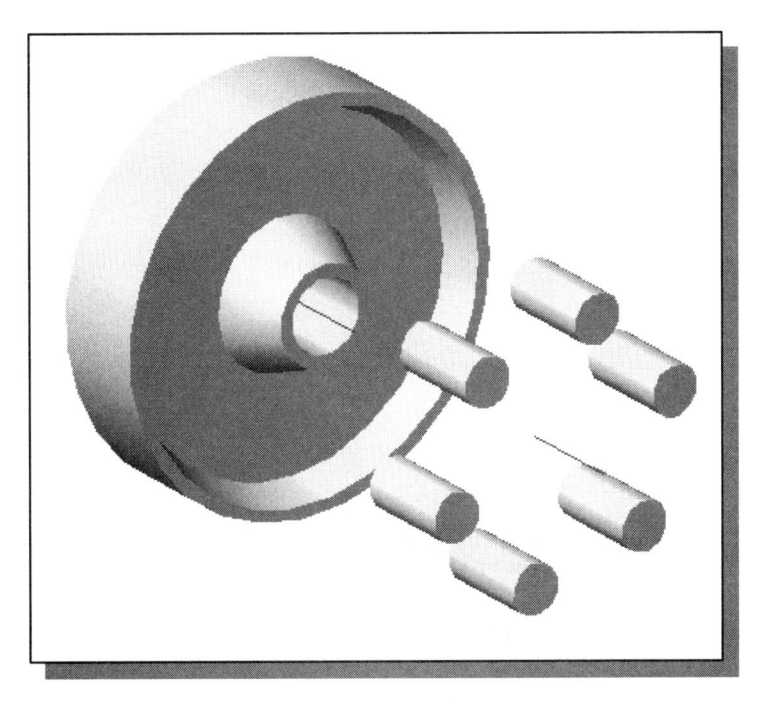

Position and Perform the Cut

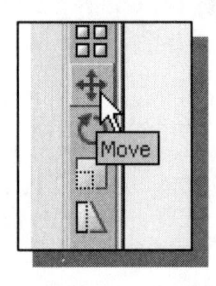

1. Select the **Move** icon in the *Draw* toolbar.

2. In the command prompt area, the message "*Select Objects:*" is displayed. Pick the six cylinders we just created.

3. Inside the graphics window, right-mouse-click to accept the selection.

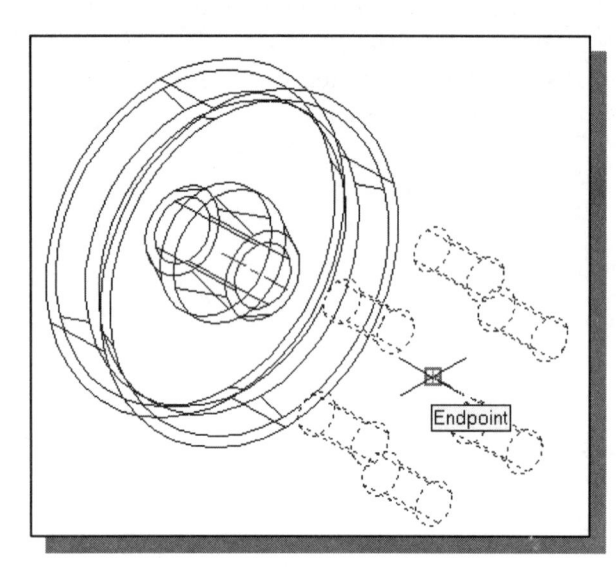

4. In the command prompt area, the message "*Specify base point or displacement:*" is displayed. Pick the **left endpoint the polar array centerline** as shown.

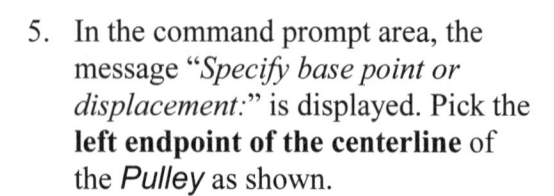

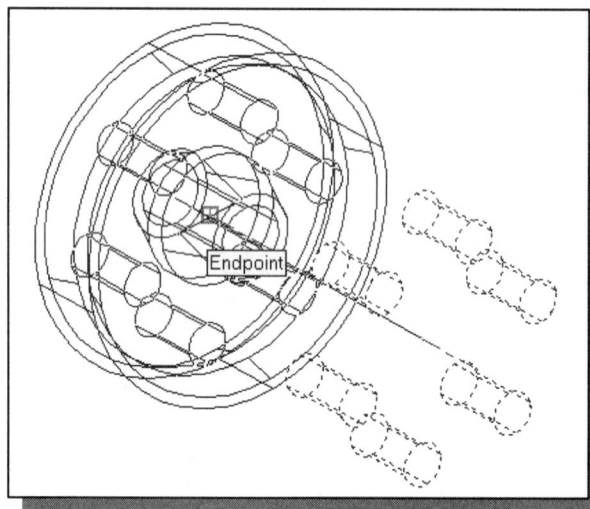

5. In the command prompt area, the message "*Specify base point or displacement:*" is displayed. Pick the **left endpoint of the centerline** of the *Pulley* as shown.

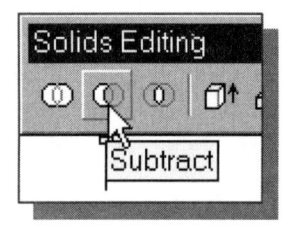

6. In the *Solids Editing* toolbar, click on the **Subtract** icon.

7. In the command prompt area, the message "*_subtract Select solids and regions to subtract from .. Select Objects:*" is displayed. Pick the *Pulley* by clicking on one of the edges.

8. Inside the graphics window, right-mouse-click to accept the selection and proceed with the **Subtract** command.

9. In the command prompt area, the message "*Select solids and regions to subtract .. Select Objects:*" is displayed.

10. Pick the six cylinder blocks by clicking on the circular edges.

11. Inside the graphics window, right-mouse-click to accept the selection and proceed with the **Subtract** command.

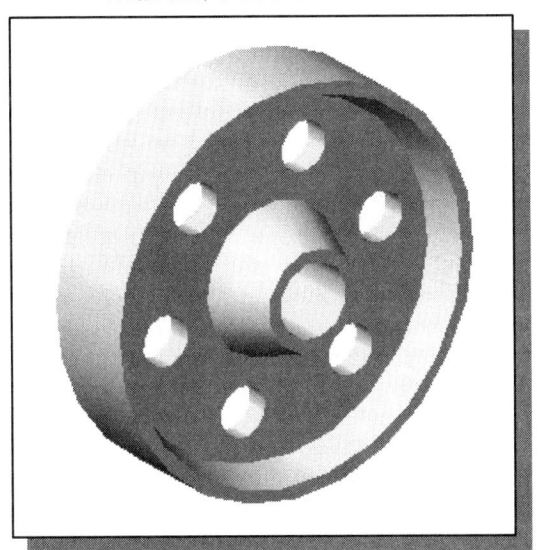

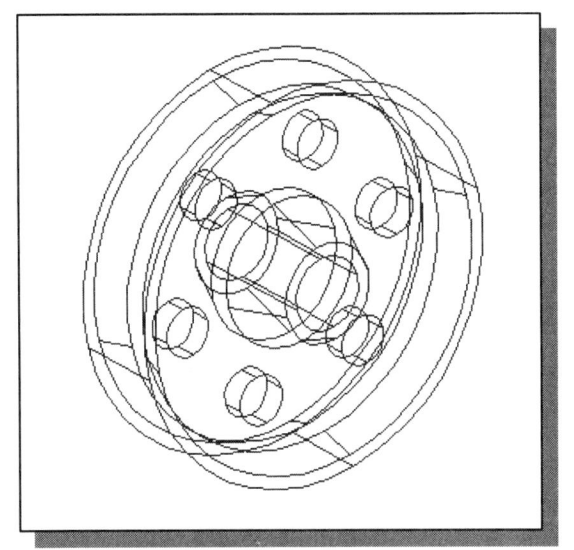

- On your own, create the 2D drawing from the 3D model as shown.

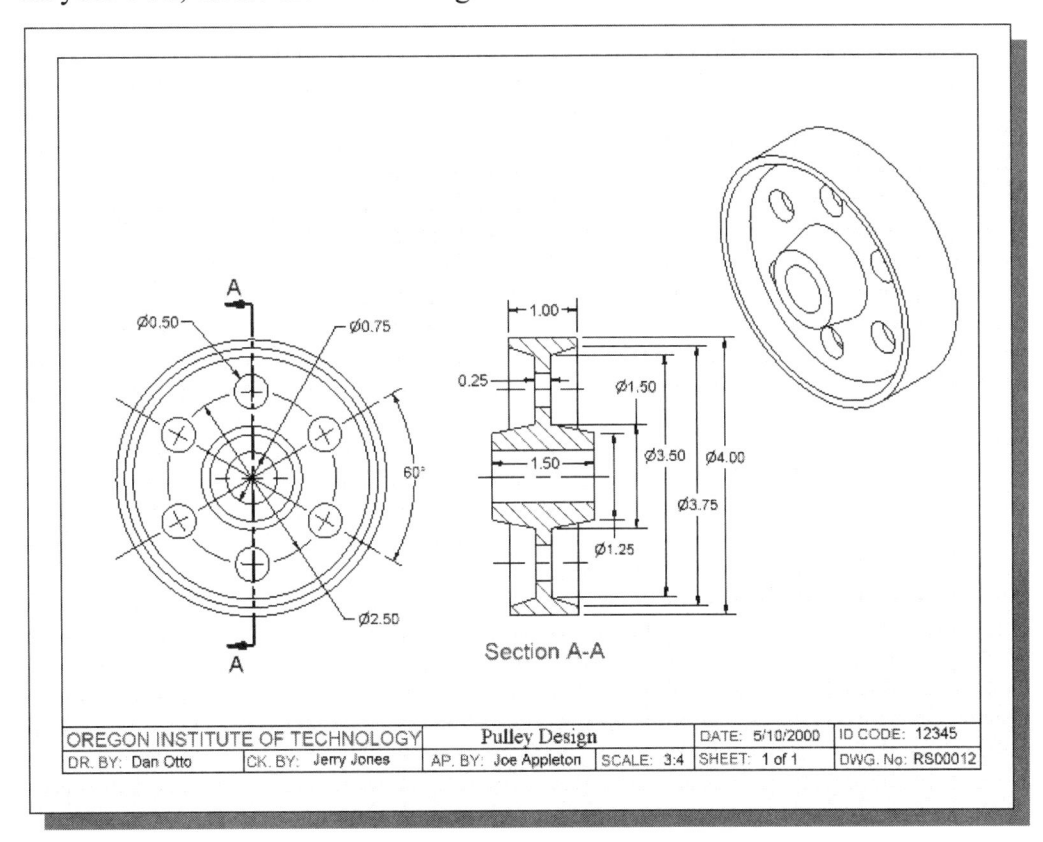

Questions:

1. List the different symmetrical features created in the *Pulley* design.

2. What are the advantages of using *regions*?

3. Describe the steps required in using the **Mirror Part** command.

4. Why is it important to identify symmetrical features in designs?

5. When and why should we use the **Array** option?

6. What is the difference between *Rectangular Array* and *Polar Array*?

7. When and why should you use the **Revolve** command in AutoCAD?

8. Identify and describe the following commands:

   (a)

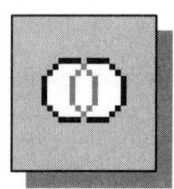

   (b)

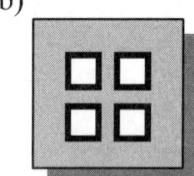

   (c)

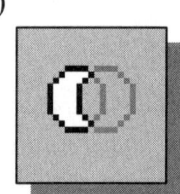

   (d)

Exercises:

1.

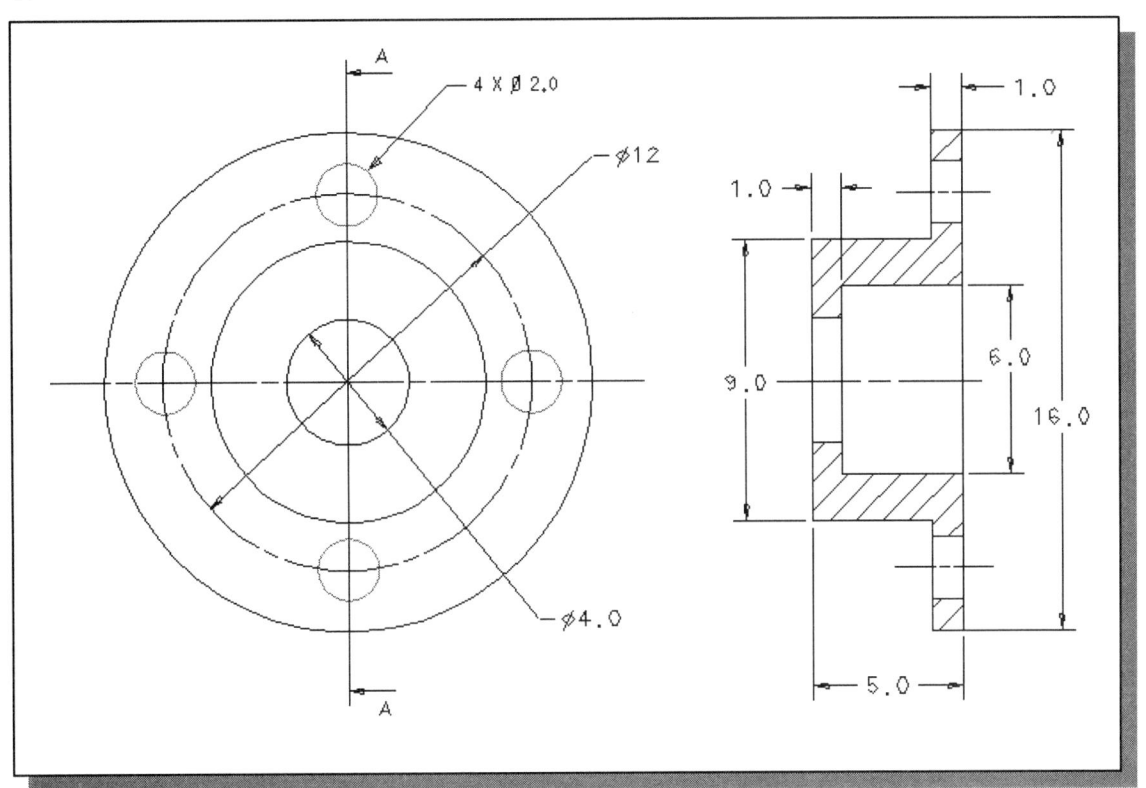

2. Plate thickness : 0.125 inch

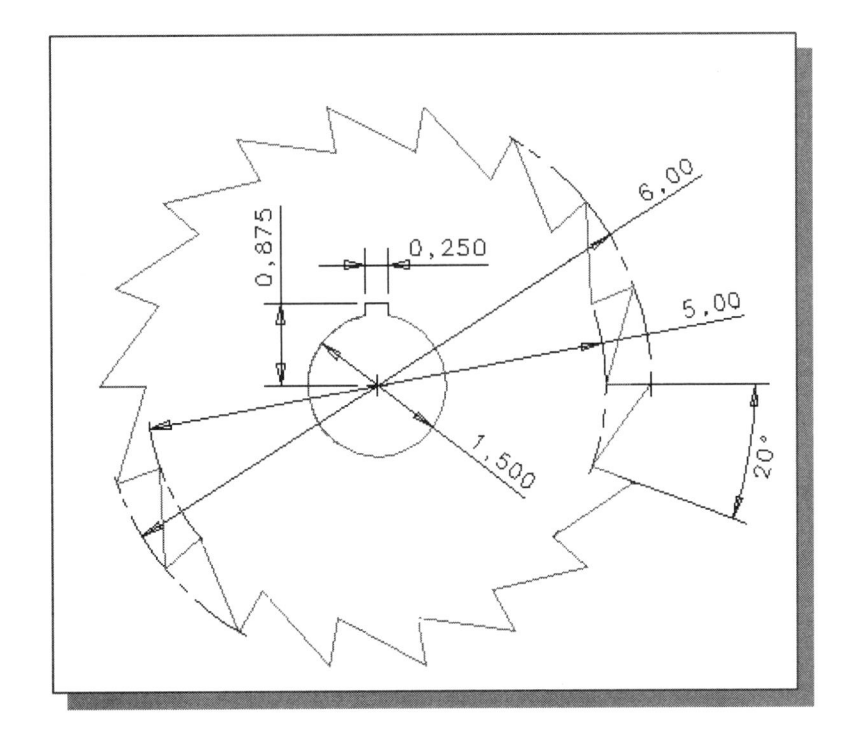

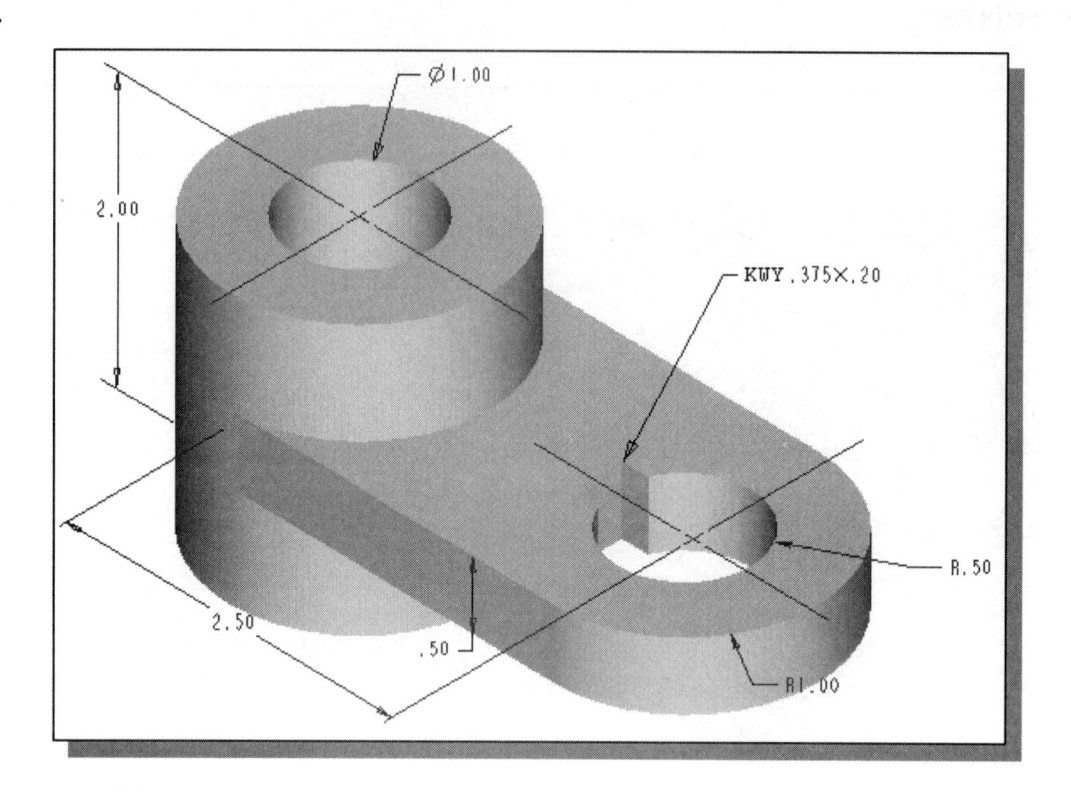

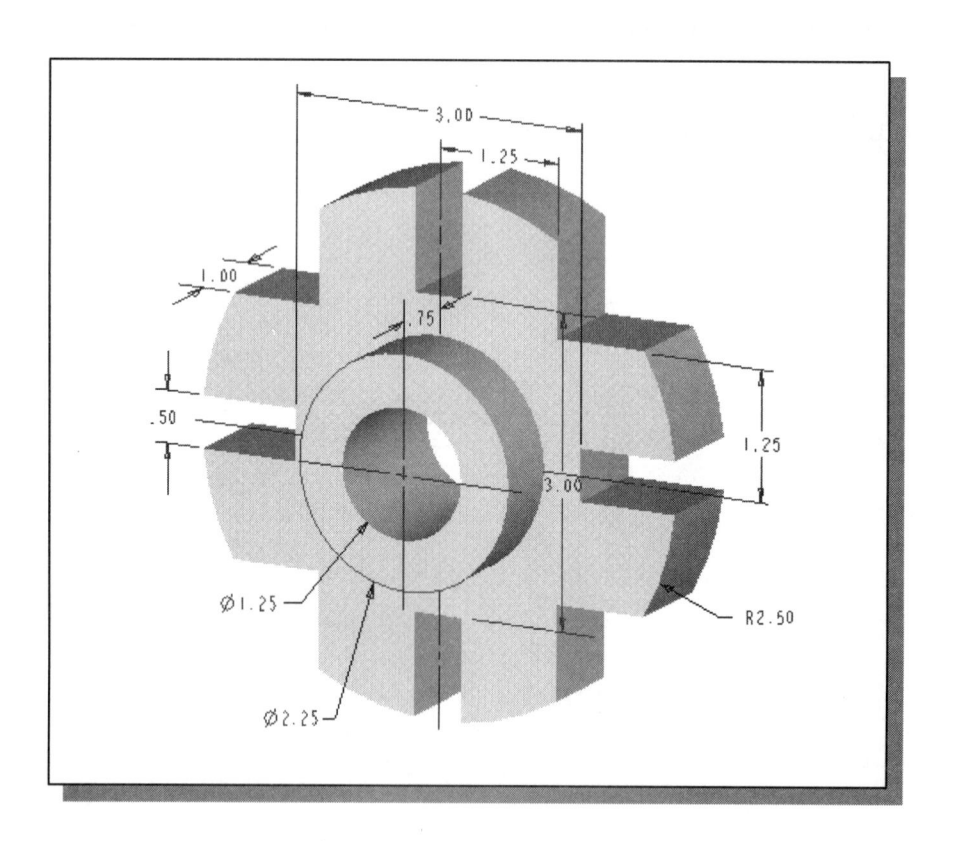

3.

4.

5.

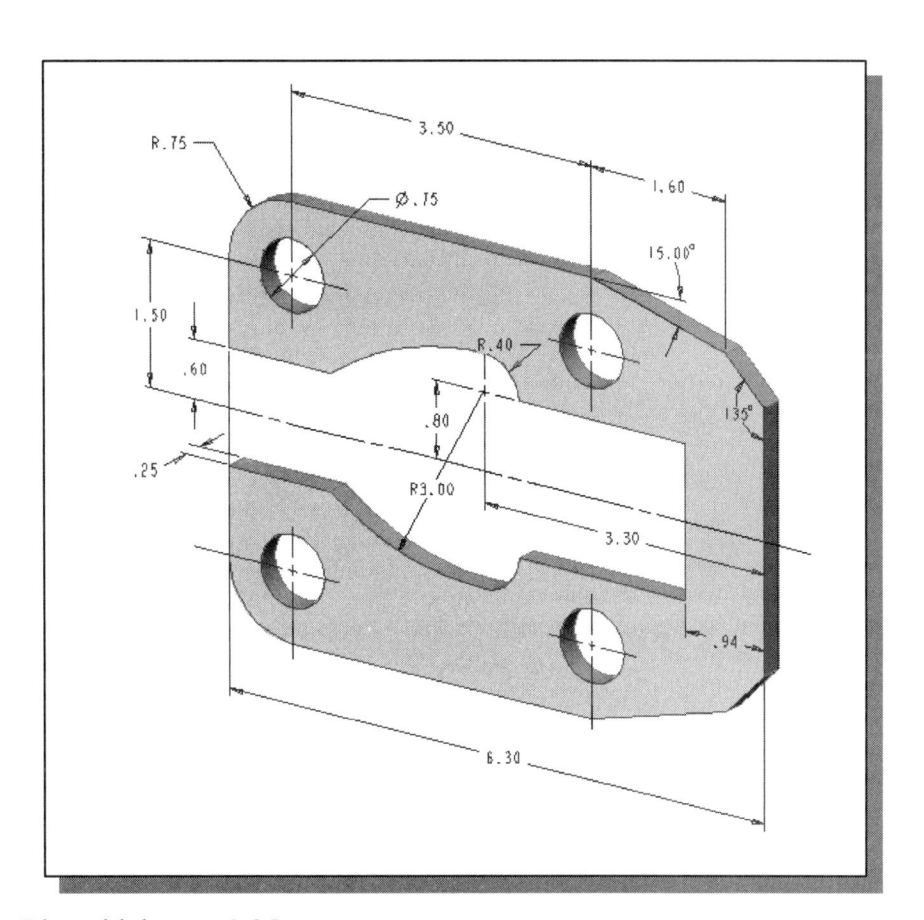

6. Plate thickness: 0.25

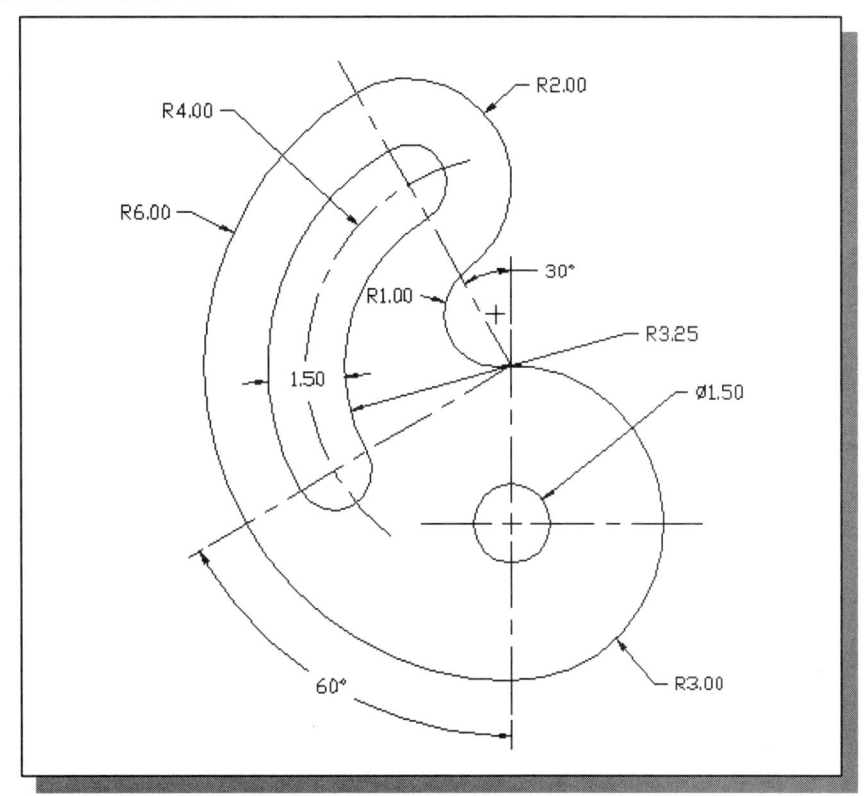

NOTES:

AutoCAD® 2006 Tutorial: 3D Modeling   10-1

Chapter 10
Advanced Modeling Tools & Techniques

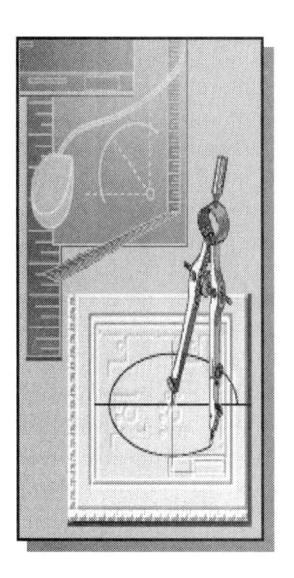

Learning Objectives

- **Using More Advanced Solid Modeling Construction Tools**
- **Create Draft Angle Features**
- **Use the 3D Rounds & Fillets Command**
- **Create Rectangular Patterns**
- **Use the Shell Command**
- **Use the Copy Faces Command**

Introduction

AutoCAD provides an assortment of three-dimensional construction tools to make the creation of solid models easier and more efficient. In this chapter, we will examine the procedures to create the *Draft Angle* feature, the *Shell* feature and also for creating three-dimensional *Rounds* and *Fillets* along edges of a solid model. These features are common characteristics of molded parts. The three-dimensional **Fillets** and the **Shell** commands will usually create complex three-dimensional spatial curves and surfaces. These commands are more sensitive to the associated geometric entities. For this reason, the *3D Fillets* and the *Shell* features are created last, after all associated solid features are created. In this chapter, we will also examine the different methods of referencing to existing surfaces.

A Thin-walled Design: *Oil Sink*

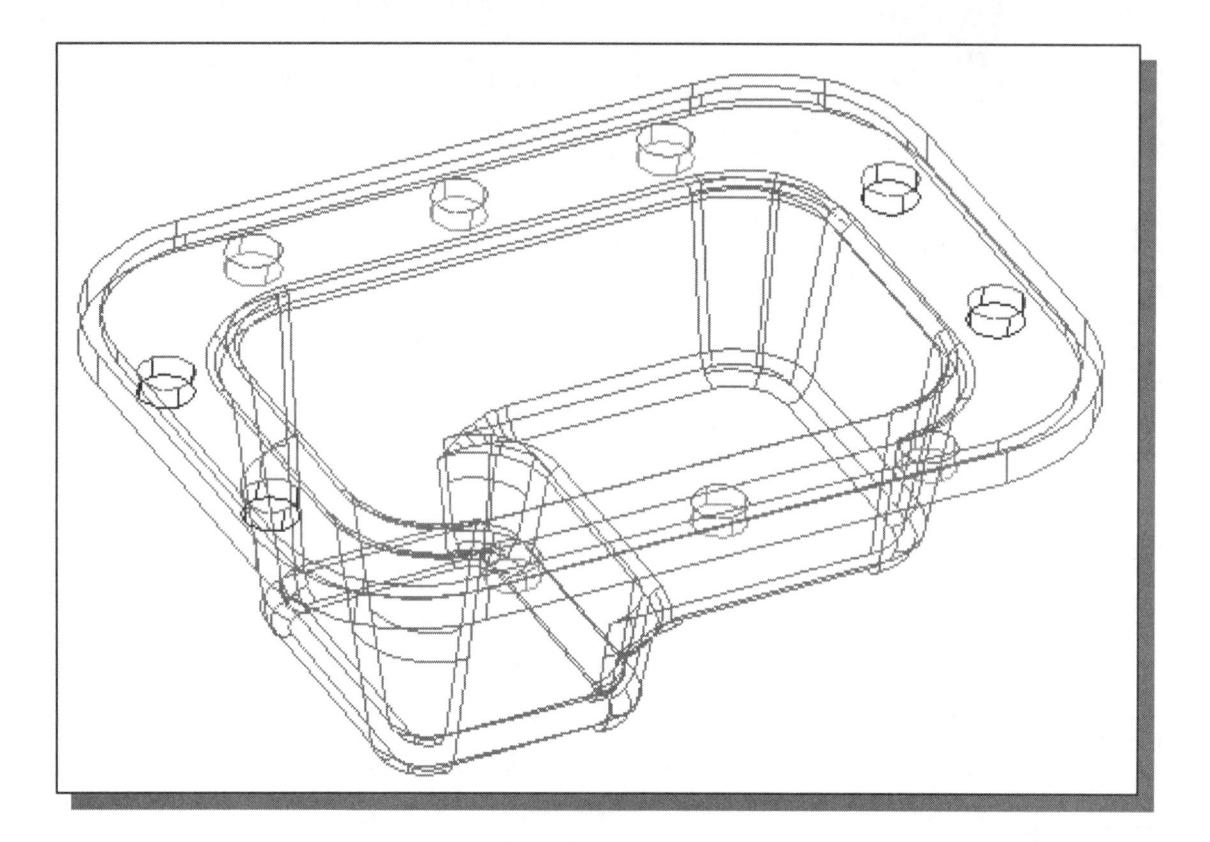

❖ Based on your knowledge of AutoCAD so far, how many features would you use to create the design? Which feature would you choose as the starting point in creating the model? What are the more difficult features involved in the design? What is your choice in arranging the order of the features? Take a few minutes to consider these questions and do preliminary planning by sketching on a piece of paper. You are also encouraged to create the design on your own prior to following through the tutorial.

Advanced Modeling Tools & Techniques   10-3

Modeling Strategy

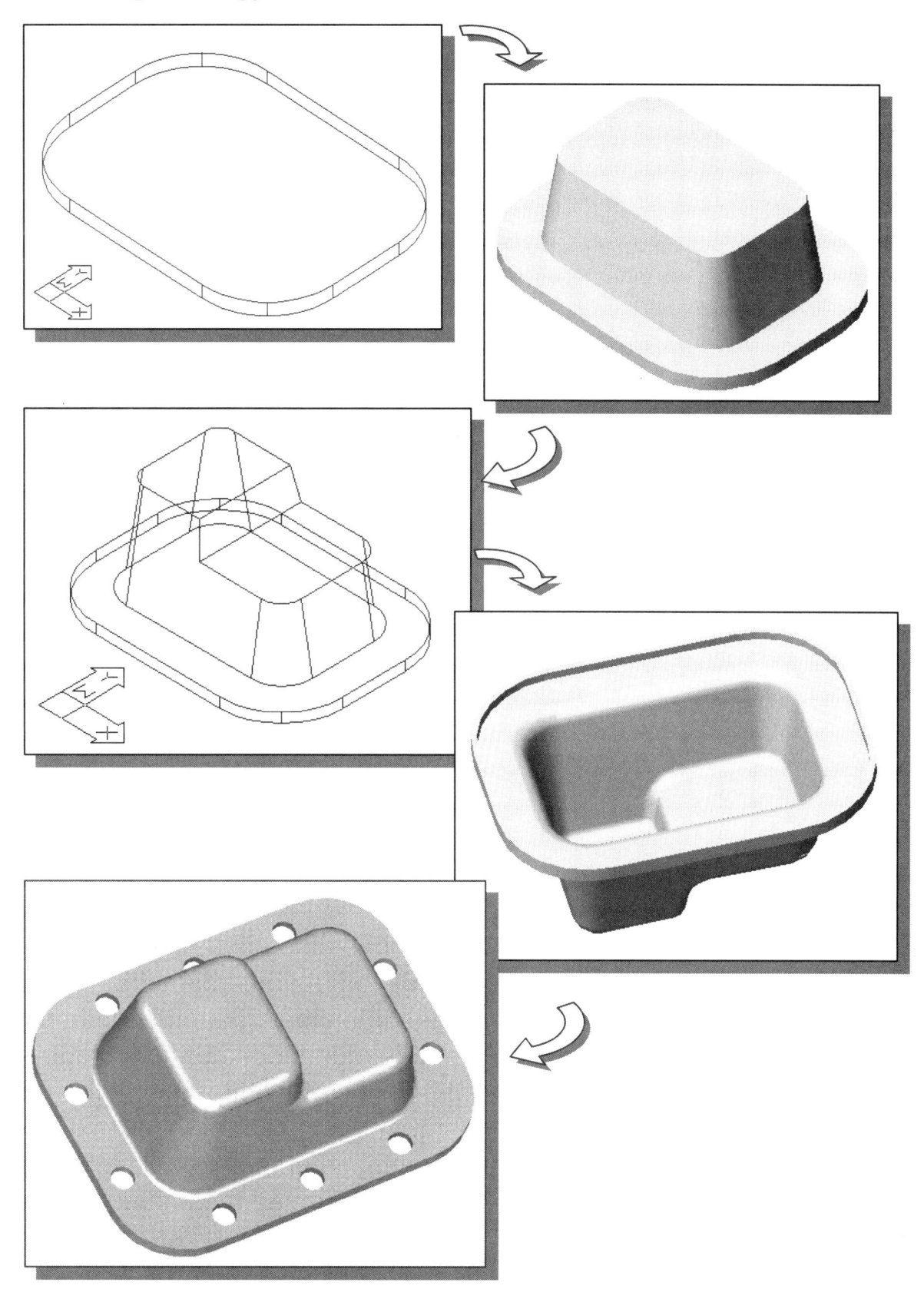

Starting Up AutoCAD® 2006

1. Select the **AutoCAD 2006** option on the *Program* menu or select the **AutoCAD 2006** icon on the *Desktop*.

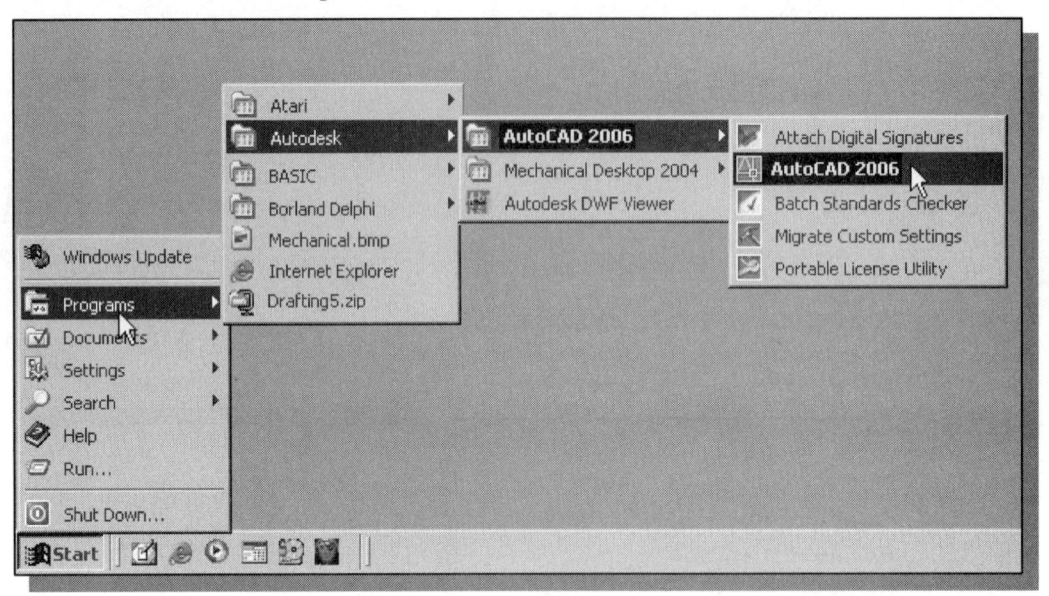

2. In the *Startup* window, select **Start from Scratch**, as shown in the figure below.

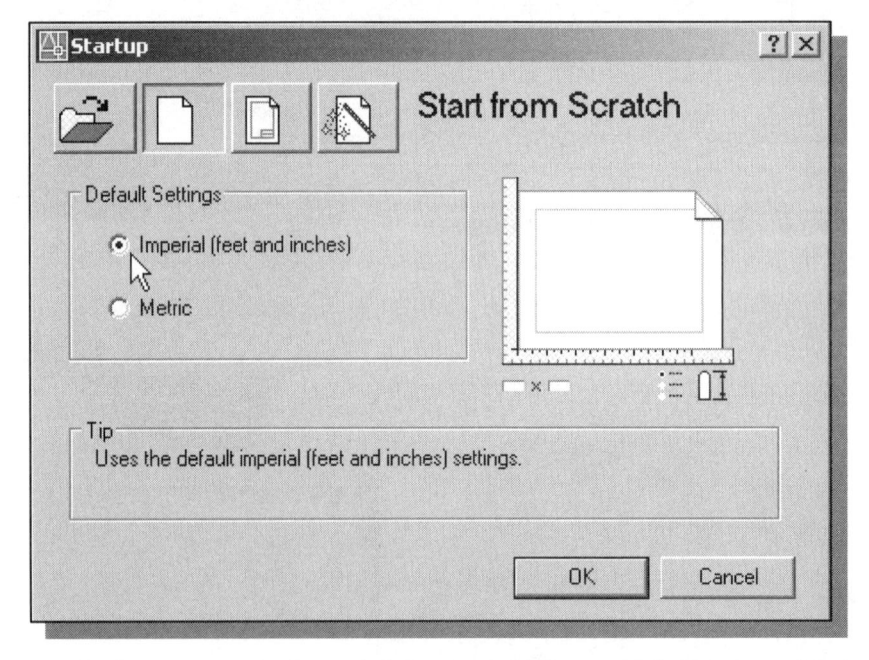

3. In the *Default Settings* section, pick **Imperial (feet and inches)** as the drawing units.

4. Pick **OK** in the *Startup* dialog box to accept the selected settings.

Layers Setup

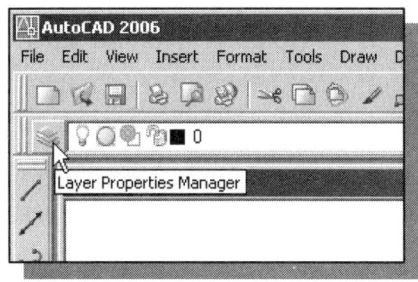

1. Pick **Layer Properties Manager** in the *Object Properties* toolbar.

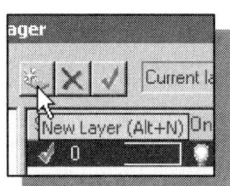

2. Click on the **New** button to create new layers.

3. Create **three new layers** with the following settings:

Layer	Color	LineType	LineWeight
Center_lines	Red	Center	Default
Dimensions	Magenta	Continuous	Default
Object_lines	Cyan	Continuous	0.30mm

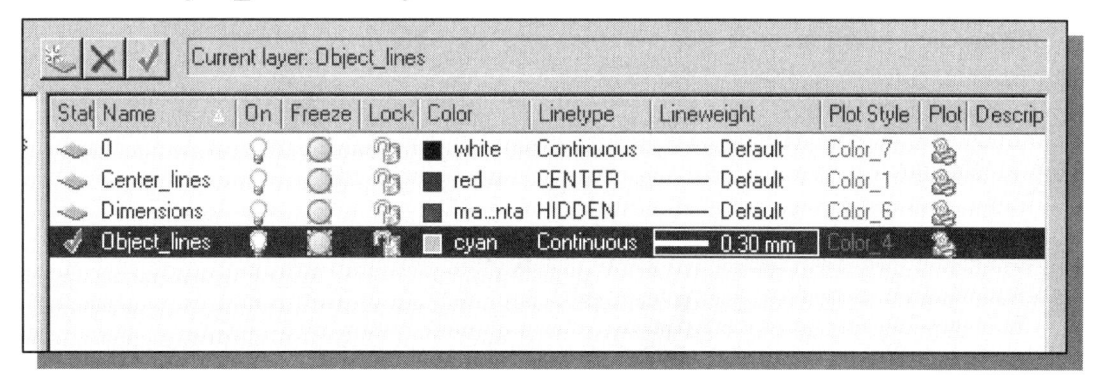

4. Highlight the layer *Object_lines* in the list of layers.

5. Click on the **Current** button to set layer *Object_lines* as the *Current Layer*.

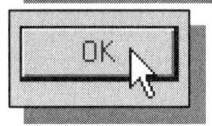

6. Click on the **OK** button to accept the settings and exit the *Layer Properties Manager* dialog box.

7. In the *Status Bar* area, reset the options and turn *ON* the *POLAR, OSNAP, OTRACK, DYN, LWT* and *MODEL* options.

The First Extruded Feature

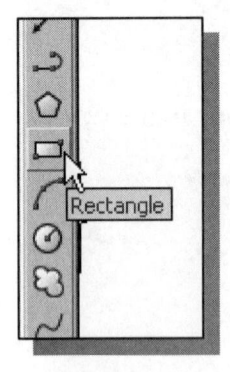

1. Select the **Rectangle** icon in the *Draw* toolbar. In the command prompt area, the message "*Specify first corner point or [Chamfer/Elevation/Fillet/Thickness/Width]:*" is displayed.

2. Place the first corner point of the rectangle near the lower left corner of the screen.

3. We will create a 16" x 12" rectangle for the main body of the sketch. Enter **@16,12 [ENTER]**.

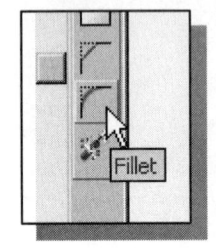

4. Select the **Fillet** command icon in the *Modify* toolbar. In the command prompt area, the message "*Select first object or [Polyline/Radius/Trim]:*" is displayed.

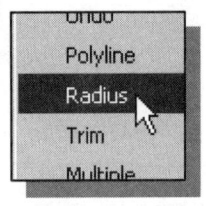

5. Inside the graphics window, right-mouse-click to activate the option menu and select the **Radius** option with the left-mouse-button to specify the radius of the fillet.

6. In the command prompt area, the message "*Specify fillet radius:*" is displayed. *Specify fillet radius:* **4.0 [ENTER]**

7. Inside the graphics window, right-mouse-click to activate the option menu and select the **Polyline** option.

8. Pick the rectangle to create four rounded corners.

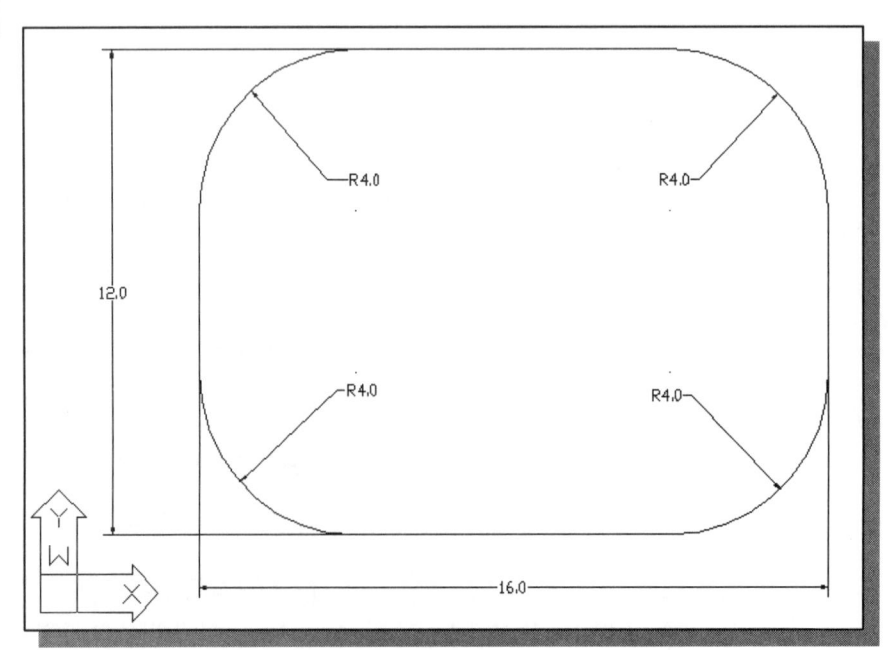

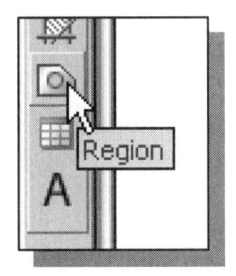

9. Select the **Region** command icon in the *Draw* toolbar.

10. In the command prompt area, the message "*Select objects:*" is displayed. Select the polyline by clicking on any segments of the 2D sketch.

11. Inside the graphics window, **right-mouse-click** to accept the selection and create a region.

12. Move the cursor to the *Standard* toolbar area and right-mouse-click once on any icon in the toolbar area to display a list of toolbar menu groups.

13. Select **Solids**, with the left-mouse-button, to display the *Solids* toolbar on the screen.

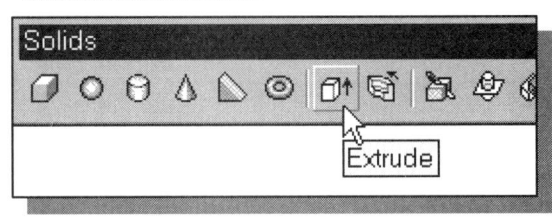

14. Select **Extrude** in the *Solids* toolbar as shown.

15. In the command prompt area, the message "*Select objects:*" is displayed. Select the region by clicking on any of the displayed segments.

16. Inside the graphics window, right-mouse-click to accept the selection and proceed with the **Extrude** command.

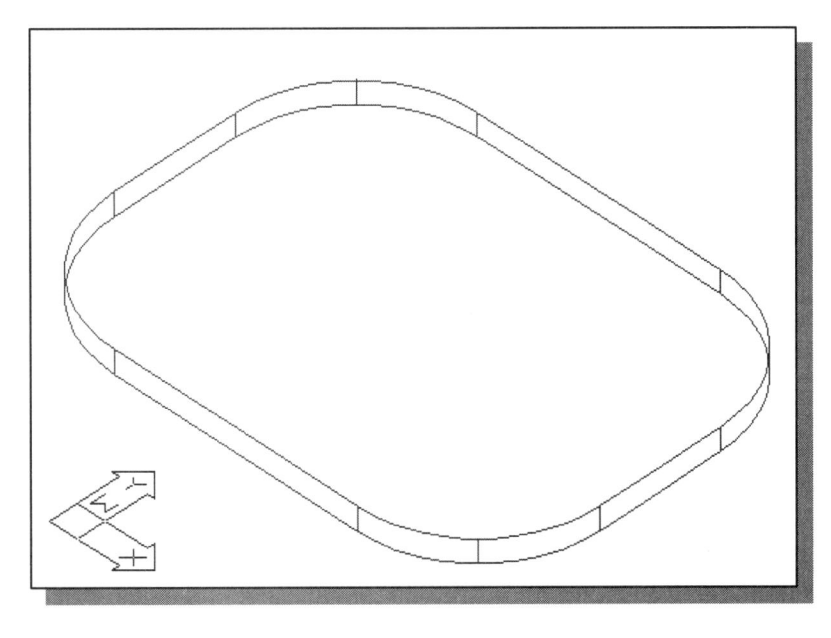

17. In the command prompt area, the message "*Specify height of extrusion or [path]:*" is displayed. Enter: **0.625** [ENTER]

18. In the command prompt area, the message "*Specify angle of taper for extrusion <0>:*" is displayed. Enter: **0.0** [ENTER]

Create an Offset Geometry from an Extracted Surface

- In AutoCAD, several options are available to create geometry from existing solid features. We will use the Copy Faces command to assist constructing geometry from the existing surfaces.

1. Move the cursor to the *Standard* toolbar area and right-click on any icon of the *Standard* toolbar to display a list of toolbar menu groups.

2. Select **Solids Editing**, with the left-mouse-button, to display the *Solids Editing* toolbar on the screen.

3. Click on the **Copy Faces** icon to activate the command.

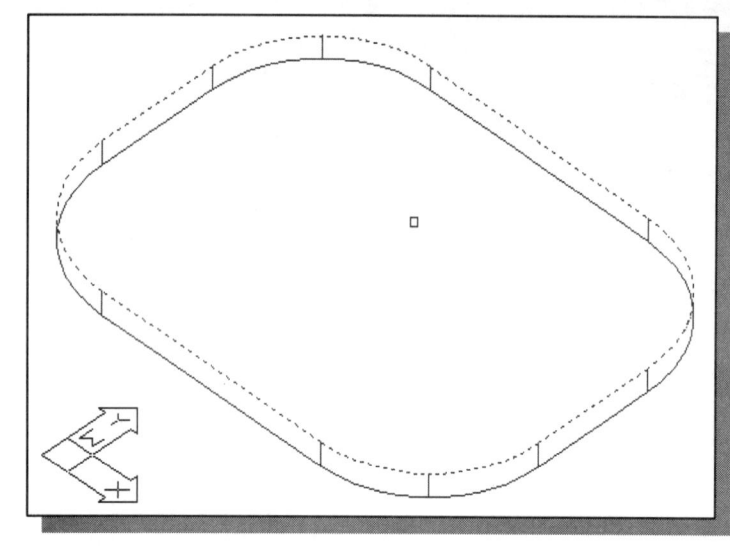

4. In the command prompt area, the message "*Select faces or [Undo/ Remove/ All]:*" is displayed. Pick the **upper surface** by clicking on the inside of the surface as shown.

5. Inside the graphics window, right-mouse-click to bring up the option menu and select **Enter** to accept the selection.

6. In the command prompt area, the message "*Specify a base point or displacement:*" is displayed. Enter: **20,0 [ENTER]**

7. Inside the graphics window, right-mouse-click to accept the entered values.

8. Inside the graphics window, right-mouse-click to bring up the option menu and select **Exit**.

9. Repeat the above step to end the **Copy Faces** command.

Advanced Modeling Tools & Techniques 10-9

10. Select the **Explode** command icon in the *Modify* toolbar. In the command prompt area, the message "*Select objects:*" is displayed.

11. Pick the *surface* we just created.

12. Inside the graphics window, right-mouse-click to end the **Explode** command.

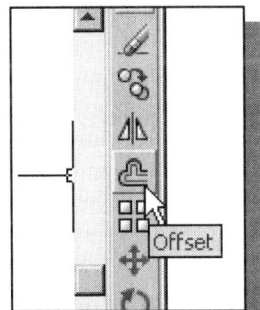

13. Select the **Offset** command icon in the *Modify* toolbar. In the command prompt area, the message "*Specify offset distance or [Through]:*" is displayed.
    Specify offset distance or [Through]: **2.0 [ENTER]**

14. On your own, create a set of entities on the inside of the exploded surface as shown.

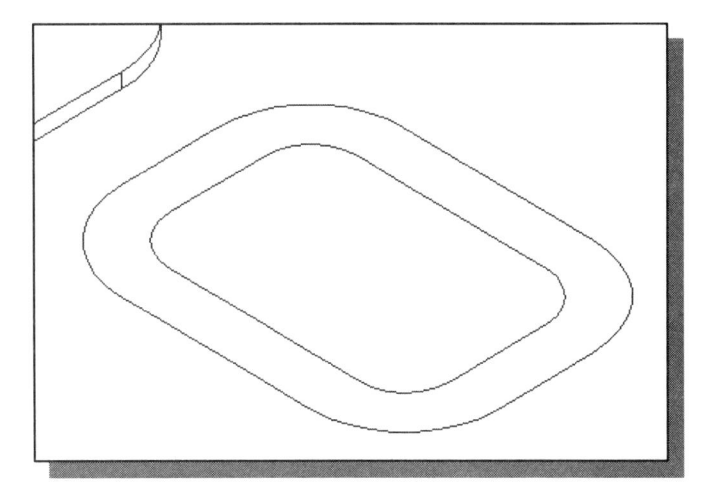

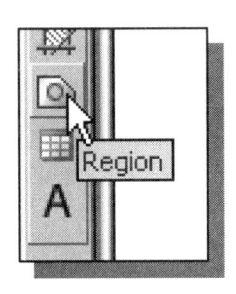

15. Select the **Region** command icon in the *Draw* toolbar.

16. In the command prompt area, the message "*Select objects:*" is displayed. Select the entities we just created with the **Offset** command (the objects inside of the *copy*).

17. Inside the graphics window, right-mouse-click to accept the selection and create a region.

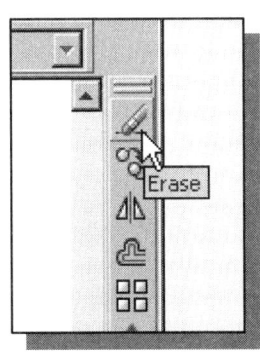

18. Select the **Erase** command icon in the *Modify* toolbar.

19. In the command prompt area, the message "*Select objects:*" is displayed. Select the outer loop entities that were created with the **Copy Faces** command.

20. Inside the graphics window, right-mouse-click to accept the selection and end the **Erase** command.

Extrude with Draft Angle

1. Select **Extrude** in the *Solids* toolbar as shown.

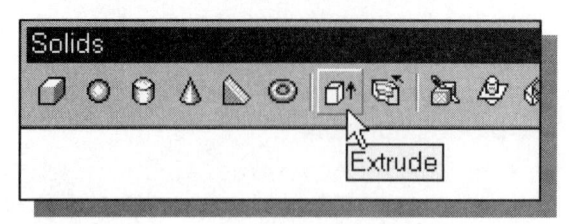

2. In the command prompt area, the message "*Select objects:*" is displayed. Select the 2D region by clicking on any of the displayed segments.

3. Inside the graphics window, right-mouse-click to accept the selection and proceed with the **Extrude** command.

4. In the command prompt area, the message "*Specify height of extrusion or [path]:*" is displayed. Enter: **4.0 [ENTER]**

5. In the command prompt area, the message "*Specify angle of taper for extrusion <0>:*" is displayed. Enter: **10.0 [ENTER]**

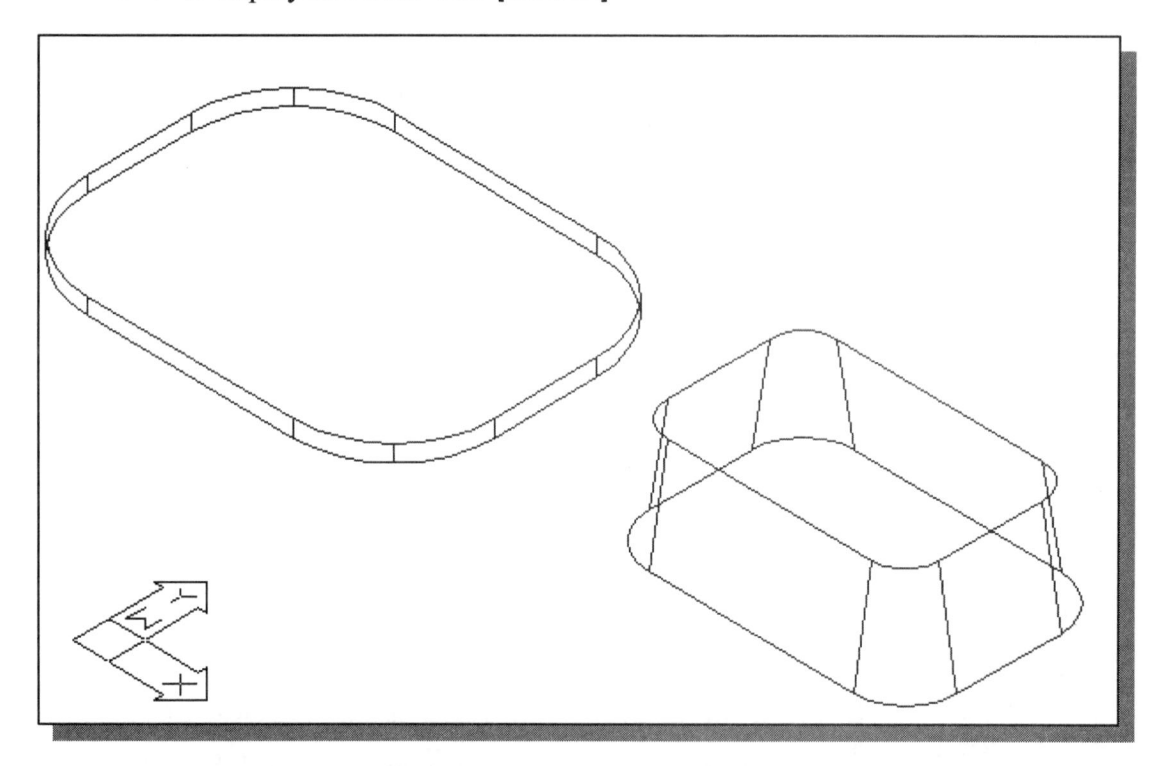

- Note that the second part is created by extruding an extracted surface from the first solid model. The ability to extract and re-use existing surface-geometry is an important aspect of solid modeling. Also note that AutoCAD allows us to create multiple parts in a single drawing file, which allows us to create assembly models under the same deign environment.

Advanced Modeling Tools & Techniques   10-11

Aligning the Parts

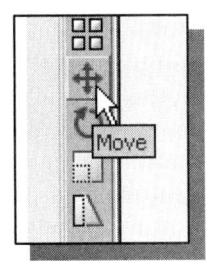

1. Select the **Move** icon in the *Draw* toolbar.

2. In the command prompt area, the message "*Select Objects:*" is displayed. Pick the *second part* we just created.

3. Inside the graphics window, right-mouse-click to accept the selection.

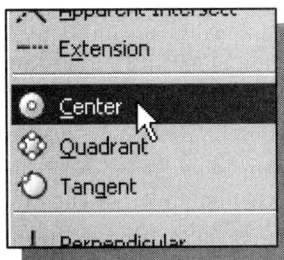

4. Inside the graphics window, hold down the [**SHIFT**] key and **right-mouse-click** once to bring up the *Object Snap* shortcut menu.

5. Select the **Center** option in popup window.

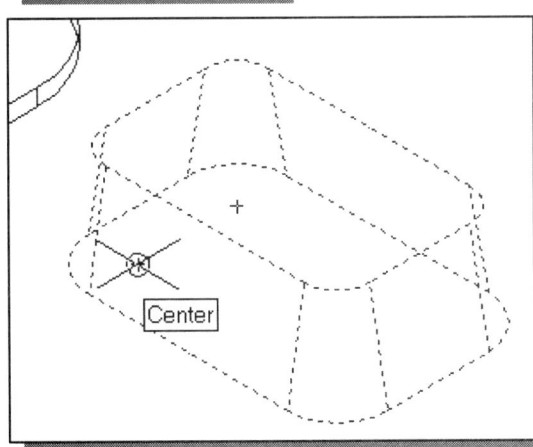

6. Pick the center of the bottom left arc as shown in the figure.

7. Inside the graphics window, hold down the [**SHIFT**] key and **right-mouse-click** once to bring up the *Object Snap* shortcut menu.

8. Select the **Center** option in popup window.

9. Pick the corresponding center point on the top surface of the solid model as shown.

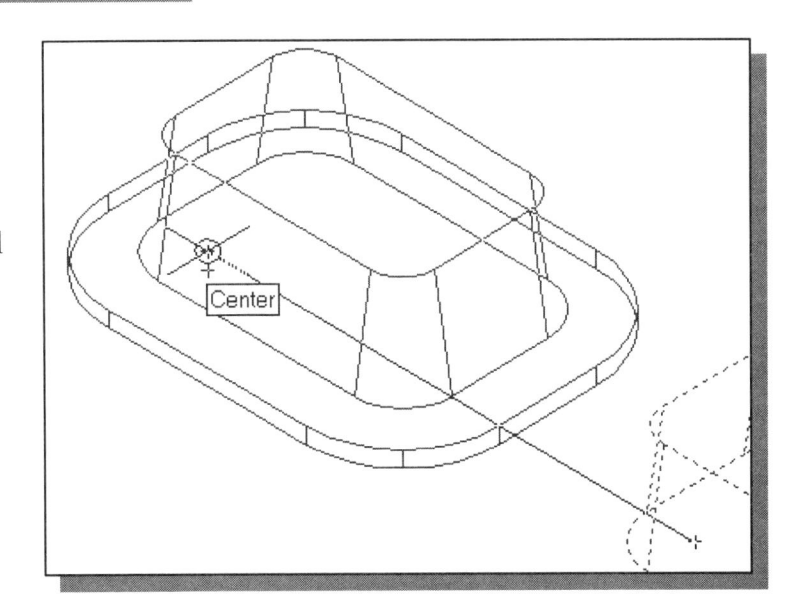

Create another Extracted Surface

1. Click on the **Copy Faces** icon to activate the command.

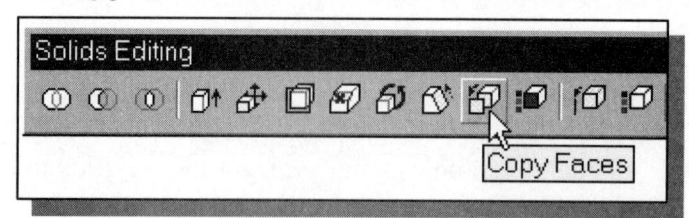

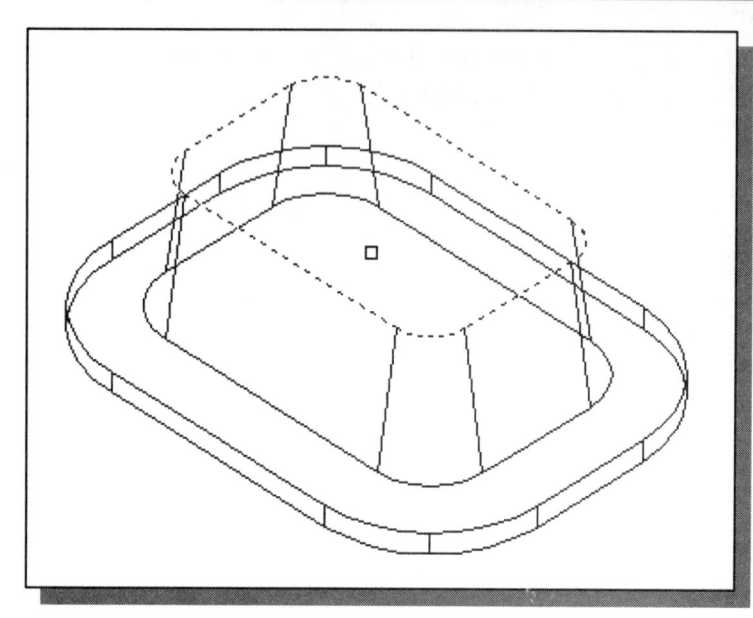

2. In the command prompt area, the message "*Select faces or [Undo/ Remove/ All]:*" is displayed. Pick the top surface of the tapered part by clicking on the inside of the surface as shown.

3. Inside the graphics window, right-mouse-click to bring up the option menu and select **Enter** to accept the selection.

4. In the command prompt area, the message "*Specify a base point or displacement:*" is displayed. Enter: **20,0 [ENTER]**

5. Inside the graphics window, right-mouse-click to accept the entered values.

6. Inside the graphics window, right-mouse-click to bring up the option menu and select **eXit**.

7. Repeat the above step to end the **Copy Faces** command.

8. Select the **Explode** command icon in the *Modify* toolbar. In the command prompt area, the message "*Select objects:*" is displayed.

9. Pick the *surface* we just created.

10. Inside the graphics window, right-mouse-click to end the **Explode** command.

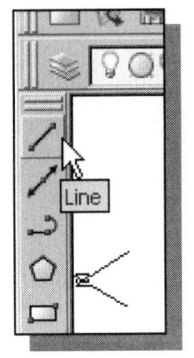

Advanced Modeling Tools & Techniques   10-13

11. Select the **Line** command icon in the *Draw* toolbar.

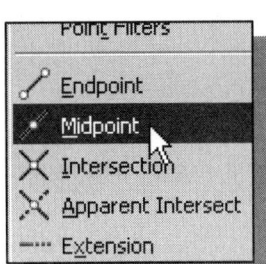

12. Inside the graphics window, hold down the [**SHIFT**] key and **right-mouse-click** once to bring up the *Object Snap* shortcut menu.

13. Select the **Midpoint** option in popup window.

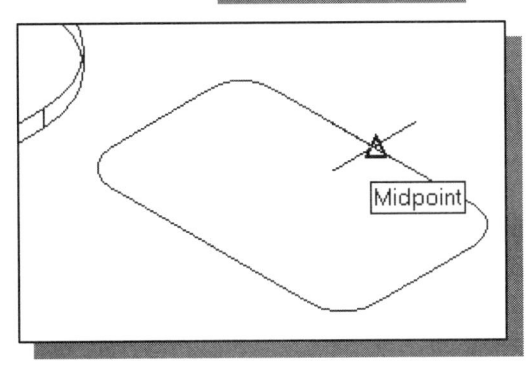

14. In the command prompt area, the message "*_line Specify first point:*" is displayed. Pick the line segment on the extracted surface as shown.

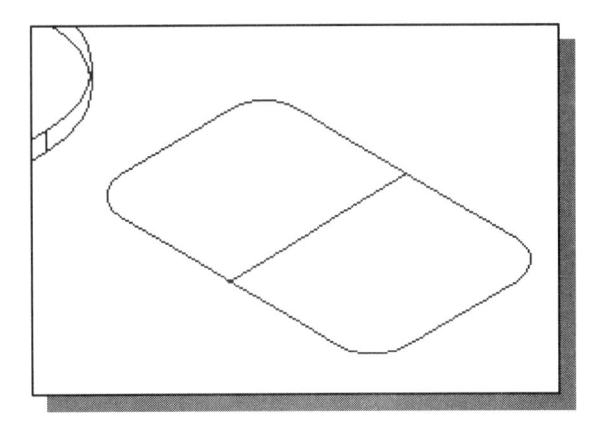

15. On your own, repeat the *Midpoint SNAP* option and create a line as shown.

16. Inside the graphics window, right-mouse-click once and select **Enter** to end the Line command.

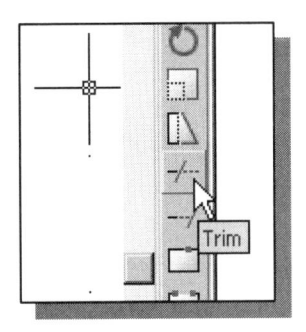

17. Select the **Trim** command icon in the *Modify* toolbar.

18. In the command prompt area, the message "*Select boundary edges... Select objects:*" is displayed. Pick the line across the center of the 2D sketch.

19. Inside the graphics window, right-mouse-click to proceed with the Trim command.

20. The message "*Select object to trim or shift-select object to extend or [Project/ Edge/ Undo]:*" is displayed in the command prompt area. Trim the two lines of the exploded geometry as shown.

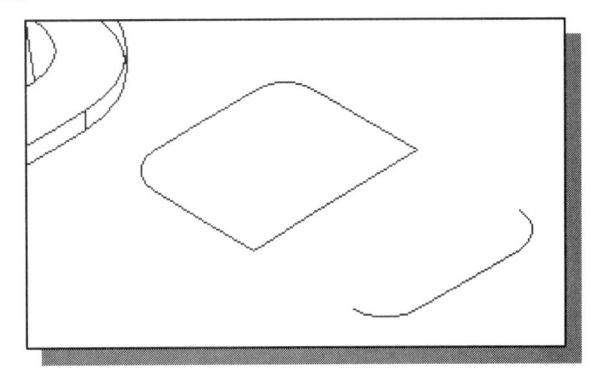

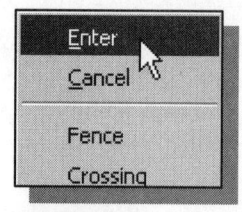

21. Inside the graphics window, right-mouse-click to activate the option menu and select **Enter** with the left-mouse-button to end the Trim command.

22. Select the **Region** command icon in the *Draw* toolbar.

23. In the command prompt area, the message "*Select objects:*" is displayed. Select the entities we just modified.

24. Inside the graphics window, right-mouse-click to accept the selection and create a region.

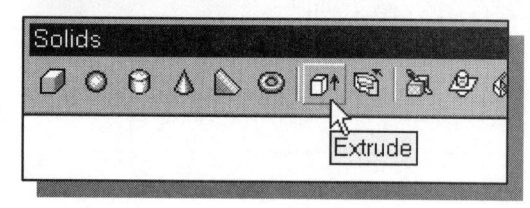

25. Select **Extrude** in the *Solids* toolbar as shown.

26. In the command prompt area, the message "*Select objects:*" is displayed. Select the 2D region by clicking on any of the displayed segments.

27. Inside the graphics window, right-mouse-click to accept the selection and proceed with the **Extrude** command.

28. In the command prompt area, the message "*Specify height of extrusion or [path]:*" is displayed. Enter: **2.0 [ENTER]**

29. In the command prompt area, the message "*Specify angle of taper for extrusion <0>:*" is displayed. Enter: **10.0 [ENTER]**

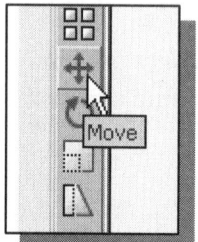

30. Select the **Move** icon in the *Modify* toolbar.

31. On your own, reposition the part on top of the other parts as shown.

❖ At this point, the three separate parts are positioned and aligned on top of each other. Note that the last part was created without applying or knowing the exact diemensions of the extracted geometry.

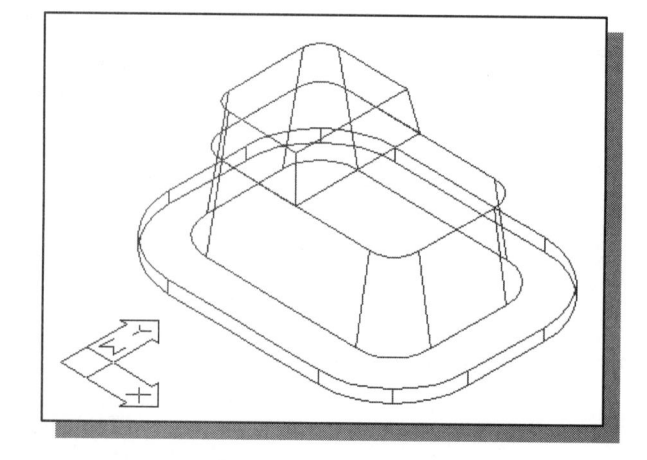

Combining Parts – Boolean UNION

1. Select **3D Orbit** in the *View* pull-down menu.
   [View] → **[3D Orbit]**

2. On your own, rotate the created models in 3D space and confirm the three parts are positioned on top of each other.

3. In the *Solids Editing* toolbar, click on the **Union** icon. In the command prompt area, the message "*Select Objects:*" is displayed.

4. Pick all the solid models by clicking on their edges.

5. Inside the graphics window, right-mouse-click to accept the selection and proceed with the **Union** command.

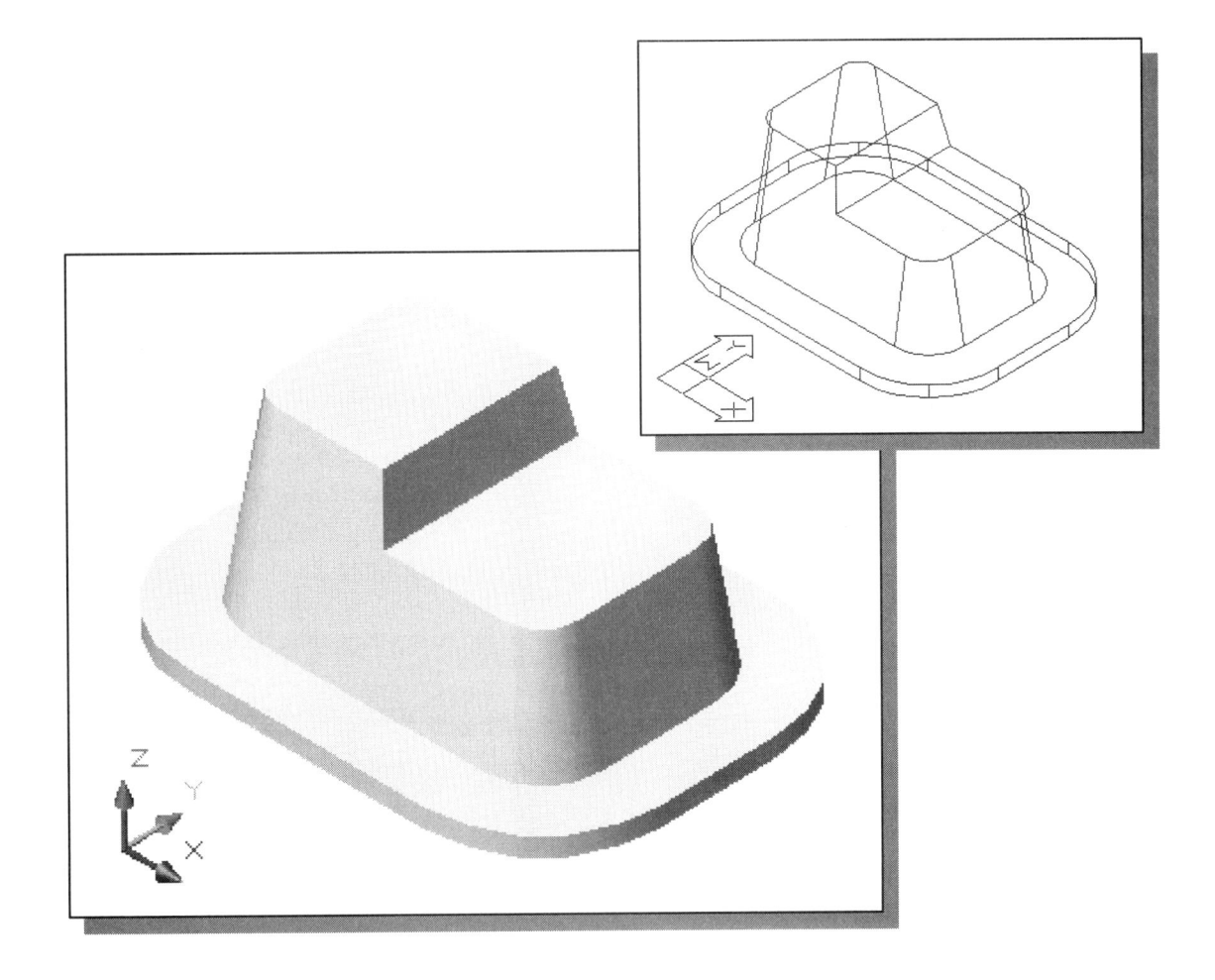

Creating 3D Rounds and Fillets

- In **AutoCAD® 2006**, the same **Fillet** command can be applied to both 2D and 3D designs.

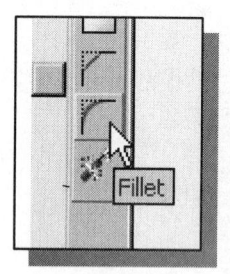

    1. Select the **Fillet** command icon in the *Modify* toolbar.

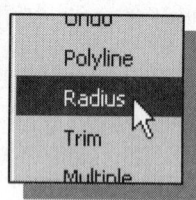

    2. In the command prompt area, the message *"Select first object or [Polyline/Radius/Trim]:"* is displayed. Inside the graphics window, right-mouse-click to activate the option menu and select the **Radius** option with the left-mouse-button to specify the radius of the fillet.

    3. In the command prompt area, the message *"Specify fillet radius:"* is displayed.
       Specify fillet radius: **0.75** **[ENTER]**

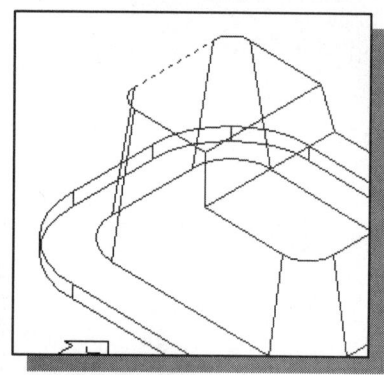

    4. Pick the **top back edge** as shown.

    5. In the command prompt area, the message *"Enter fillet radius <0.75>:"* is displayed. Right-mouse-click to accept the displayed value.

    6. In the command prompt area, the message *"Select an edge or [Chain/Radius]:"* is displayed. Right-mouse-click to activate the option menu and select the **Chain** option.

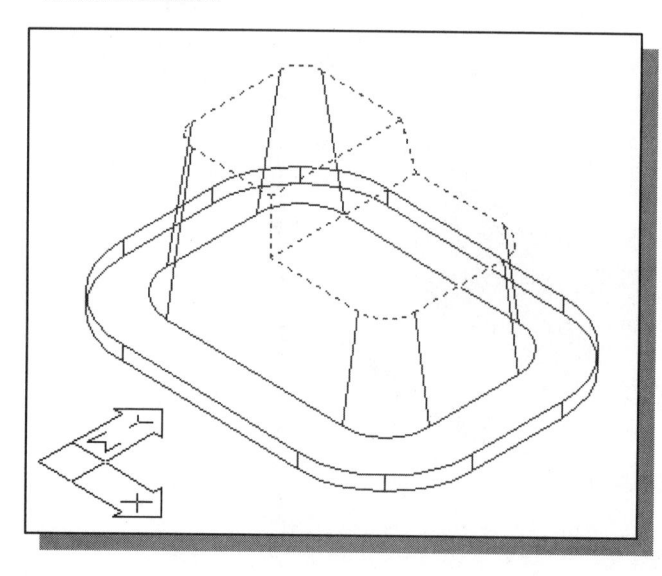

    7. In the command prompt area, the message *"Select an edge or [Chain/Radius]:"* is displayed. Select the adjacent curve on the top surface.

    8. On your own, select the adjacent edges on the top surfaces as shown in the figure.

    9. Press the **[ENTER]** key or the right-mouse-button and select **Enter** to accept the selection.

Advanced Modeling Tools & Techniques   10-17

- The **Fillet** command can be used to create complex 3D curves and surfaces as shown in this example.

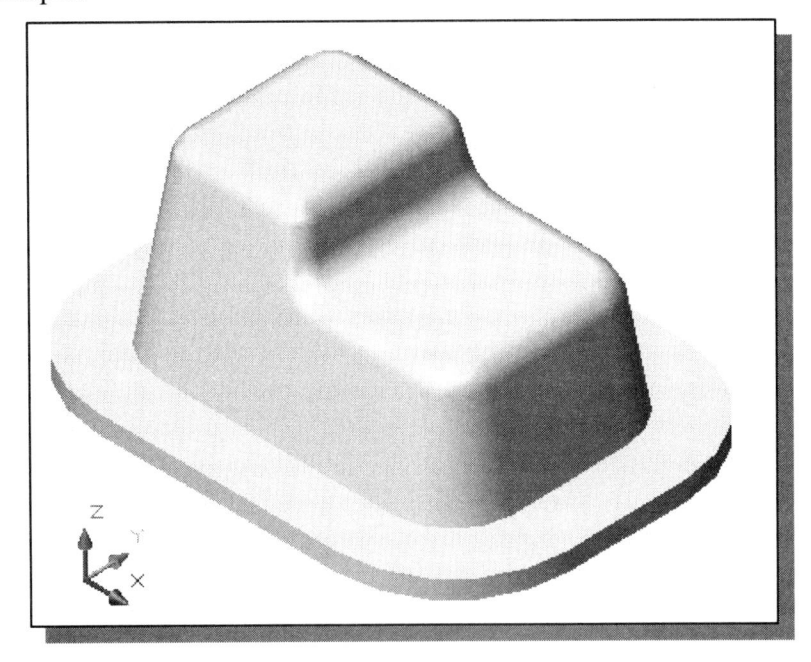

- On your own, create a *chained fillet* of **0.5 radius** at the bottom edges as shown.

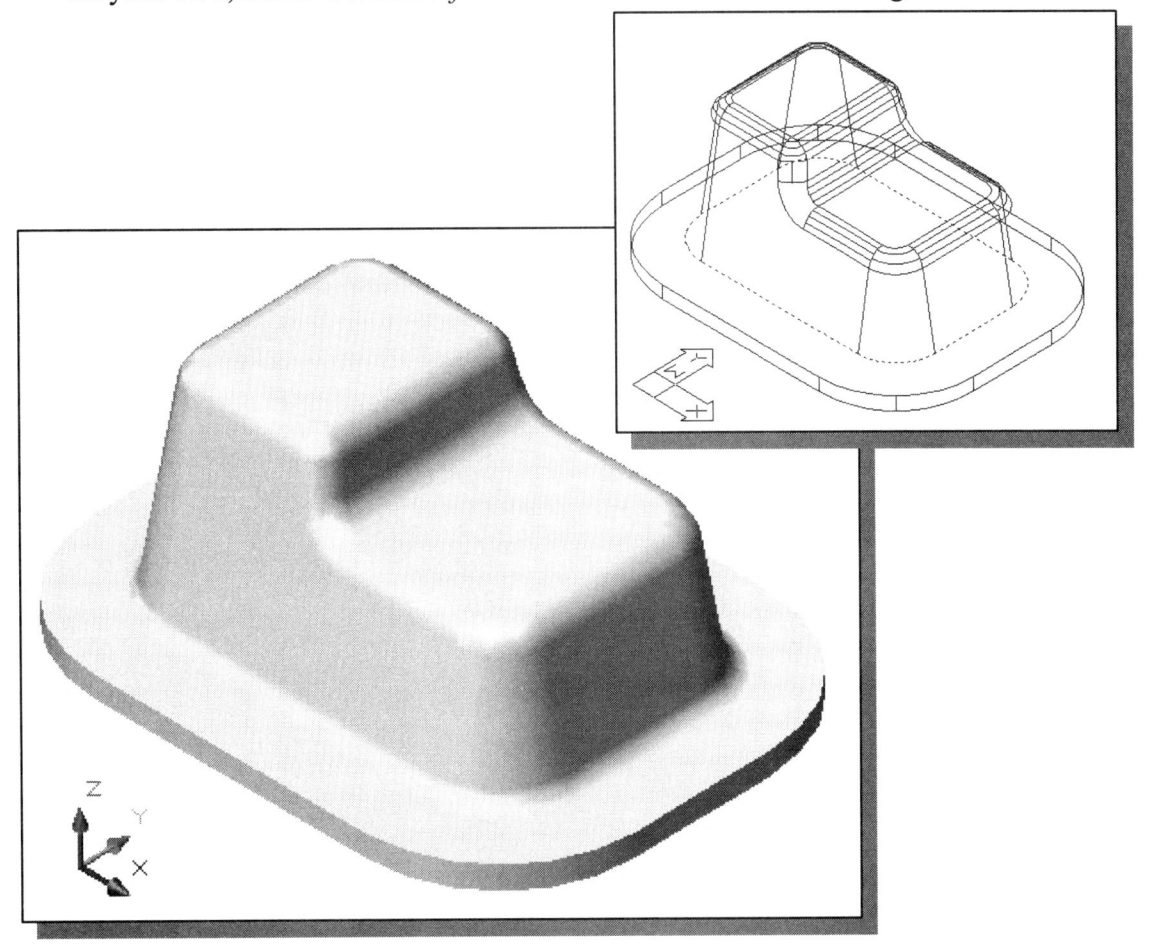

Creating a Shell Feature

- The **Shell** command can be used to create new faces by offsetting existing ones inside or outside of their original positions.

1. Select **3D Orbit** in the *View* pull-down menu.
   [View] → [3D Orbit]

2. On your own, rotate the created models in 3D space so that we are viewing the bottom flat face of the model as shown in the figure below.

3. Choose **Shell** in the *Solids Editing* toolbar.

4. The message "*Select a 3D solid:*" is displayed in the command prompt area. Pick the top surface of the 3D solid. Note the entire solid is selected.

5. In the command prompt area, the message "*Remove faces or [Undo /Add /All]:*" is displayed. Select the top flat side (as oriented in the figure below) of the *Oil Sink* model. (AutoCAD will remove this surface before creating the shell feature.)

6. Inside the graphics window, right-mouse-click to accept the selection and proceed with the **Shell** command.

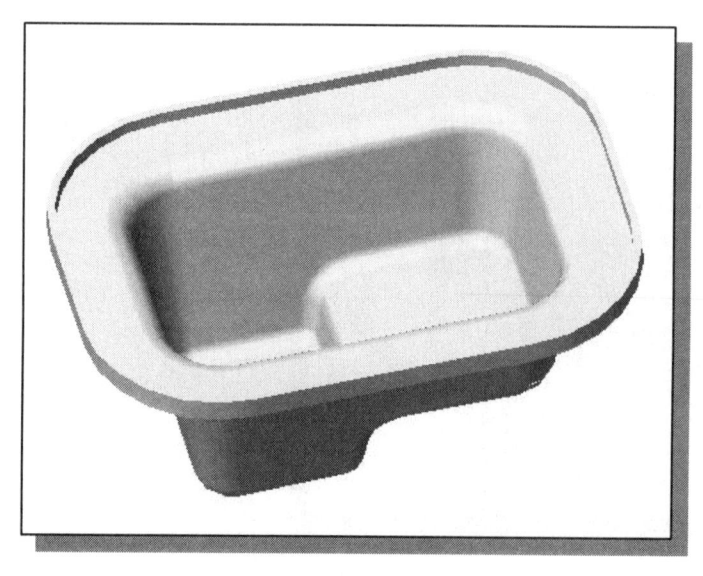

7. In the command prompt area, the message "*Enter the shell offset distance:*" is displayed. Set the thickness to **0.25**.

8. Press the [**ENTER**] key or the **middle-mouse-button** to bring up the option menu and select **Exit** to continue.

9. Repeat the above step and **Exit** the Shell command.

Creating a Rectangular Array Cut Feature

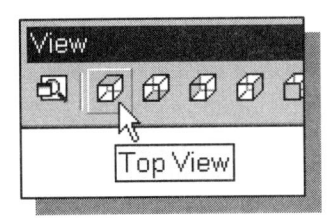

1. In the *View* toolbar, select the **Top View** option to reset the display to the XY plane of the coordinate system.

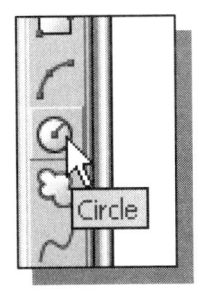

2. Select the **Circle** icon in the *Draw* toolbar.

3. On your own, create a circle (diameter **0.75**) that is toward the right side of the solid model. Note that, in AutoCAD, *circles* are created as valid *regions*.

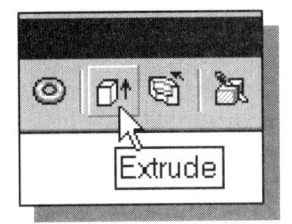

4. On your own, extrude the region into a **1"** high cylinder. (Taper angle set to **0**.)

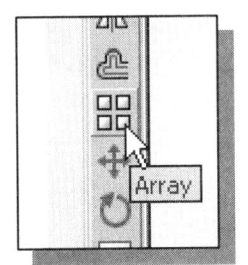

5. Select **Array** in the *Modify* toolbar as shown.

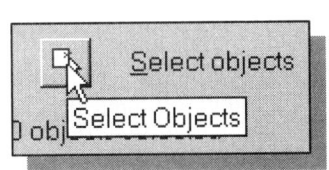

6. Click on the **Select Objects** button.

7. Select the cylinder.

8. Inside the graphics window, right-mouse-click to accept the selection and return to the Array command.

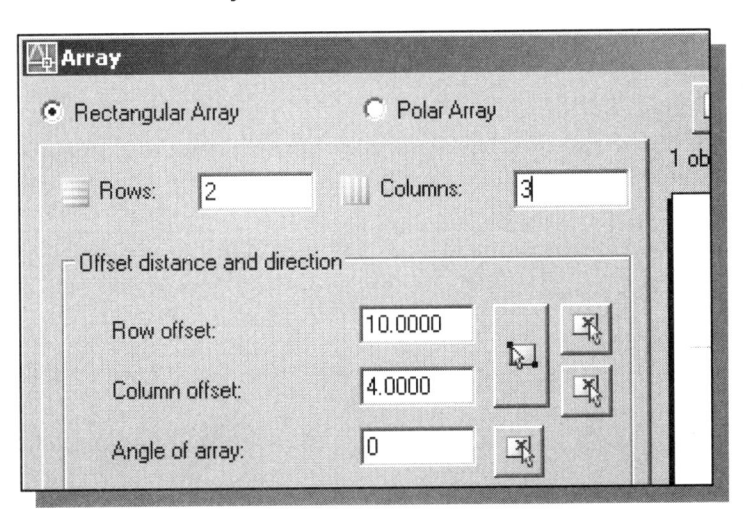

9. In the *Array* window, fill in the items as shown: *Rectangular array*, **2** *Rows* with **10"** *Row offset* and **3** *Columns* with **4"** *Column offset*.

10. Click on the **OK** button to accept the settings.

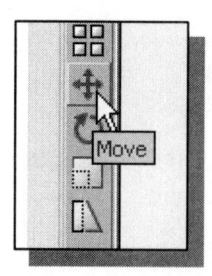

11. Select the **Move** icon in the *Draw* toolbar.

12. On your own, reposition the six cylinders as shown in the figure below. Make sure the cylinders pass through the *Oil Pan* model before proceeding to the next step.

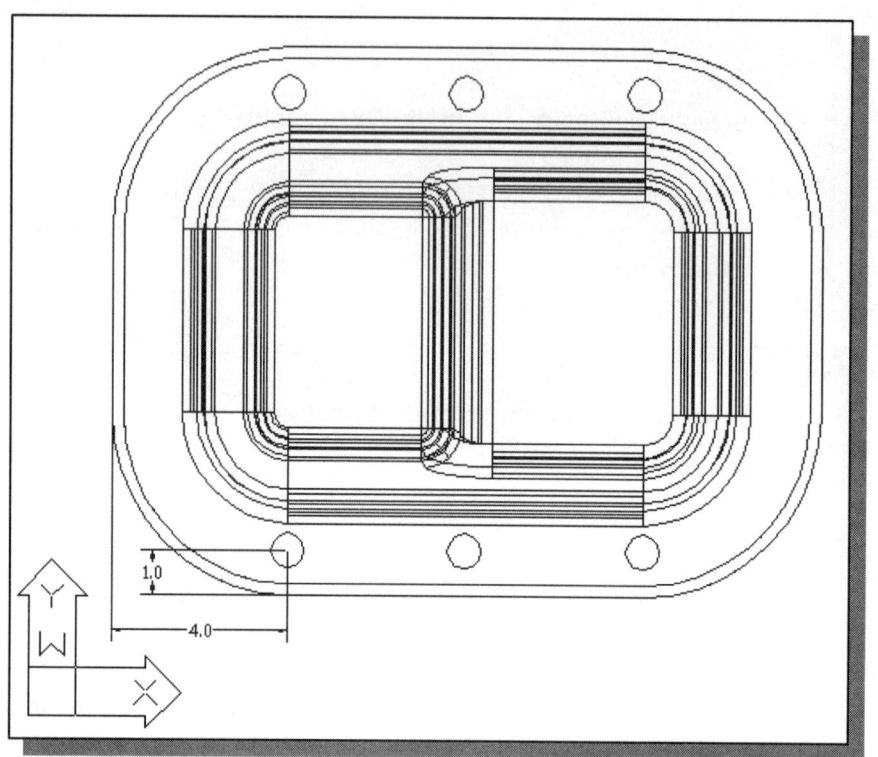

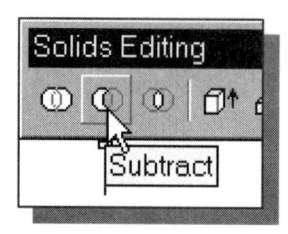

13. In the *Solids Editing* toolbar, click on the **Subtract** icon.

14. In the command prompt area, the message "*_subtract Select solids and regions to subtract from .. Select Objects:*" is displayed. Pick the *Oil Pan* model by clicking on one of the edges.

15. Inside the graphics window, right-mouse-click to accept the selection and proceed with the **Subtract** command.

16. In the command prompt area, the message "*Select solids and regions to subtract .. Select Objects:*" is displayed.

17. Pick the six cylinder blocks by clicking on the circular edges.

18. Inside the graphics window, right-mouse-click to accept the selection and proceed with the **Subtract** command.

Creating Another Rectangular Array Cut Feature

1. On your own, repeat the above steps and create another rectangular array as shown (circular hole: diameter **0.75**, *Rectangular array*, **2** *Rows* with **4"** *Row offset* and **2** *Columns* with **14"** *Column offset*.)

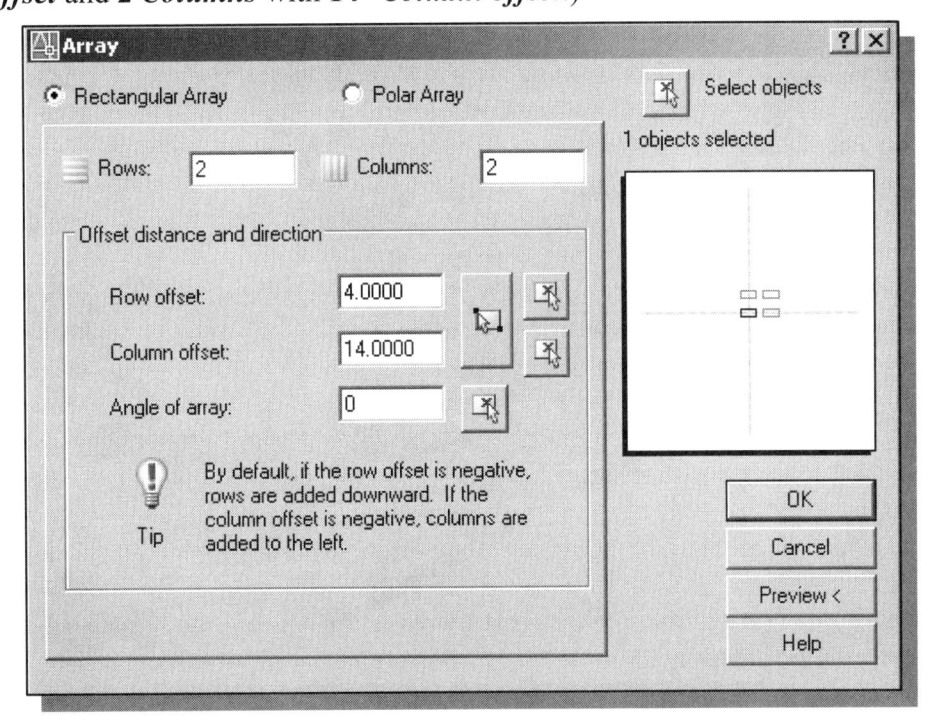

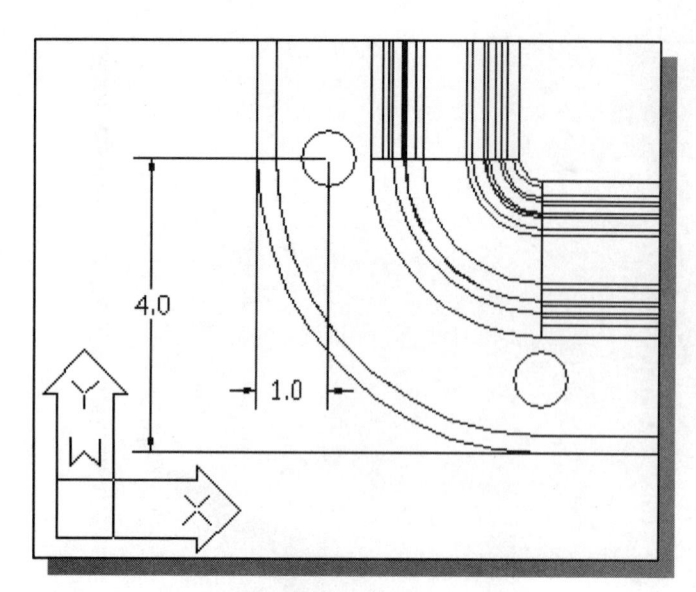

2. On your own, reposition the rectangular array so that the lower left cylinder is located at 4.0 inches away from the bottom edge of the base feature and 1.0 inch away from the left edge of the *Oil Pan* model.

3. Complete the model with the **Subtract** command.

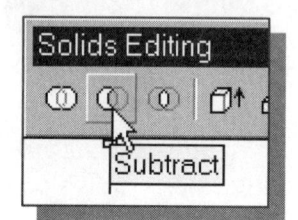

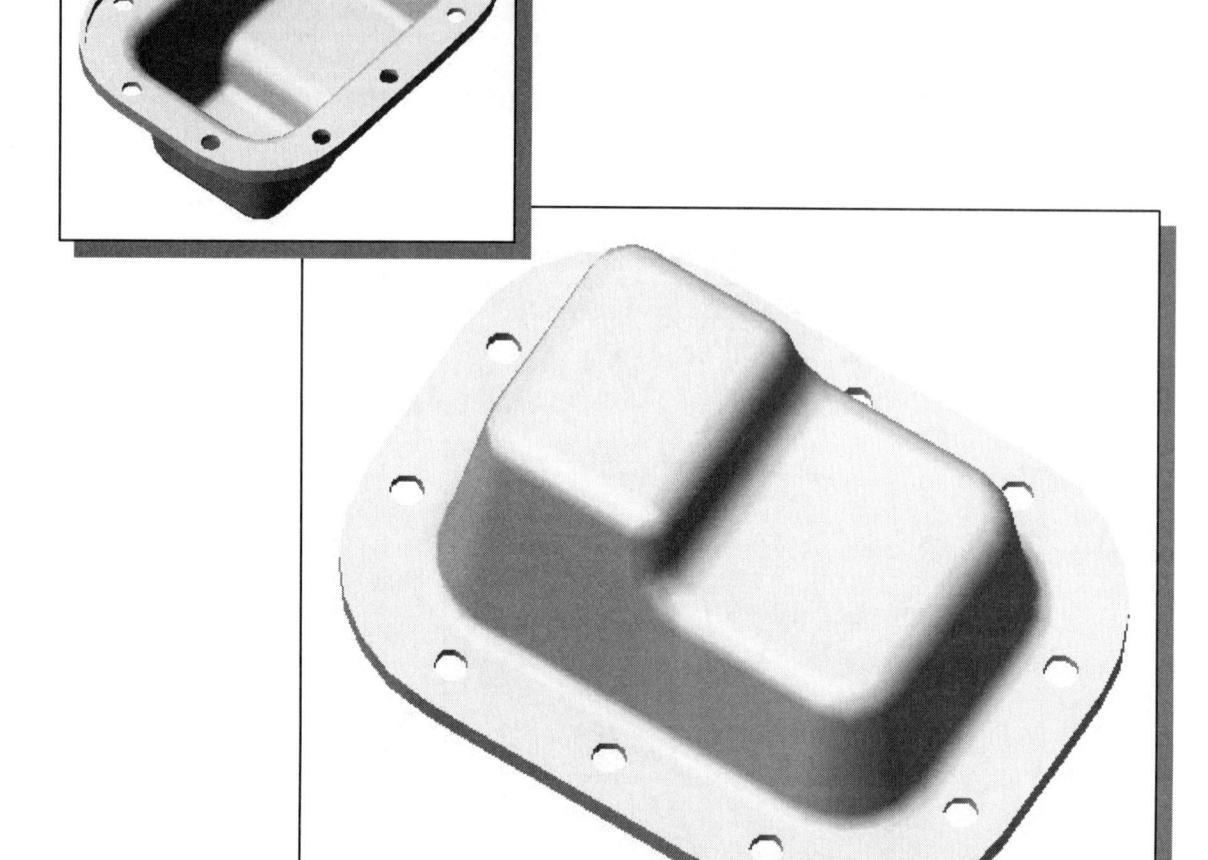

Conclusion

Design includes all activities involved from the original concept to the finished product. Design is the process by which products are created and modified. For many years designers sought ways to describe and analyze three-dimensional designs without building physical models. With the advancements in computer technology, the creation of computer solid models on computers offers a wide range of benefits. Solid models are easier to interpret and can be easily altered. Solid models can also be analyzed using finite element analysis software, and simulation of real-life loads can be applied to the models and the results graphically displayed.

Throughout this text, various modeling techniques have been presented. Mastering these techniques will enable you to create intelligent and flexible solid models. The goal is to make use of the tools provided by AutoCAD and to successfully perform solid models of the product. In many instances, only a single approach to the modeling tasks was presented; you are encouraged to repeat all of the chapters and develop different ways of thinking in accomplishing the same tasks. We have only scratched the surface of AutoCAD's functionality. The more time you spend using the system, the easier it will be to perform solid modeling with AutoCAD.

Questions:

1. Describe the method to create *Draft Angle* features in AutoCAD.

2. List and describe the differences between *Cut with a Pattern* and *Cut Each One Individually*?

3. Which command is used to create 3D rounded corners in AutoCAD?

4. What elements are required to define a *Rectangular Array* in AutoCAD?

5. How do we create a thin-walled part in AutoCAD? What elements are required to perform the operation?

6. Identify the following commands:

   (a)

   (b)

   (c)

   (d)

Exercises:

1. Dimensions are in inches.

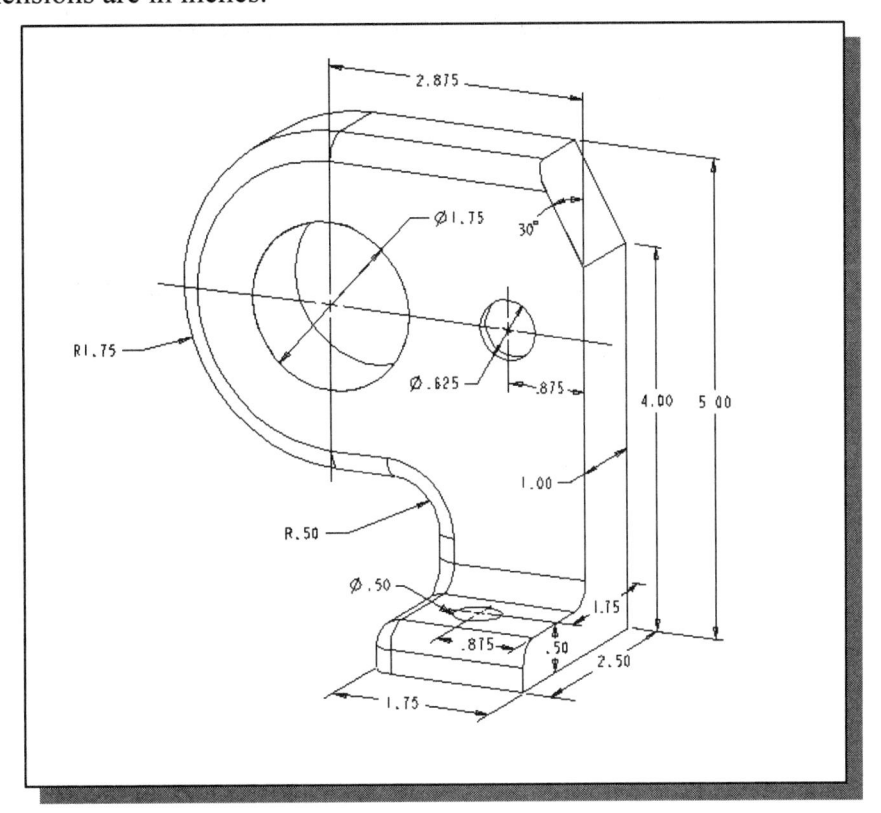

2. Dimensions are in millimeters.

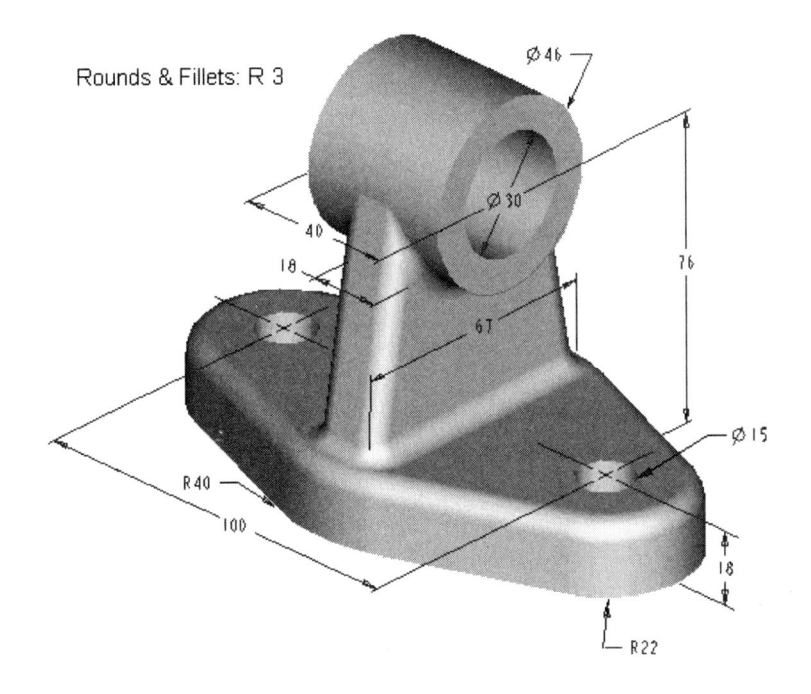

Rounds & Fillets: R 3

10-26　AutoCAD® 2006 Tutorial: 3D Modeling

3. **Leveling Assembly** (Create a set of detail and assembly drawings. All dimensions are in mm.)

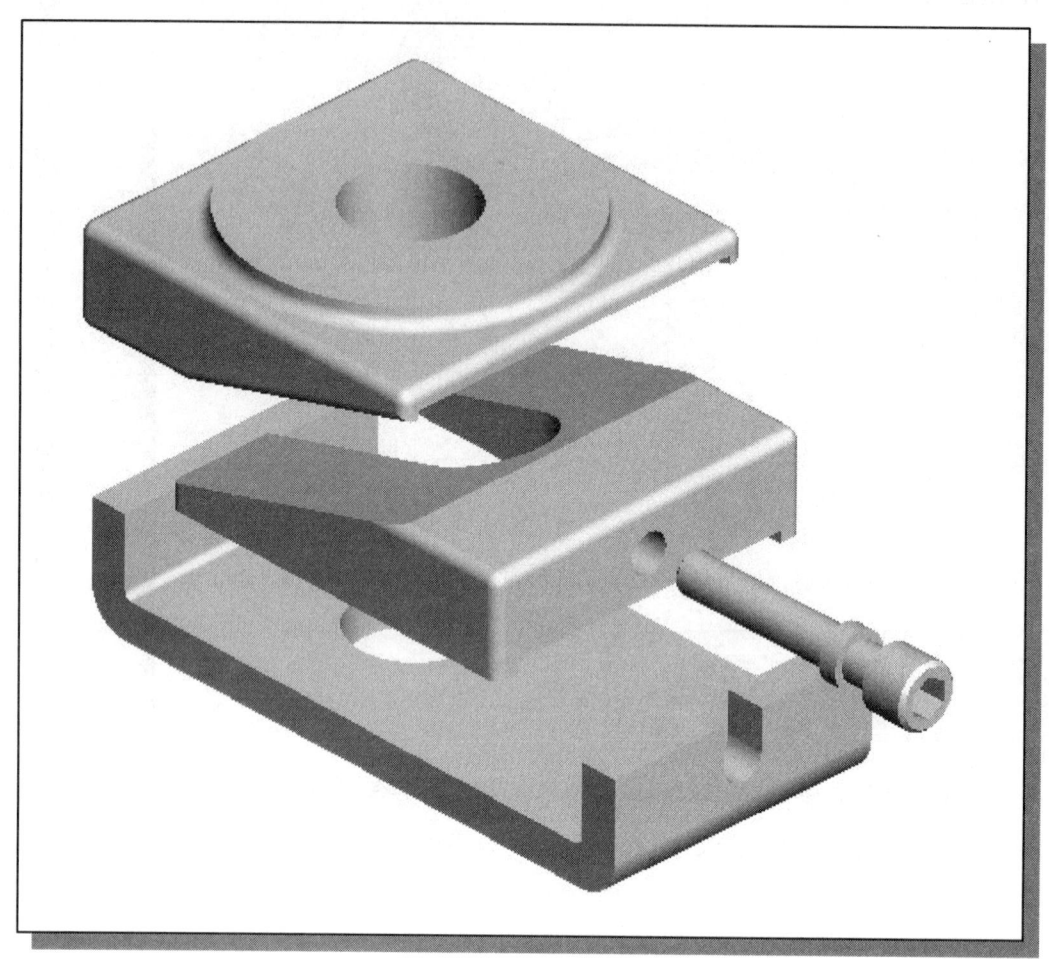

(a) Base Plate

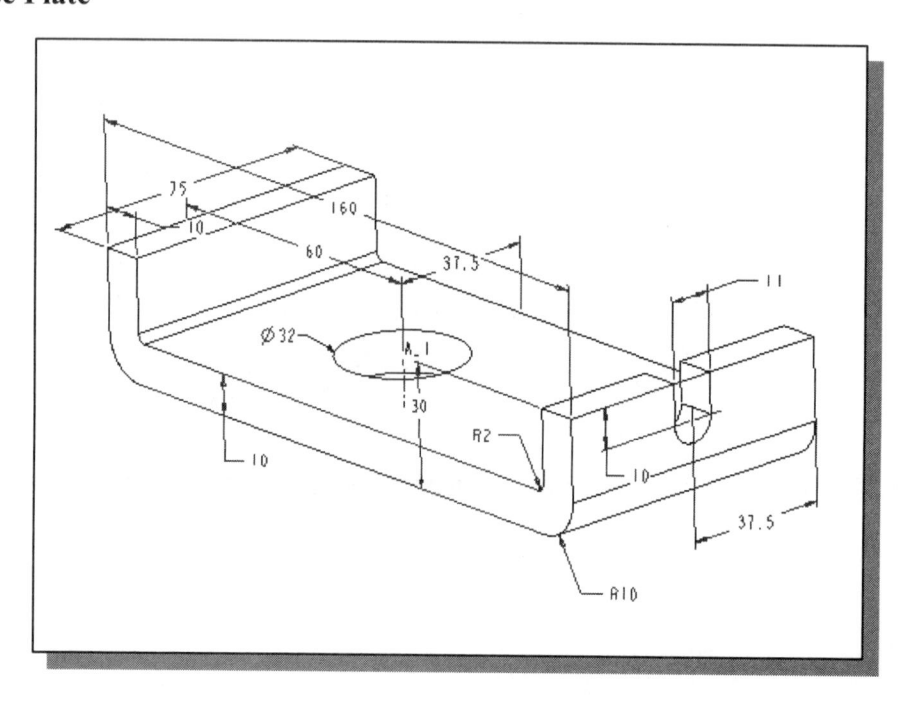

(b) **Sliding Block** (Rounds & Fillets: R3)

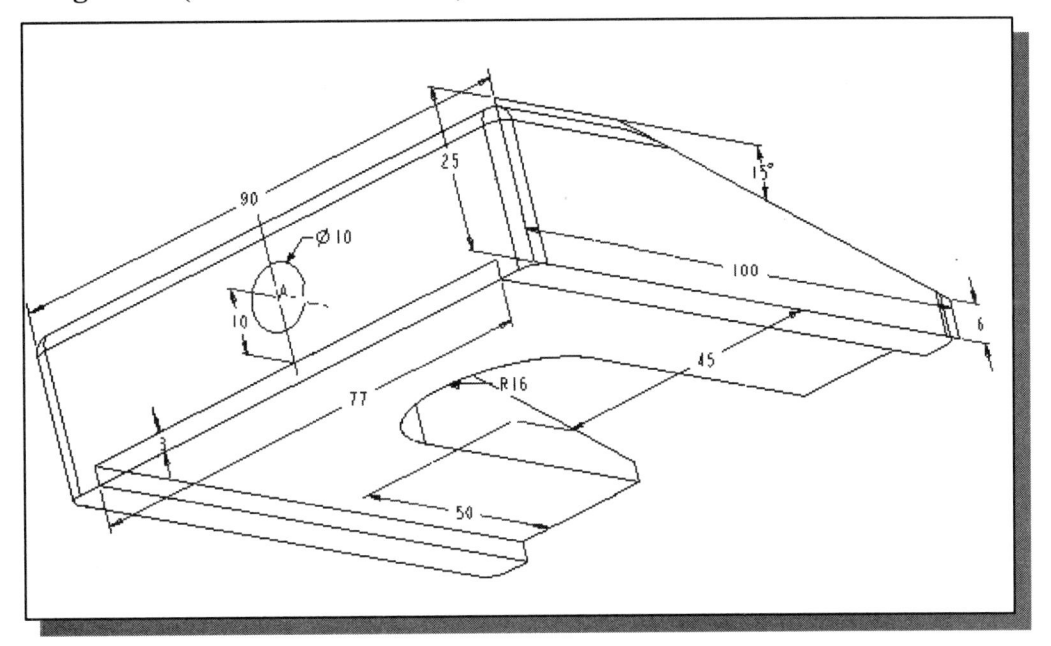

(c) **Lifting Block** (Rounds & Fillets: R3)

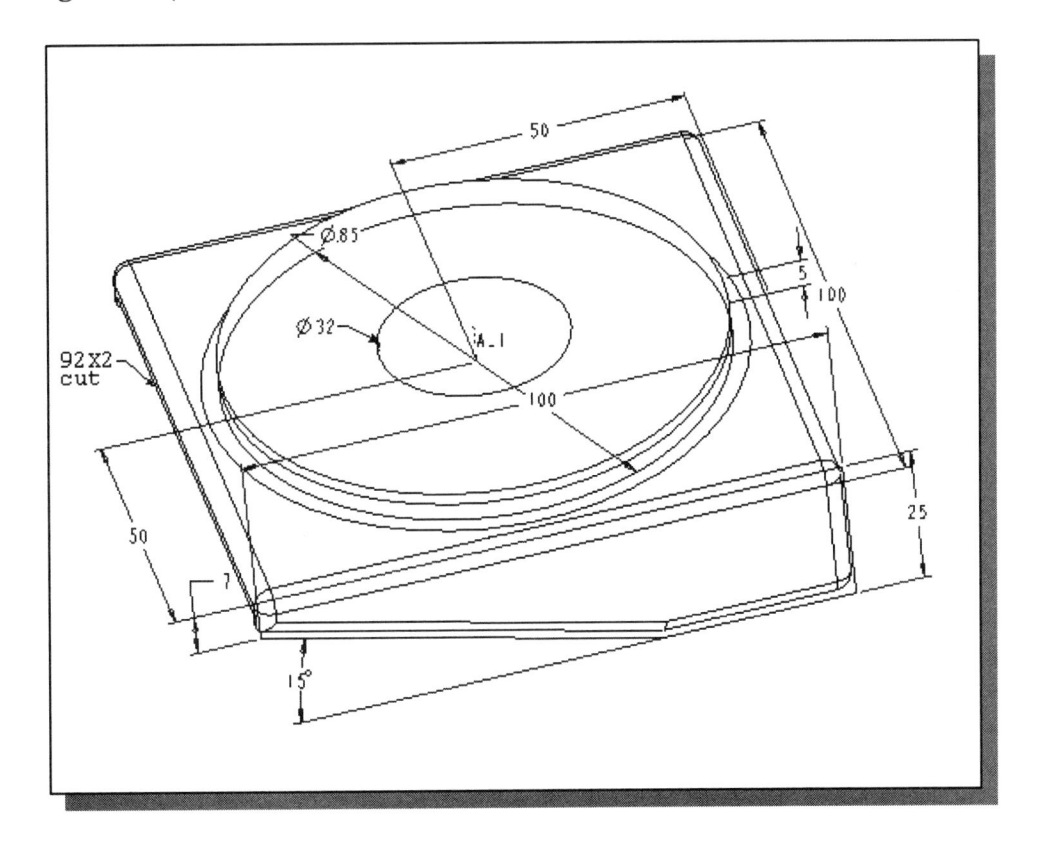

(d) **Adjusting Screw** (M10 × 1.5)

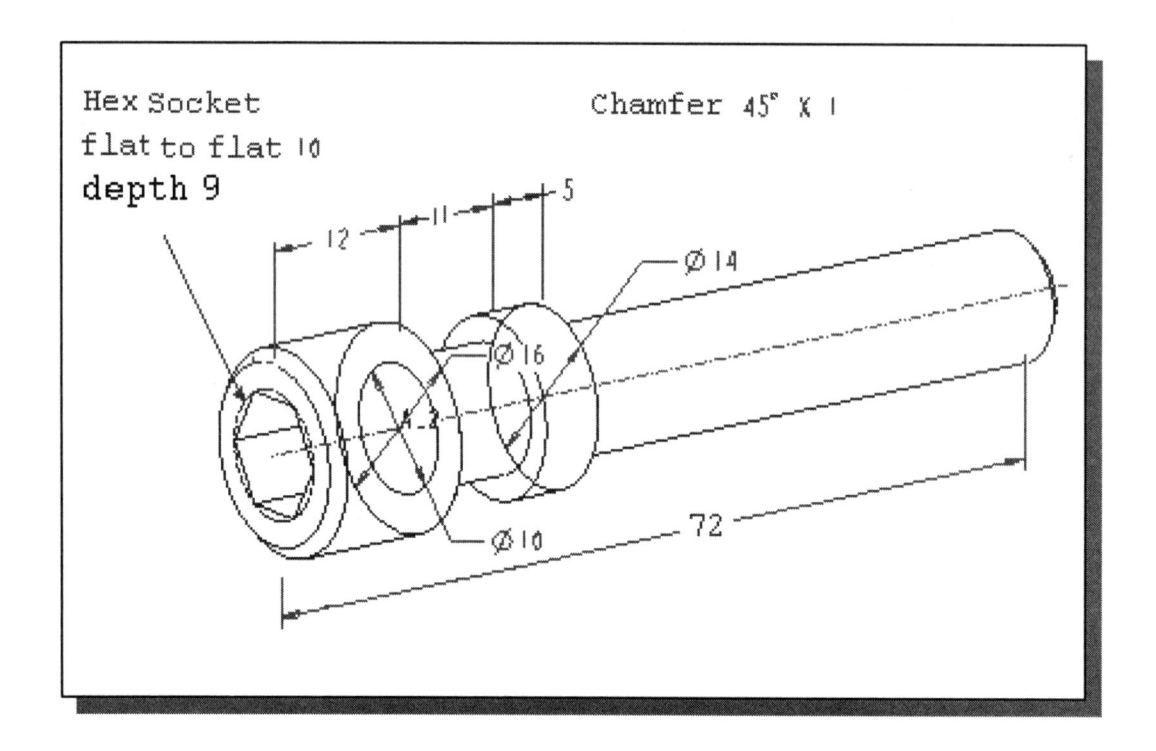

Hex Socket
flat to flat 10
depth 9

Chamfer 45° X 1

12
11
5
Ø14
Ø16
Ø10
72

INDEX

A
Additional Resources, 1-11
Arcball, 3-17
Array, 3D, 9-16

B
Binary Tree, 6-4
Block, solids, 6-7
Boolean Intersect, 6-2
Boolean Operations, 6-2
Boundary representation (B-rep), 7-2
Box, Solid, 6-7
Box Method, 3-2
Break, 5-21
Buttons, Mouse, 1-10

C
CAD, 1-2
CAE, 1-2
Canceling commands, 1-11
Cartesian coordinate system, 2-13
CIRCLE command, 3-21
Circular Pattern, 9-16
Close dialog box, 2-16
Close window, 4-3
Combine Parts, 9-14
Command Prompt, 1-8
Command Line area, 1-9
Computer Geometric Modeling, 1-2
Construction Line, 8-18
Constructive solid geometry, 6-2
Coordinate systems, 2-13
Coordinates, 2-13
Copy Faces, 10-8
Copy Object, 3-7
Create Drawings, 1-7
CSG, 6-2
Current UCS, 4-20
Cursor Coordinates display, 1-8
Cylinder, Solid, 6-9

D
Display Locked, 8-18
Draft Angle, 10-10
Drafting Settings, 2-5, 4-8
Draw Toolbar, 1-10
Drawing Layout, 8-2
Drawing Limits, 2-5
Drawing Views, Scale, 8-17
DVIEW, 8-16

E
Edge, hidden, 5-14
Endpoint, snap, 2-12
Erase, 2-14
Esc, 1-11
EXIT, 1-12
Explode, 3-19
Extract faces, 10-8
Extend, 4-23
Extrude, 7-2, 7-15
Extrude, Draft Angle, 10-10

F
Face, 3D, 5-2, 5-10, 5-12
FILLET, 10-6
Fillet, 3D, 10-16, 10-17
Flat Shaded, 5-8
Folder create new, 1-13
Format, pull-down, 2-4
4 viewports, 4-14
From, Snap, 3-11
Front View, 4-23

G
Global space, 5-15
Graphics Window, 1-8, 1-9
GRID Interval setup, 2-7
GRIP, editing, 4-23
Gouraud Shaded, 5-8

H
Help, 1-11, 1-12
Help Line, 1-8
Hide, 2-19
Hidden, 5-8, 5-11

I
Inquiry, 7-17
INTERSECT, 4-9
Invisible Edge, 5-14
Isometric Views, 2-16

J
JOIN, 6-3

L
Layer Properties Manager, 2-20
Layout, 8-2
Layout Tab, 8-4
Limits, 2-5
LINE command, 3-8
Line, close, 7-9
Local Coordinate System (LCS), 2-15

M
Mass Properties, 7-17
Meshes, 5-2
Midpoint, snap, 2-10
Mirror, 4-16, 7-14
Mirror 3D, 9-13
Model Space, 8-4
Model Tab, 8-11
Model Mode, 8-4
Modify toolbar, 1-8, 1-10
Mouse Buttons, 1-10
Move, 6-17
Move UCS, 2-21
Multiview drawing, 8-2

N
Named view, 4-20
New UCS, 4-18
New UCS, Face, 7-18

O
Object Properties toolbar, 1-8, 1-9
Object Snap, 2-10
Object Snap Settings, 4-8
Object Snap Toolbar, 3-10
OFFSET, 2-9
On-Line Help, 1-11
1 Viewport, 4-17
Open a Drawing, 5-4, 5-16
Orbit, 3D, 3-18

P
Page Setup, 8-4
Pan Realtime, 8-8
Paper Space, 8-2
Plan View, 4-20
Polar Angle settings, 4-8
Polar Tracking, 4-8
Polygonal Viewport, 8-13
Polyline, 2-8
Predefined surface models, 5-3, 5-18
Primitive Solids, 6-3
Properties, 2-18
Pull-down menus, 1-8, 1-9

Q
Qnew, 3-5
Quadrant, Snap to, 3-23
Quick Select, 4-15
Quick Setup, 2-3
Quit, 1-12

R
RECTANGLE command, 2-23
RECTANGULAR Array, 10-19
Reference area setup, 2-4
Regen, 2-19
Region, 5-2, 7-10
Region, Subtract, 9-3
Repeat, 3-9
Restore Down, 3-5
Revolve, 7-2, 9-12
Revolved Feature, 9-2

Revolved surfaces, 5-3, 5-20
ROTATE, 7-6
Rotate Faces, 6-16
Rotate 3D, 7-12
Rotate UCS, 2-21
Ruled surfaces, 5-3, 5-20, 5-24

S
SAVE AS, 2-14
Scale Drawing Views, 8-17
Screen Layout, 1-8
SE Isometric, 2-16
Selection window, 2-15
Settings, 4-8
Shade, 2-27
Shade toolbar, 5-8
Shaded Solids, 6-14
SHELL, 10-18
Show Startup dialog box, 3-4
Single Line Help, 1-8
Snap from, 2-12
Snap Interval Setup, 2-7
Snap to Endpoint, 2-12
Snap to Midpoint, 2-10
Snap to Quadrant, 3-23
Solids Editing, 6-10
Solid Modelers, 1-3
Solid, 2D, 5-2
Solids toolbar, 6-7
SOLPROF, 8-20
Standard scale property, 8-17
Standard Toolbar, 1-9, 1-10
Start From Scratch, 3-5
Startup options, 3-4
Startup window, 3-5
Status Bar, 1-9
Stretch, 4-24
Subtract, 6-4, 6-12
Surface Modelers, 1-3
Surface Modeling commands, 5-20
Surfaces Toolbars, 5-5
SW Isometric View, 4-24
Symmetrical Features, 9-2

T
Tabulated surfaces, 5-3, 5-20
Thickness, 2-2, 2-18
Thickness option, Rectangle, 2-23
3 Point, UCS, 4-18, 5-5
3D Face, 5-10
3D Orbit, 3-18
3D Rounds and Fillets, 10-16
3D View, 2-16
3D Wireframe, 2-23, 5-8
Thin Shell, 10-18
Title Block, 8-2
Toolbars, 1-9, 1-10
Tool Palettes, 1-8, 2-6
Tools, pull-down, 2-6
Top view, 4-14
Trim, 2-13
Trim, Project View option, 3-16
2½D solid modeling approach, 2-18
2D Extrusion method, 3-2

U
Undo, 2-14
Union, 6-10
Units Setup, 2-4
UCS, 2-2, 4-7, 5-5
UCS, 3 point, 5-5
UCS Icon, 4-19
UCS II, 4-7
UCS Toolbar, 4-7
User Coordinate System (UCS), 2-2
User Coordinate System, 2-15

V
View, Hide, 2-17
View, 3D Views, 3-6
View, pull-down, 2-16
View, Regen, 2-17
View, Shade, 2-25
View Toolbar, 3-18
Viewports, 4-14
Viewport Scale, 8-17

W

Wireframe, 2D, 5-8
Wireframe, 3D, 2-27
Wireframe Modeler, 1-3
Wizard, quick setup, 2-3
World Coordinate System, 2-13
World space, 2-15
World UCS, 5-9

X

X Axis Rotate UCS, 2-25
XY-Plane, 3-6

Y

YZ-Plane, 2-13

Z

Zoom All, 2-6, 3-8
ZX-Plane, 2-13

Notes:

Notes:

Notes:

Notes: